Towards a Biosemiotic Theoretical Biology

Vienna Series in Theoretical Biology

Gerd B Müller, editor-in-chief

Thomas Pradeu and Katrin Schäfer, associate editors

Convergent Evolution: Limited Forms Most Beautiful, by George McGhee, 2011

From Groups to Individuals, edited by Frédéric Bouchard and Philippe Huneman, 2013

Developing Scaffolds in Evolution, Culture, and Cognition, edited by Linnda R. Caporael, James Griesemer, and William C. Wimsatt, 2013

Multicellularity: Origins and Evolution, edited by Karl J. Niklas and Stuart A. Newman, 2016

Vivarium: Experimental, Quantitative, and Theoretical Biology at Vienna's Biologische Versuchsanstalt, edited by Gerd B. Müller, 2017

Landscapes of Collectivity in the Life Sciences, edited by Snait B. Gissis, Ehud Lamm, and Ayelet Shavit, 2017

Rethinking Human Evolution, edited by Jeffrey H. Schwartz, 2018

Convergent Evolution in Stone-Tool Technology, edited by Michael J. O'Brien, Briggs Buchanan, and Metin I. Erin, 2018

Evolutionary Causation: Biological and Philosophical Reflections, edited by Tobias Uller and Kevin N. Lala, 2019

Convergent Evolution on Earth: Lessons for the Search for Extraterrestrial Life, by George McGhee, 2019

Contingency and Convergence: Toward a Cosmic Biology of Body and Mind, by Russell Powell, 2020

How Molecular Forces and Rotating Planets Create Life, Jan Spitzer, 2021

Rethinking Cancer: A New Understanding for the Post-Genomics Era, edited by Bernhard Strauss, Marta Bertolaso, Ingemar Ernberg, and Mina J. Bissell, 2021

Levels of Organization in the Biological Sciences, edited by Daniel S. Brooks, James DiFrisco, and William C. Wimsatt, 2021

The Convergent Evolution of Agriculture in Humans and Insects, edited by Ted R. Schultz, Richard Gawne, and Peter N. Peregrine, 2022

Evolvability: A Unifying Concept in Evolutionary Biology?, edited by Thomas F. Hansen, David Houle, Mihaela Pavlicev, and Christophe Pélabon, 2023

Evolution "On Purpose": Teleonomy in Living systems, edited by Peter A. Corning, Stuart A. Kauffman, Denis Noble, James A. Shapiro, Richard I. Vane-Wright, and Addy Pross, 2023

Properties of Life: Toward a Theory of Organismic Biology, Bernd Rosslenbroich, 2023

The Evolution of Techniques: Rigidity and Flexibility in Use, Transmission, and Innovation, edited by Martin Charbonneau, 2024

The Organism-Environment Pairing: A Historical and Philosophical Re-Appraisal, Alejandro Fábregas-Tejeda, 2026

Towards a Biosemiotic Theoretical Biology: Sign Processes and Meaning-Making in Living Systems, edited by Kalevi Kull and Donald Favareau, 2026

Towards a Biosemiotic Theoretical Biology

Sign Processes and Meaning-Making in Living Systems

edited by Kalevi Kull and Donald Favareau

The MIT Press
Cambridge, Massachusetts
London, England

The MIT Press
Massachusetts Institute of Technology
77 Massachusetts Avenue
Cambridge, MA 02139
mitpress.mit.edu

The MIT Press would like to thank the anonymous peer reviewers who provided comments on drafts of this book. The generous work of academic experts is essential for establishing the authority and quality of our publications. We acknowledge with gratitude the contributions of these otherwise uncredited readers.

This book was set in Times New Roman by Westchester Publishing Services. Printed and bound in the United States of America.

Library of Congress Cataloging-in-Publication Data

Names: Kull, Kalevi editor | Favareau, Donald editor
Title: Towards a biosemiotic theoretical biology : sign processes and
meaning-making in living systems / edited by Kalevi Kull and Donald Favareau.
Description: Cambridge, Massachusetts : The MIT Press, [2026] | Series:
Vienna series in theoretical biology | Includes bibliographical references and index.
Identifiers: LCCN 2025032814 (print) | LCCN 2025032815 (ebook) |
ISBN 9780262053617 paperback | ISBN 9780262053624 pdf |
ISBN 9780262053631 epub
Subjects: LCSH: Biology—Semiotics | Signs and symbols | Organisms
Classification: LCC QH331 .T658 2026 (print) | LCC QH331 (ebook)
LC record available at https://lccn.loc.gov/2025032814
LC ebook record available at https://lccn.loc.gov/2025032815

EU Authorised Representative: Easy Access System Europe, Mustamäe tee 50, 10621 Tallinn, Estonia | Email: gpsr.requests@easproject.com

Contents

III DRAFTS

IV ESSAYS

Series Foreword

Biology is a leading science in this century. As in all other sciences, progress in biology depends on the interrelations between empirical research, theory building, modeling, and societal context. But whereas molecular and experimental biology have evolved dramatically in recent years, generating a flood of highly detailed data, the integration of these results into useful theoretical frameworks has lagged behind. Driven largely by pragmatic and technical considerations, research in biology continues to be less guided by theory than seems indicated. By promoting the formulation and discussion of new theoretical concepts in the biosciences, this series intends to help fill important gaps in our understanding of some of the major open questions of biology, such as the origin and organization of organismal form, the relationship between development and evolution, and the biological bases of cognition and mind. Theoretical biology has important roots in the experimental tradition of early twentieth-century Vienna. Paul Weiss and Ludwig von Bertalanffy were among the first to use the term theoretical biology in its modern sense. In their understanding the subject was not limited to mathematical formalization, as is often the case today, but extended to the conceptual foundations of biology. It is this commitment to a comprehensive and cross-disciplinary integration of theoretical concepts that the Vienna Series intends to emphasize. Today, theoretical biology has genetic, developmental, and evolutionary components, the central connective themes in modern biology, but it also includes relevant aspects of computational or systems biology and extends to the naturalistic philosophy of sciences. The Vienna Series grew out of theory-oriented workshops organized by the KLI, an international institute for the advanced study of natural complex systems. The KLI fosters research projects, workshops, book projects, and the journal Biological Theory, all devoted to aspects of theoretical biology, with an emphasis on—but not restriction to—integrating the developmental, evolutionary, and cognitive sciences. The series editors welcome suggestions for book projects in these domains.

Gerd B. Müller, Thomas Pradeu, Katrin Schäfer

Preface

Semiotics is the scientific study of sign processes and meaning-making—neither of which turn out to be unique to human language use, once they are properly defined and understood from the biological perspective. Accordingly, *biosemiotics* is the study of prelinguistic and nonlinguistic sign processes and meaning-making, as they manifest in different forms throughout the entire realm of the living. Moreover, to the extent that meaning-making, non-anthropomorphically understood, appears to be one of the fundamental attributes of living systems, the biosemiotic study of sign processes in life processes is fundamental to the whole endeavor of biology.

Yet for reasons that are detailed in several of the selections in this volume, the role of meaning-making processes (*semiosis*) and their associated phenomena (such as *contextuality, interpretation, cognition, decision-making,* and *choice*) in explaining the behavior, dynamics, and development of living beings had not yet become a central focus in the biology of the twentieth century. Research in this direction has been slowly but persistently growing, however, since then.

Still, the problems in this area are theoretically rather complex, requiring the development of models and scientifically responsible frameworks for such subtle and not at all easily explicated phenomena as are agency, anticipation, arbitrariness, categorization, context, freedom, intentionality, meaning, motivation, need, interpretation, wholeness, biological space, and time—all attributes of living systems that have not yet received their rightful place in most treatises of general biology, yet without which, a proper understanding of living being is unattainable.

Theoretical biology, however, can already boast of several influential achievements from the twentieth century that have pointed toward the importance of studying the semiotic processes of living systems. These include: the monograph *Theoretische Biologie* by Jakob von Uexküll, and his life's work studying animal perception and action at his institute of umwelt research and Naples biological station; the symposia and volumes *Towards a Theoretical Biology* as organized by Conrad Hal Waddington; the deep analysis of living beings as anticipatory modelling systems by Robert Rosen; and the direct application of semiotic theory in biology by Thomas Sebeok, Jesper Hoffmeyer, Terrence Deacon, and several others. Similarly, the first quarter of the twenty-first century has seen the epigenetic turn in biology, a growing coalescence of biologists moving away from neo-Darwinism and towards a more extended evolutionary synthesis, and a remarkable growth in the research agenda of biosemiotics *per se*, marked by the regular annual international Gatherings in Biosemiotics conferences, the journal and book series on biosemiotics, and the application of semiotics in many areas of biology.

We are honored to now add to this burgeoning research trajectory, the present volume—in which we have collected a variety of exciting new approaches to the examination and analysis of meaning-making in living systems, written by some of the most forward-thinking biologists of the present time. In so doing, we are actively paying homage to the symposia and related volumes of the *Towards a Theoretical Biology* meetings organized by Conrad Hal Waddington from 1966 to 1969 in Bellagio, Italy, and we are glad to have one of the original participants of these historic meetings, Stuart Kauffman, with us to mark the continuity between Waddington's project and our own.

As part of that homage, too, we have reproduced Waddington's titling scheme in the organization of his volumes. Thus, in part I, "Prolegomena," our introductory chapter first reviews some of the major inflection points in the development of theoretical biology, before embarking upon a detailed discussion on the importance of semiotic phenomena—such as *semiosis, meaning, umwelt, communication,* and *categorization*—for living systems, and offering our own most recent thinking on how to informatively incorporate such phenomena within a biological framework. In the two chapters that follow, Erik Peterson surfaces some earlier unnoticed connections of Conrad Waddington to semiotics, while Charles Wolfe compares and contrasts the biosemiotic explorations of Jakob von Uexküll, Kurt Goldstein, and Georges Canguilhem.

Part II, "Sketches," which brings together a wide range of approaches to semiotic emergence and evolution, consists of five chapters. There, Raymond Noble and Denis Noble propose a possible explanation for organisms' choice-making; Scott F. Gilbert describes the role of communication in the life of holobionts; Alexei Sharov provides a theoretical review of a semiotic theory of agency; Stuart Kauffman examines the intertwined evolution of agency, function, semantics, and meaning; and Matt Kalkman and Terrence Deacon demonstrate the importance of the semiotic concept of *abduction* for the theory of biological evolution.

Part III, "Drafts," deals with some outstanding conceptual problems in biosemiotics and comprises four chapters. Jana Švorcová and Karel Kleisner discuss the often under-analyzed role of *habit* in evolution; Richard I. Vane-Wright examines the semiotic processes that actualize the potential of Batesian mimicry; Arantza Etxeberria and David Cortés-García expand the concept of evolutionary reproduction as the reproduction of *umwelt*; and the section concludes with a chapter by Tom Froese, Georgii Karelin, and Takashi Ikegami, who examine the conceptualization and testability of the phenomenon of end-directedness in living beings.

Part IV, "Essays," also comprises four chapters. There, Henrik Nielsen, Mariana Vitti Rodrigues, and Claus Emmeche apply Peircean semiotic concepts to analyze peptide recognition in the cell; Federico Vega advances a new theoretical synthesis joining Robert Rosen's relational modeling with a Peircean approach to the modeling of biosemiotic processes; Daniel C. Mayer-Foulkes examines the primary logical or pre-logical structure of organic sense-making, and the entire volume concludes with Anton Markoš's call for "taking a life stance" in the biological sciences—and beyond.

As insightful and important as all these individual contributions "towards" the development of a biosemiotic theoretical biology are, however, a completed theory of the origins and workings of semiotic processes in living systems is still very much a work in development—for, indeed, semiotic processes and their related phenomena are so ubiquitous and ineliminable in the world of the living that their study will not be exhausted in the century to come. But by pursuing these lines of inquiry presently, however, the thinkers who are represented here are collectively leading us to the day that biology will at last no longer be said to be "meaning-blind."

We thank them all for the Waddingtonian spirit in which they have engaged this work.

Kalevi Kull
Donald Favareau

I PROLEGOMENA

1 Theoretical Biology on Its Way Towards Biosemiotics: History and Current Theory

Kalevi Kull and Donald Favareau

The study of the nature and workings of meaning-making remains critically underdeveloped. In our recent publication, "Semiotics in General Biology" (Kull and Favareau, 2022), we give an overview of the history and main concepts of the study of sign processes (semiotics) as they have been applied in general biology thus far, and we indicate the ways in which the application of a more explicitly semiotic understanding of life may aid in the development of general biology in the future.[1] This article, together with a lengthy description of the history of biosemiotics by Favareau (2010, pp.1-77), would serve well as an informative contextualizing introduction to the current collection of essays.

The aim of our paper here is to introduce our understanding of semiotic processes and their role in living systems, while avoiding an elaborated terminology of semiotics and without repeating the former material. Accordingly, the first half of the paper (section 1.1) reviews some historical moments and trends in the formation of theoretical biology—noting parallel developments in the history of semiotics—that led to the emergence of overt interdisciplinary connections being made between biology and semiotics toward the end of the twentieth century. In the second and more theoretical half of the paper (section 1.2), we describe some key biological processes that are at the same time semiotic processes, and we argue for the development of a theoretical biology that can do full justice to both aspects of these biosemiotic processes in explaining the organization and behavior of living beings. Accordingly, the concepts presented in this latter section form the basis of a potential theoretical biology that can finally incorporate the reality of meaning-making and other semiotic processes in its understanding and modeling of living systems.

1.1 Prior Theoretical Biology That Incorporates Elements of Semiotics

In the following sketch of selected moments in the history of theoretical biology, we describe the some of the earlier moves made towards incorporating semiotics into (or as applied to) biology.

1.1.1 Disciplinary Beginnings: Theoretical Biology

Manfred Laubichler (2007) notes that the history of theoretical biology should be an important part of theoretical biology itself, given that the existence of different views, models, and theories in it are at all times absolutely necessary, according to Allison Shaw (2022). Yet, as historians of science have recently noted, "The history of theoretical biology is almost completely unexplored" (Böhm et al., 2022).

Indeed, a wide range of scientifically discussed theorizing has existed in the field of wildlife research since biology became a science—that is, more or less since the eighteenth century. However, theoretical biology as a scientific discipline did not appear under its own name until the end of the nineteenth century.

Aside from some occasional early mentioning of theoretical biology (e.g., by von Paula Gruithuisen (1823, p. 55)), one of the first instances of the expression "theoretical biology" found in print in English comes from the philosopher Chauncey Wright (1830–1875), a colleague of, and intellectual influence upon, both the American philosopher and semiotician Charles Sanders Peirce and the psychologist William James. In his monograph entitled *Darwinism: Being an Examination of Mr. St. George Mivart's 'Genesis of Species,'* Wright analyzes Mivart's important criticism of Darwin's theory of species evolution by natural selection and speaks of "the great additions to our knowledge of natural history, and of general physiology, or theoretical biology, which the discussion [of such ideas] has produced" (Wright, 1871, pp. 3, 7).

According to Mivart's alternative evolutionary theory (Mivart, 1871), Wright notes, organisms are potentially more or less suited to different environments, and evolution is primarily concerned with the transitions of organisms to modified forms in response to changes in the environment, while remaining permanently intact and relatively stable as species. It is worth noting that Mivart's ideas have certain similarities with arguments now being made under the aegis of the modern extended theory of evolution (Pigliucci and Müller, 2010).

Appropriately, one of the first books to have the word "theoretical biology" in its title was Alfred Russel Wallace's (1891) *Natural Selection and Tropical Nature: Essays on Descriptive and Theoretical Biology*. In the same last decade of the nineteenth century, theoretical biology appeared as a subject in universities, and theoretical biology as a discipline started to take shape. A booklet by zoologist Robert Bigelow (1863–1955), entitled *Syllabus of Lectures in Theoretical Biology* (Bigelow, 1896), provides an overview of the subject of theoretical biology as taught to fourth-year biology students over two semesters by Bigelow at the Massachusetts Institute of Technology.[2]

Theoretical biology was also being discussed within academia at this time—for example, at Bryn Mawr College in Pennsylvania, first in the course entitled "Advanced Biology: Lectures on Special and Theoretical Topics"[3] and later as a separate course entitled "Theoretical Biology," taught by Thomas Hunt Morgan (who also offered a course entitled "Advanced Theoretical Zoology").[4] There was also a course entitled "Historical and Theoretical Biology" at the University of Georgia, taught by John Pendleton Campbell (1863–1918)[5]—and these were not the only ones. For example, the content description of the course "Theoretical Biology" given at Adelphi College in New York in 1898 reads as follows: "The history of Biological Discovery. The Examination of philosophical problems of Biology: such as heredity, variations, animal metamorphoses, the origin of the individual, instinct and intelligence. Readings and discussions of Darwin, Spencer, Romanes, Weismann, and similar authorities. Senior Year, three hours a week."[6]

The first known monograph entitled *Theoretical Biology* was published in German in 1901, written by the botanist Johannes Reinke (Reinke, 1901; 2nd edn. 1911). There he considered morphogenesis and symbiosis—in the latter case, using the term *consortium*—to be the central life processes, and his understanding of the role of the organism in gene regulation—for which he used the term *dominant*—is a foreshadowing of the current view (e.g., Wissemann, 2006).

Not long after, Jakob von Uexküll's important monograph *Theoretische Biologie* was printed (Uexküll, 1920, 1928), about which we will have much to say later. The first book series on theoretical biology (*Abhandlungen zur theoretischen Biologie*) was founded by Julius Schaxel in 1919, and thirty volumes were published between that year and 1931. Soon afterward, a monograph on theoretical biology was published by the Austrian biologist Ludwig von Bertalanffy (1932, 1942), and, in 1935, one by the Hungarian biologist Ervin Bauer.

The opposition between vitalism and mechanism in the first years of the twentieth century gave rise to the organism-based theory of biology, *organicism*,[7] which importantly prepared the ground for addressing the mechanisms of meaning-making in biology. This was a holistic trend that supported *emergentism* (i.e., the appearance of new properties and irreducible rules in the organism—properties that individual parts of the organism alone may not possess). At the same time, organicism was also a structuralist trend that sought to formalize the general concepts of biology. Analysis on the population level was not enough for doing so because the basic processes of life lie in the physiology of organisms; it was understood that it is only at the organism level that the key to the phenomenon of both agency and biological integrity should lie.

Descriptions of organismal processes using mathematical models were therefore developed (Bertalanffy, 1932, 1942; Rashevsky, 1938) in the effort to capture the emergent and holistic nature of such processes. Doing so was thought to be a sound systemic approach, as mathematics is a good tool to help uncover subtle patterns and not always immediately noticeable systemic effects. In England, the Theoretical Biology Club, uniting organismic biological theorists, was founded in the 1930s (Peterson, 2016), and in the Netherlands, in Leiden, both the Institute of Theoretical Biology and the journal *Acta Biotheoretica* were founded in 1935 (Hemerik and Reydon, 2005; Reydon et al., 2005).

Thus, in the last decades of the nineteenth century and the first decades of the twentieth century, theoretical biology took shape under its own name—on the one hand, as German-languaged and nonmathematical, in the tradition of continental philosophy, where its important predecessors were Goethe and Schelling—and on the other hand, as English-languaged and mathematical, with an emphasis on biophysics and mathematical statistics. Of course, there was and is a lot of mixing between each of these two broad approaches, ofttimes to the consternation of the practitioners of each. The relationship between the qualitative and the quantitative characteristics in living systems constantly creates problems for the development of sound theorizing in the life sciences. Empirical replicability requires precise definitions and mathematics, while accurate understanding often demands the acceptance of unquantifiable and overlapping concepts.

So how to clearly describe phenomena and objects that themselves have blurry boundaries, as is the case in the realm of life? We have a biology that can very rigorously account for the biophysical aspects of organismic beings; we now need to augment that with one that can just as rigorously account for its subjective, experiential, perceptual, and communicative aspects—which is to say, its sign processing, or semiotic, side—before we can have a fully explanatory account of living beings. It was toward this end that the research project of biosemiotics would be eventually developed.

In the first half of the nineteenth century, however, semiotics as such was mainly known through its usage in medicine, as detailed descriptions of signs of pathologies and health. The rise of interest in the study of sign processes of all kinds appeared in the later decades of that century through the works of philosophers such as Charles Sanders Peirce, Victoria Lady Welby, Edmund Husserl, Gottlob Frege, and Gustav Teichmüller, despite being explicitly called for much earlier in the concluding argument of John Locke's *Essay Concerning Human Understanding* (1690).[8] In the beginning of the twentieth century, the pioneering structuralist analyses of linguist Ferdinand de Saussure (1857–1913) gave rise to the semiotic study of language—then under the name of semiology.

The more direct connections between biology and semiotics took place only in the early years of the twentieth century, especially due to the works of Jakob von Uexküll (1928, 1940) from the side of biology (while not using the term *semiotics*) and those of Charles Morris (1925, 1937) from the side of philosophy of science and semiotics. However, this connection between the disciplines of biology and semiotics remained almost unnoticed until the 1960s.

1.1.2 Waddington's Symposia and the Development of Biosemiotics

The mid-twentieth century saw the appearance of biocybernetics and a rapid development of mathematical modeling in biology. In 1961, the *Journal of Theoretical Biology* was established. The genetic code was discovered in the same decade, and neo-Darwinian theory became formalized based on population genetics. At the same time, however, ethology flourished (for example, von Frisch, 1957; Goodall, 1990; Griffin, 1962; Lorenz, 1970; Tinbergen, 1972). Peter Marler (1961), who noticed semiotic aspects in animal behavior, published a logical analysis of animal communication in the first volume of the *Journal of Theoretical Biology*. Yet developmental biology was not yet properly integrated into the Modern Synthesis.

This was the context within which Conrad Hal Waddington arranged the four "Towards a Theoretical Biology" symposia, which took place at Villa Serbelloni in Bellagio on the shores of Lake Como in Italy in 1966, 1967, 1968, and 1969. Each symposium had seventeen to twenty-one participants, comprised of the leading thinkers in theoretical biology, along with some of his students from Edinburgh, for a total of forty-three researchers participating across the four meetings. Most were from the US and the UK (including eight from Edinburgh), one was from France, and one was from Switzerland. Besides Waddington himself, five participated in all four meetings: René Thom, Howard Pattee, Brian Goodwin, Jack Cowan, and Christopher Longuet-Higgins. John Maynard Smith and Lewis Wolpert participated in three, and all the others in either two or just one event.

We wish to take some time to explain why these symposia and their proceedings (Waddington 1968, 1969, 1970, 1972) were so important in the context of connecting biology and semiotics, even if the term *semiotics* was most probably not used there.

It is important to remember that Waddington and several of the other participants in these symposia referred to an extended concept of *language* as the foundation for a future theoretical biology. And it is equally important to remember in this context that a very influential linguist and semiotician of these times, Thomas Sebeok, made a clear distinction between *language as a specifically human faculty*, as studied by linguistics, and *all other sign systems* that are used by living beings of many other species and studied by semiotics in general.[9] Thus, we can interpret Waddington's (1972, p. 289) claim that "it is language . . . that I suggest may become a paradigm for the theory of General Biology," as actually referring to what Sebeok distinguishes as the broader category of semiotics—which for Waddington, at that time, mostly meant semiotics in its structuralist version.[10]

Moreover, several key participants of these symposia made significant contributions to what would later be called biosemiotics in their work, and among these, we must surely count Howard Pattee, Rene Thom, Brian Goodwin, Stuart Kauffman, Susanne Langer, and Marjorie Grene. Importantly, too, as demonstrated by Erik Peterson in his chapter for the current volume, Waddington had a lifelong friendship with Gregory Bateson (1904–1980), one of the acknowledged pioneers of biosemiotics, whose efforts to naturalize epistemology in biology (Bateson, 1972, 1979) was a formative influence on the thought of both Jesper Hoffmeyer and Terrence Deacon.

The subsequent influence of the Waddington Symposia on the development of biosemiotics has been significant, moreover, and to illustrate that, we will highlight just one representative manifestation of it here. This was the Winter School in Theoretical Biology in Viitna, Estonia, followed by the conference "Biology and Linguistics" in 1978 in Tartu, which brought together a remarkable group of semioticians and biologists and became an important starting point for biosemiotic research in Estonia and in Russia—and which later would become global with the joining together of the Copenhagen "Semiotics of Nature" group and the Prague group of biosemiotics (Favareau, 2010, pp. 50–51). The meeting was organized by three groups of young Russian and Estonian biologists, independently informed in 1970s by the *Towards a Theoretical Biology* volumes, and several of the original participants to these Tartu-Moscow meetings have gone on to work in and develop the field of biosemiotics.[11]

As participant Alexei Sharov (then of Moscow University) notes, "The influence [of the Waddington volumes] was enormous. We read and discussed every chapter in this book. . . . At that time we did not know about semiotics and talked about coding and 'language' as a metaphor. But we seriously considered the chapters by Pattee and Thom, who both were close to biosemiotic thinking, but from the physical point of view. It was [Julij] Shreider who taught us the elements of semiotics ([in the traditions of] Frege and Saussure, not Peirce). Waddington's cybernetic doctrine was an intermediate stepping-stone for biologists to approach biosemiotics."

Sergey Chebanov (St. Petersburg University) likewise recalls, "For me, the first volume of '*Towards a Theoretical Biology*' (which was published in Russian in 1970) is the book that had the greatest influence [and] significance for me, because I found in it the largest number of like-minded people. The emergence of [the Winter School] seminar at the university in 1972, and [Alexander] Levich's group at Moscow State University was also connected with it."

Kalevi Kull (University of Tartu) reports that "the Tartu Group in Theoretical Biology came up the idea to have an honorary member, and we proposed to Waddington that he become it. He agreed, and sent to us volumes 2, 3 and 4 of *Towards a Theoretical Biology*, in the spring of 1975. In the autumn of 1976, we admitted another honorary member—Robert Rosen, after which no more honorary members were elected. Since 1975 we have been convening annual meetings which we call the Spring Schools in Theoretical Biology. In each meeting we have a display table showcasing the relevant literature, and the *Towards a Theoretical Biology* volumes have been exhibited every year, without exception."

Meanwhile, in Copenhagen, the biosemiotic research agenda began coalescing around biochemist Jesper Hoffmeyer and the Danish "Semiotics of Nature" study group in the 1980s.

Group member Frederik Stjernfelt recalls that "Waddington was a reference in the Danish group, but not very often. I discovered him before the biosemiotics movement of the 1990s when I wrote, in the mid-1980s, my treatise on René Thom and Jean Petitot," while biosemiotician Claus Emmeche reports that

> Jesper [Hoffmeyer] had the volumes in his office, I remember that I also had a look in them when I did my Ph.D. work there (as a part of this, I was especially interested in Brian Goodwin's work). And I am sure he [Jesper] read them. . . . It is very likely that Jesper could have been much inspired by the open curiosity and the style of non-reductionist approaches to (especially developmental) biology that many of the contributions expressed. In that sense (but here I am guessing), these volumes helped Jesper to formulate his critique of reductionism in Neo-Darwinist thinking that he formulated a long time before [he started developing] biosemiotics, namely in a series of five articles in *Naturkampen*, 1977 to 1979, that later came out as a little book in Danish in 1980 called *Evolution, Økologi, Historie: Neodarwinismens Krise*.[12] In that book, he refers to works by R. C. Lewontin and Waddington. I think that the "spirit" expressed in the volumes—gathering so many different approaches and allowing for a combination of speculation, modelling, and philosophical/theoretical approaches—has had some distant impact for the contemporary biosemiotic research activities.[13]

Anton Markoš from the Prague biosemiotics group recalls, "In 1980, I got acquainted with [Waddington] through his book (or better—collected papers) *Evolution of an Evolutionist* (1975), became very enthusiastic, and have followed this trend since: i.e. [the study of] influencing development through external stimuli, plus subsequent genetic assimilation of traits."

It should be added that in addition to the inspiring work of Waddington, another factor in the development of contemporary biosemiotics was the well-developed non-Darwinian and anti-reductionist traditions in biology—well known in Eastern Europe—as established already by Karl Ernst von Baer and Lev Berg. Meanwhile, the 1970s was a period of high popularity for the study of general semiotics, which had considerable interdisciplinary influence. There the differences between French and East-European structuralism were extensively documented (Sériot, 2014). Together, all of the above conditions were particularly fertile ground for the development of biosemiotic thought in general.

Perhaps the strongest move toward connecting semiotics and biology, however, was made by the polymath and linguist Thomas Albert Sebeok (1920–2001), who, after an intense immersion into the extant literature on animal communication systems, founded first the discipline of *zoösemiotics*—"a discipline within which the science of signs intersects with ethology, devoted to the scientific study of signalling behaviour in and across animal species" (1963)—and later, the even more encompassing research agenda of *biosemiotics*, the study of all sign processes in life processes (Sebeok and Umiker-Sebeok, 1992).

Independently during these decades, also, ethologist Peter Marler (1961), psychologist Friedrich Rothschild (1968), semiotician Yuri Stepanov (1971), molecular biologist Marcel

Florkin (1974), communicologist Günter Bentele (1984), immunologists Eli Sercarz et al. (1988), bacteriologist Sorin Sonea (1992), embryologist Marcello Barbieri (1985), and others each made attempts to incorporate the understandings and frameworks from various schools of sign study into their analyses of biology.

As recounted extensively elsewhere (Favareau, 2010; Sebeok, 1998, 2001), Sebeok and his colleague Thure von Uexküll, the medical doctor son of Jakob von Uexküll, spent the 1970s up until the time of Sebeok's death in 2001 bringing together biologists and scientists of all kinds—including mathematician René Thom (1975), oncologist Giorgio Prodi (2021 [1977]), zoologist Heini Hediger (1974), ecologist Kalevi Kull (1993), and molecular biologist Jesper Hoffmeyer (1996)—into dialogue with the leading semioticians of the day, such as John Deely (2002) and Umberto Eco (1999), to develop what Kull and Hoffmeyer, with many other like-minded colleagues, would later go on to institutionalize as the contemporary research field of biosemiotics.

In the early 1990s, then, several groups of biosemioticians from around the world, including the four groups mentioned above, were linked together thanks to the efforts of Thomas Sebeok and Thure von Uexküll, first through contributing to the collected volume entitled *Biosemiotics* (Sebeok and Umiker-Sebeok, 1992) and afterward by a growing number of meetings and linked publications, drawing in even more such groups and independent researchers.

Among the most important publications on biosemiotics from the 1990s, the books by Hoffmeyer (1996) and Deacon (1997), in particular, have had a wide and enduring influence. Synergistically, Robert Rosen's (1988, 1991, 1999) work on anticipatory systems, as well as the epigenetic turn in biology (Jablonka and Ginsburg, 2022; Jablonka and Lamb 2005; Noble, 2021; Sinha, 2006; West-Eberhard, 2003) have had a strengthening effect on the development of biosemiotics.

Indeed, the extent of biosemiotic research has grown remarkably in the twenty-first century, helped in no small part by the international annual Gatherings in Biosemiotics conferences, now in their twenty-fifth year, the journal *Biosemiotics*, the *Springer Book Series in Biosemiotics*, the International Society for Biosemiotic Studies, and the teaching of biosemiotics in a number of scientifically well-regarded universities around the world.

1.1.3 Biosemiotics as a Part of Biology, Defined

Having now concluded our brief overview of the history of biosemiotics as a discipline, it is relevant here to define again what we mean by the term *biosemiotics* itself.

Joining together the definitions of semiotics (the study of sign processes and meaning-making) and biology (the study of living systems), biosemiotics is defined as the study of sign processes and meaning-making in living systems—and, as such, biosemiotics includes, broadly, the study of biocommunication, codes, and interpretation processes in living systems. Occasionally, biosemiotics has also been defined as the study of what and how organisms "know;" the study of the subjective worlds (or *umwelten*) of organisms; the study of prehuman and nonhuman epistemology; the study of the natural history of meaning; an investigation into the ontology of knowledge; and/or a scientific inquiry into the nature of organismic freedom.

However it is defined, biosemioticians recognize that the processes and phenomena that biosemiotics focuses on—meaning-making, communication, interpretation, freedom, choice, coding, umwelt, arbitrariness, representation, normativity, knowing—are all essential attributes of life, and that their scientific study, including in development and evolution, must become an inevitable part of general biology.[14]

1.2 Semiotic Processes in Life

In the following section, we introduce some central biosemiotic concepts—semiosis, habit, semiotic scaffolding, umwelt, modeling system, and meaningful communication—that reflect phenomena that we feel must be satisfactorily accounted for, whether using these exact terms or not, in the development of a more fully explanatory theoretical biology. In the concluding section, we sketch the role of semiosis in evolution.

1.2.1 Semiosis, or Meaning-Making, in Living Systems

Unlike as is the case with nonliving systems, in the case of living systems, a condition of *experiential indeterminacy* often emerges as the organism finds itself confronted by at least two or more *possibilities* for its immediate-next action. Do note that what we are concerned with here are not just "possibilities" from the perspective of an outside observer (which can be described through probability distributions based on a stochastic model), but the detection of the presence of actual alternative possibilities at a given moment by and for the organism itself.

Under such conditions, the very concept of *possibility* cannot be singular, as that would entail just one way to behave, in which case such behavior is predetermined, or at least stochastically determined. Rather, the internal break of determinacy that is the forking of two or more equally available alternatives means that a genuine measure of *freedom* appears in the life of the organism (and in some cases, for the cell) when left to choose between these alternatives. Again, what we are referring to is not merely the emergence of different behaviors in response to different conditions, but the situation in which two or more different possibilities are simultaneously available and can be freely chosen.

The process that takes place under such conditions of multiple possibility is thus one of *interpretation*—by which we mean a process that is not uniquely determined by initial conditions, and in which distinguishing between alternative simultaneous possibilities, and choosing to act upon only one of them, is manifested. For it is in the process of such distinguishing simultaneous available trajectories that such possibilities become related *as* alternative opportunities, entering into what is thus called a *semiotic relation.*

A semiotic (i.e., a meaning- or sign-based) relation does not function in interaction in the same way as does brute physical interaction; it takes the form, rather, of a kind of logical relation. By way of an analogy, just as the concept of a "gift" does not refer to any specific singular object preexisting in its essential "giftness," but instead refers to any imaginable object *to the extent that it is involved in* a three-way interaction between a person who at that moment is taking on the role of a "gift giver" and another the role of a "gift receiver," so too do the non-essentially "semiotic" things of the world (such as molecules, sound waves, odors, and visual displays) become *used as* particular "sign vehicles" denoting possible "sign objects" *in the context* of a living being's need to coordinate its sensations with action.

In the simplest case, then, such semiotic relations are primordially logical, or prelogical, in the sense that they do not rely on systems of notation or linguistic proposition. It is in this sense that Charles Peirce identified semiotics with the kind of logic that must be operative in "any intelligence that is capable of learning from experience" (1897). For as Karl Popper (1999, 2014) and others have pointed out, organisms are problem-solvers and meaning-makers, and their behavior is based on the *interpretation* of a situation, however simple the task might be, more often than it is on their being literally pushed and pulled around by the power of external forces, as is the case with all inanimate matter. In this way, "nature comes to the organism in the form of questions," wrote Jakob von Uexküll (1940, p. 25), which we may also understand as "in the form of simultaneous possibilities" that must be actively registered as such and selectively acted upon.

Crucially, for such possibilities to *be* alternative possibilities, they have to be simultaneously available. If they are sequential, they are thus not alternatives, and this necessary simultaneity of alternative possibilities implies that semiotic relations are, like logical ones, in a certain sense atemporal. Due to the processual nature of living beings, however, such relations become *instantiated* in a given moment, which, in the framework of physical time, involve a certain duration. This is why we have argued that semiosis must be explained in its full process-*and*-relation duality: we must scientifically account for both the manifestation of its temporal physiological processes and, at the same time, the semiotic nature of the specific atemporal logical relations that are being manifested by the co-appearance of those physiological processes (Kull and Favareau, 2023, pp. 20–21).

The actual real-world manifestation of any given *semiotic* (meaning- or sign-based) relation is called *semiosis*. In the Peircean sense of the sign, semiosis is a synonym for *meaning-making* or *interpretation*. Semiosis is where meaning—the activity whereby something is used or taken to signify something other than itself—arises. Semiosis is thus the act of meaning-making *as such*—the putting into formally triadic relationships that which on its own is not—which is why semiosis is so fundamentally related to the phenomenon of indeterminacy.

Here we see an important difference between physics and semiotics. One of the principles followed by physics is the principle of noncontradiction—stating that while the laws of nature may give rise to effects that are in some sense the *inverse* of one another (as in the case of condensation and evaporation, for example), no one given law of nature *contradicts* any another (and in fact, we see that many of the same laws govern both cases, as in our example). There are therefore no contradictions *as such* in the inevitable workings of physics in nature, including on the quantum level.

Contradictions—relationships and entailments that can be considered both A and not A at the same time—can and must exist in the realm of semiotics, however, due to its freedom in establishing nondeterministic relations. Indeed, one of the defining characteristics of a "sign" is in its intrinsic non-self-identity: the word is not the thing, and the map is not the territory. Accordingly, such intrinsic disunities appear in the realm of semiotic relations as the *useable elements of an extended logic* rather than as the barriers to action that are physical impossibilities. This also explains why "meaning" (the nondeterministic assigning of significance) is excluded from strictly physicalist models—for in the non-animate and strictly noncontradictory realm, there is no choosing from among alternative possibilities, and therefore, the notion of "freedom" is neither useable nor needed. The recognition of multiple available possibilities, by contrast, creates the conditions for free choice, and the active negotiation of such indeterminacy constitutes the semiotic, or meaning-making, *agency* of the organism.[15]

Accordingly, in the works on the biological origin of semiosis (e.g., Di Paolo, 2018; Markoš, 2002; Pattee, 1969; Rocha, 2001; Rosen, 1991; Sharov and Tønnessen, 2021), a number of other particularly agentive features of semioisis have been studied—among them: *organismic autonomy*, *anticipation*, *intentionality*, *communication*, *representation*, and *normativity*. Importantly, Deacon's (2011) model of *interacting and generative constraints* sheds light on the preconditions for semiosis and its major forms (i.e., iconic, indexical, and symbolic; qualitative, quantitative, and habitual).

Finally, it should be said that the physical mechanisms instantiating semiosis—and especially the origin of *subjective time*, which seems somewhat counterintuitive from the physicalist point of view—are only in the very earliest stages of scientific investigation, despite the concept's considerable elaboration in many of the works briefly noted above. It will be the work of the biosemiotic theoretical biology that we are proposing with this volume to further investigate and articulate these mechanisms.[16]

1.2.2 Habit as the Scaffolding of Semiotic Relations

We have seen that semiosis is essentially a process of actively and nondeterministically *relating* one thing with another and that such relations of non-self-identity can be arbitrarily established, in principle, infinitely (as Chomsky, quoting Humboldt, noted similarly about linguistic relations).[17] In actual practice, however, the freedom for establishing such relations in the nonlinguistic realm is always to some extent constrained, in the same sense that it is enabled, by an organism's body and its circumstances.

Moreover, semiosis—to the extent that it leaves traces of its "relating" in the form of biological structures and association patterns—is therefore also a powerful platform for phylogenetic and ontogenetic *learning*. Such "traces" can be modifications in the organization of the organism and/or in its environment, and both act subsequently as constraints upon, as well as the possibility space for, further semiosis.

Such facilitative constraints for the development of subsequent semioses were termed *semiotic scaffolding* by Hoffmeyer (2015a, 2015b). This his colleague in biosemiotics, Claus Emmeche, defines as the interlocking of a number of *enabling processes of sign activity* unfolding at several levels of organization, "focusing energy flow and agency of the system or subsystem upon a constrained repertoire of possibilities," thereby guiding the system's behavior to follow a more definite sequence of events (Emmeche, 2015, p. 275). Such scaffolding "canalizes further behavior," writes Kull, and in so doing, becomes "the frame for habits" (Kull, 2012, p. 228).[18]

In this way, semiotic scaffolding provides a whole interpretative landscape for the behavior of living beings, similar to Waddington's epigenetic landscape in some ways, but differing from it in that the agent actively navigates the choice points, and not vice versa. This organismic semiotic activity thus modifies its own conditions, and through this recursivity, its repertoire of agent-object relations obtains certain *regularities*, often "canalizing" into repeatable *tendencies* and *habits*. Such semiotic canalization is not entirely deterministic, however, since a habit is still a habit, and not a "law"—that is, it is not so fixed that it does not allow the manifestation of alternative behavior, but simply requires that such alternative (nonhabitualized) ones may entail more effort to realize.

1.2.3 Umwelt

Umwelt is a concept, introduced and developed stepwise in the works of physiologist and theoretical biologist Jakob von Uexküll (1928, 1932, 1940), that describes the individual's experienced "model" of the world—and that delineates the semiotic space of perception and action of an organism. It is based on the observation that individuals of different species have different capacities for distinguishing and categorizing sensations. Umwelt is thus the subjective and *experiential* world of a living being—the environment as known by an organism, through which its possibilities are presented to it, and on the basis of which its choices for action must be made.

Umwelt as such, in fact, arises from the copresence of the possibilities distinguished through the act of semiosis. The simplest semiotic space, or umwelt, would then consist of one semiotic relation that is realized only in the presence of one discernable binary difference. Richer umwelten can be multimodal and include a variety of perceived differences used by the organism in the organization of its behavior (hence the concept of the umwelt as a kind of "organizer" of perception-action joinings).

Also, while the active manifestation of umwelt is fundamentally momentary (since its difference-making presupposes the copresence of two or more relata), *its experienced "present"* as such is continuous in time. Therefore, in addition to its umwelt as experienced in the moment, an organism's ever-aggregating umwelt is also temporally extended throughout of the whole lifespan of the organism, in what is termed by Uexküll its *umwelt tunnel*.

Communally, the sharing of umwelt relations between organisms is most commonly based on the similarity of the distinctions and categorizations made by each of them. In the case of human umwelt sharing, this primarily takes the form of a shared language and set of cultural conceptions. A more important prerequisite of umwelt sharing for virtually all other living systems is the similarity of their sensorimotor systems, where the physiological mechanism that is responsible for its semiosis, or meaning-making, is the *functional cycle*, as described by Uexküll. This is where the temporal physiological processes of perception and action are meaningfully joined together, so as to realize the atemporal semiotic relations by which an organism can come to "know" and respond "intelligently" in its world.[19]

In order to understand organisms' behavior, then, the study of their umwelten is unavoidable. Accordingly, the concept of umwelt—as distinguished from the mere physicality of the surrounding environment (*Umgebung*)—has been refined through a whole series of qualifying concepts (see table 1.1; for another version of this table, see Magnus et al. 2026; see also Tønnessen et al., 2016).

Uexküll primarily described the umwelten of animals, both vertebrate and invertebrate, but he also applied the concept to unicellular organisms such as *Amoeba*, *Paramecium*, and *Didinium*. By extension, we can, at least hypothetically, suppose the existence of umwelt-like relations organizing at least some of the cellular interactions within a multicellular organism.

1.2.4 Living Beings as Modeling Systems

The evolutionary adaptation of auto-reproductive systems, as described by neo-Darwinian theory, can certainly occur even if such auto-reproductive systems have no knowledge of the environment, nor even any embodiment or sentience, such as in computer simulations.

Table 1.1
Concepts related to *umwelt* (modified: Magnus et al. 2026)

Term	Definition	References
Momentary umwelt	The space of the subjective present, experienced as an individual moment in time. This is the umwelt in the "now" or "specious present," when meaning is created and experienced.	Deely, 2001; Uexküll, 1928, p. 70
Core umwelt	The primary space of differences that can be detected by the organism	Brentari, 2015; Tønnessen, 2015
Mediated umwelt	Umwelt that uses remembered associations of objects for its categorizations	Brentari, 2015; Tønnessen, 2015
Conceptual umwelt	Umwelt that uses human language for introducing and indicating differences	Brentari, 2015; Tønnessen, 2015
Umwelt tunnel	An individual's subjective world throughout its ontogeny. This includes the panopoly of momentary umwelten throughout a life cycle.	Uexküll, 1928, p. 70
Functional cycle, functional circle	The cyclic connection between parts of the environment (object), their perception and interpretation by an organism (subject), and the subject's behavioral activity (action) toward the object. If this process is consecutive (diachronic), then we term it as a *cycle*, whereas if it is taking place in the phenomenal present of animal synchronicality, then we term it as a *circle*.	Brentari, 2015; Uexküll, 1928, p. 100ff
Umwelt transition	A lasting, systematic change within the life cycle of a living system, considered from an ontogenetic (individual), phylogenetic (population-, species-), or cultural perspective, from one typical appearance of its umwelt to another	Tønnessen, 2009, p. 49
Umwelt reversion	A process whereby an organism or a population of organisms returns to a previous state (configuration) of umwelt organization after experiencing a considerable umwelt alteration	Mäekivi, 2024, p. 254; Magnus and Mäekivi, 2023, p. 249
Semiotic alignment, cooperative action	Views communication as a dynamic process of joint semiosis, resulting in the ever-ongoing alignment and co-construction of the interactants' umwelten	Cerrone and Mäekivi, 2021; Goodwin, 2018; Ribó, 2019
Umwelt-net	The network of organisms' meaningful relations and communications in its local ecosystem	
Umweb	The network of aligned umwelten. While momentary and synchronic in its individual relations, the network of aligned umwelten as a whole is extended in time, being distributed and diachronic.	Kull, 2023; Machtyl, 2024
Umscape	A concept that generalizes the former concepts of *soundscape* (which includes *geophony*, *biophony*, and *anthropophony*); likewise, *sensescape*, *smellscape*, *tastescape*, and *visual landscape*	cf. Krause, 2012; Yong, 2022
Eco-field	Spatial configuration carrier of a specific meaning perceived when a specific living function is activated: the field of affordances	Farina and Belgrano, 2006
Affordances	The qualitative features that the environment offers the animal and what it provides or furnishes, either for good or ill; the properties and qualities of the environment *for* its inhabitants	Campbell et al., 2019; Gibson, 1979
Semiotic scaffolding	A structure that stabilizes and delimits organisms' decision-making in behavior and interpretation	Hoffmeyer, 2015a, 2015b
Semiotic niche	The world of sensory cues around the organism (or species) that the organism must necessarily interpret skillfully in order to live well	Hoffmeyer, 2008b

Term	Definition	References
Semiotic fitting	An agent's capacity for making and preserving local semiotic bonds; the agent's functional or communicational match with its surrounding	Kull, 2023
Consortium	A network of organisms at a place, persistently connected to each other by sign relations. Consortium is a local functional unit of an ecosystem, and the semiotic core structure of multispecies community.	Masing, 1981
Semiosphere	The whole space in which meaning is created and can exist, analogous to the concept of the biosphere	Lotman, 2005 [1984]; Hoffmeyer, 1996
Ecosemiosphere	The entire complex of semiosis in an ecosystem, including the involvement of human cultural semiosis and the cultural semiosphere	Maran, 2021

In contrast, although the umwelt is non-coterminus with the environment, it is still intimately related to it, much in the way that an organism's cognitive map has some relation to the spatial coordinates of the landscape. It is for this reason that Thomas Sebeok described umwelt as a "model" (Sebeok and Danesi, 2000). Yet because it is the organism's meaningful world, umwelt is not so much an internally represented "guiding pattern" that it *consults*—it is, rather, the omnipresent context and condition in which the entirety of its life is at every moment *experienced.* The meaning-making of semiosis is therefore best described as "representation with reference"—but not necessarily "depiction"—and *sentience* as "responsiveness to normative conditions," with normative conditions being determined *in the context of* the experiential umwelt.

Accordingly, there cannot be even the most primitive form of "knowing" without semiotic relations. The transformation of a pattern through mechanical coding does not give meaning to the pattern, as in our computer simulation example noted above—just as converting an alphabetic text into Morse code is not in itself an act of meaning-making. Such coding often does play an important role as a *component* in the semiotic process, however, to the extent that it conveys a pattern from one substance to another—for example, from nucleic acid sequences to amino acid sequences, or from the sequences of keystrokes to the sequences of letters on a computer screen, which may make that sequence available to genuinely interpretable (semiotic) processes on some higher level of sign-processing by a living agent.

Importantly, when an organism's acts of semiosis establish a relation, and that relation becomes scaffolded through learning (which in turn allows it to become repeatable), then the type of learning mechanism involved will be responsible for which relations can be habituated or fixed. There are two simple mechanisms of learning that are widely available among organisms. The simplest one is *imprinting*, which is based on the fixation of whatever objects fit within the organism's recognition window. The other is *association*, which is based on the fixation of the links between the objects and events that are spatiotemporally close within the umwelt.

Thus, the relations fixed by imprinting reflect relative likenesses, while the relations fixed by association reflect correlatedness. The former are called *iconic* relations, and the latter are called *indexical* relations. Indexical relations constitute the associations made between iconic relations (Deacon, 1997; Peirce, 1904), and together these two types of semiosis build sign relations and sign systems, mapping the likenesses and correlatedness of the environment, as experienced within the context of the organism's perceptual and action capabilities, into the choice-enabling arena of the organism's umwelt.

There exist, in addition, several more complex mechanisms of learning that arise from the establishment of iconic and indexical (and, in the case of humans, *symbolic*) relations, and these tend to be based on either *imitation*—seen almost exclusively in vertebrates—or *arbitrary naming*, which is the signature semiotic modality of humans. With the latter comes the capacity for purposively counterfactual sign manipulation, or *imagination*, which allows for the establishment of relations that are independent from the currently copresent features and physical restraints of the environment.

On this profoundly important difference between human and all other living beings' sign systems, we recommend all readers to consult Deacon (1997), which will also clarify what biosemioticians are and are *not* referring to when they use what might commonly be misperceived as "anthropomorphic terminology" such as *sign*, *meaning*, *habit*, *interpretation*, and *choice*. Anthroposemiosis—the species-specific human use of signs—enters almost not at all into the models of theoretical biology that we are proposing for the understanding of organismic organization and interaction.

1.2.5 Meaningful Communication: Sign Systems, but Not Yet Language

Indeed, an important root of the discipline of zoosemiotics is the study of nonverbal communication. Thomas Sebeok, who introduced the concept of zoosemiotics in the 1960s, first applied it to the study of animal communication, including ritual behavior and the aesthetic aspects of communication.

Since semiotics originated, among elsewhere, from general linguistics, the concept of language is sometimes used very broadly by its practitioners in fields outside of biosemiotics. Sebeok, however, made a clear distinction between human language and the sign systems of other animals. Human language has double articulation, conjunctions that are used universally, a heavy reliance on the generative meaning-making power of recursivity, and syntactic structure used as a scaffolding for the ordering and interpretation of communicative meaning—each of these are rare in the sign systems of other animals and have never yet been discovered to have occurred all together outside of human language.

The uniqueness of human language as our species' primary sign system for knowing and communicating was further strengthened by Deacon's work, which explains the distinctiveness of *symbolic reference* compared to the indexical and iconic relations out of which it emerges, but in some sense transcends and supplants, as our primary tool for knowing and acting in the world (Deacon, 1997). Our capacity for symbolic reference is the reason for our almost unlimited ability to freely build new signs and combine them—immune, as they are, from the real-world limitations and consequences by which all other species' sign systems are delimited. Accordingly, both the referential and the combinatorial potentials in the communication systems of other animals are remarkably more constrained, tying them closer, in a profound sense, to the immediacy of their perception-action joinings, and the consequences thereof, in real time.

As should be apparent from all of the foregoing discussion, then, both for human and for nonhuman organisms, something takes on the characteristics of a "sign" only when and because it is actively interpretated as such by some living system. Biologists often call the various molecules that move between cells to effect a change in the recipient's metabolism *signaling molecules*, and the various vibrations and radiations that influence the behavior of organisms are also often referred to as *signals*. But no molecule or oscillation pattern has the property of "sign-hood" (or signification) in itself. It is only the process of interpretation that assigns (or realizes) such things as "signs." Analogously, no series of figures on paper or across a screen is a "text" outside the reading process of a living agent that assigns those figures meanings.

Uexküll (1940: 1) spoke of the biology of his day's "meaning blindness" (*Bedeutungsblind*) and developed his umwelt theory to illuminate for scientists the important processes of meaning-making in living systems. Its lack of full adoption in our own time has resulted in a corollary "blindness to semiotic process" that conceptualizes "information" as being self-inherent in a given pattern—whether genetic, linguistic, perceptual, or digital.

Doing so, as we have seen above, entails a "fallacy of misplaced concreteness" that has kept biology unable to transcend its merely physical and chemical level of explanation and understanding. Communications theorist Sut Jhally puts it well when he reminds us that "*fetishism* consists of seeing the meaning of things as an inherent part of their physical existence—when in fact, that meaning is only created by their integration into a system of relations" (1987, p. 29).

Biosemiotics, like Uexküll's umwelt theory, was founded explicitly to bring these systems of relations into the explanatory framework of biology (Hoffmeyer, 1996; Kull et al., 2009).

1.2.6 Semiotic Evolution

As we have noted earlier, semiotic relations that are initially rather arbitrary—in the technical sense of not being the result of physical necessity—may turn into *habits* in the course of repetition and learning. Yet these habitual connections can further become so simplified and overdetermined that the possibility of choosing between alternative responses all but, or entirely, disappears—that is, they become instantiated as hard-wired biological connections, many of which can become heritable in the form of "fixed action patterns." Despite this gaining of automaticity, the relationships thus formed remain to carry through their lineage an arbitrary bond—that is, the bond that required semiosis and learning for its fixation.

Moreover, the fact that some features of organisms that are acquired through learning can be inherited has been demonstrated through the discovery of the mechanisms of epigenetic inheritance—which means that at least some innovations that are introduced by semiosis can and do influence evolution. An obvious modification that may be a result of organisms' choice (i.e., semiosis) is a functional change, and this may not be very rare when a whole population chooses a new habitat or starts to adopt a new source for food, which can lead to physiological changes and new habits, some of which can be inherited (Deacon, 2015; Jablonka and Ginsburg, 2022; Kull, 2023; Noble, 2021; Pagni and Simanke, 2021; Walsh, 2015; West-Eberhard, 2003).

The possible role of functional change as an initiator of evolutionary change was described by Gould and Vrba (1982) as *exaptation*, which was meant to help explain the *punctualist evolution of species* (Eldredge and Gould, 1972). A more explicit demonstration of the role of semiotic processes in the formation of species has been described through the *recognition concept of species* (Paterson, 1993; Sharov and Kull, 2022).

1.3 Concluding Perspectives

Theory loves unsolved problems, and a catalog of currently unsolved problems in a field can often be useful in predicting future research directions and trends. Biology, happily, always has an abundance of unsolved problems. For example, "one of the difficult aspects of biology, especially with respect to physical insights," write Longo and Montévil, "is the understanding of organisms, and by extension the implications of what it means for an object of knowledge to be a part of an organism" (2014, p. xi). Stated somewhat differently, the question is the very one that we have been examining here: What are the necessary and sufficient conditions for an organism to "know" something?

Propitiously, relevant assessments for the future have been expressed by Rosslenbroich (2023, p. 52), who writes, "Something like a 'new frontier of biology' (Henning and Scarfe 2013) seems to be developing, emphasizing the importance of organization, integration, and regulation. This also includes such areas as systems biology, epigenetics, biosemiotics, emergence theory, niche construction, extended evolutionary synthesis, and many more. These are still not mainstream but have a growing influence."

Our hope for the present volume, like the Waddington volumes that inspired it, is that it will likewise serve as yet another milestone on the path towards the development of a theoretical biology that can take into account the ineliminable semiotic processes of living beings.

Acknowledgments

We thank Gerd Müller for his support of our studies. Kalevi Kull's work on this project was also supported by the EU Horisont project 101084220, *Coevolvers*. Donald Favareau is grateful for the time that he was allowed to work on this text during his appointment as a visiting professor at Deep Springs College in California.

Notes

1. See Henning and Scarfe, 2013.

2. *Massachusetts Institute of Technology, Boston. Twenty-Ninth Annual Catalogue of the Officers and Students, with a Statement of the Courses of Instruction and a Register of the Alumni. 1893–1894* (p. 43), 1894, University Press, John Wilson and Son.

3. *Bryn Mawr College: Program. 1887–88* (p. 44), 1887, Sherman & Co.

4. *Program: Bryn Mawr College* (p. 149), 1898, Sherman & Co. Also, *Bryn Mawr College: Academic Year—1899–1900* (p. 175), 1899, Sherman & Co.

5. *Announcement of the University of Georgia with a Catalogue of the Officers and Students* (p. 29, 56), 1898, The Foote & Davies Company.

6. *Third Annual Catalog of Adelphi College and Thirtieth Annual Catalog of Adelphi Academy: 1898–1899* (p. 25), 1899, Brooklyn.

7. See Herring and Radick (2019); Nicholson and Gawne (2015).

8. "Science may be divided into three sorts. All that can fall within the compass of human understanding, being either, First, the nature of things, as they are in themselves, their relations, and their manner of operation: or, Secondly, that which man himself ought to do, as a rational and voluntary agent, for the attainment of any end, especially happiness: or, Thirdly, the ways and means whereby the knowledge of both the one and the other of these is attained and communicated; I think science may be divided properly into these three sorts: *Physica, Practica,* [and] *Semeiotike* . . . the [latter's] business whereof is to consider the nature of signs the mind makes use of for the understanding of things, or conveying its knowledge to others" (Locke, 1690, pp. 717–718).

9. It can be mentioned that there exist some general regularities in language that share commonalities with other sign systems (see, for example, Di Paolo et al., 2018; Semple et al., 2022), though it would be a critical mistake to use human language as the paradigm with which to understand any other species' sign systems, as argued extensively in Deacon (1997) and Deely (2002).

10. In semiotics, the structuralist approach is one that follows the Saussurean tradition. In biology, it has been applied, for example, by the Osaka group (Goodwin et al., 1989).

11. What follows are responses to a query from the authors made on July 11, 2024.

12. *Evolution, Ecology, History: The Crisis of Neo-Darwinism.*

13. Pers. comm. (Letter from CE to KK, July 11, 2024)

14. Biosemiotics is also greatly informed by the philosophy of biology, where some of these features of life are addressed using frameworks not commonly employed in mainstream biology journals.

15. As we have noted previously, "the use of the terms 'non-deterministic' and 'free choice' here does not imply that such choosing is not motivated, constrained, or free from the reliance on [the organism's immediately manifested biological condition] or any other influencing factors. It means simply, rather, that the choice is not the product of a deterministic algorithm or its equivalent" (Favareau and Kull, 2024, p. 45).

16. It should be noted here, too, that there exist several models of semiosis, among which the Saussurean binary model and the Peircean triadic model are most well known. In biology, the pentadic model of Morris and the circular model of Uexküll have received much attention, with the latter forming the basis on neuroscientist Joaquin Fuster's (2003) notion of the "cognit" network model of cortical representation. A comparative review of the various models of semiosis was published by Krampen (1997), and among more recent circular models of meaning-making, one could notice the one proposed by Édeline and Klinkenberg (2024), which employs the biologically inspired concepts of anasemiosis and catasemiosis. Needless to say, we are unable to delve into the discussion of all such models here, though the interested reader may wish to consult Pelkey (2022) for a state-of-the-art overview.

17. "We should also bear in mind Wilhelm von Humboldt's now oft-quoted aphorism that language involves [the] infinite use of finite means" (Chomsky, 2016, p. 8).

18. See also Favareau (2015).

19. In this regard, the reader is directed to Augustyn (2023) for an informative comparison between Jakob von Uexküll's *Funktionskreis*, Viktor von Weizsäcker's *Gestaltkreis*, and Thure von Uexküll's *Situationskreis* models of perception-action cycles.

References

Augustyn, P. 2023. Funktionskreis, Gestaltkreis, and Situationskreis in the context of integrated medicine. *Semiotica, 253*, 23–50.

Auletta, G. 2011. *Cognitive biology: Dealing with information from bacteria to minds.* Oxford University Press.

Barbieri, M. 1985. *The semantic theory of evolution.* Harwood Academic.

Bateson, G. 1972. *Steps to an ecology of mind: Collected essays in anthropology, psychiatry, evolution, and epistemology.* Chandler Pub. Co.

Bateson, G. 1979. *Mind and nature: A necessary unity.* Dutton Press.

Bechtel, W., and Bich, L. 2021. Grounding cognition: Heterarchical control mechanisms in biology. *Philosophical Transactions of the Royal Society B, 376*, 20190751, 1–11.

Bentele, G. 1984. *Zeichen und Entwicklung: Vorüberlegungen zu einer genetischen Semiotik.* G. Narr.

Bertalanffy, L. von 1932. *Theoretische biologie.* Borntraeger.

Bertalanffy, L. von 1942. *Theoretische biologie.* Borntraeger.

Bigelow, R. P. 1896. *Syllabus of lectures in theoretical biology: Prepared for the use of students in the Massachusetts Institute of Technology.* Massachusetts Institute of Technology.

Böhm, A., Reiners-Selbach, S., Baedke, J., Fábregas-Tejeda, A., Straetmanns, V., and Nicholson, D. J. 2022. What was theoretical biology? A topic-modelling analysis of a multilingual corpus of monographs and journals, 1901–1971. *8. Jahrestagung des verbands digital humanities im deutschsprachigen raum 2022: Potsdam, Germany.*

Brentari, C. 2015. *Jakob von Uexküll: The discovery of the umwelt between biosemiotics and theoretical biology.* Springer.

Campbell, C., Olteanu, A., and Kull, K. 2019. Learning and knowing as semiosis: Extending the conceptual apparatus of semiotics. *Sign Systems Studies, 47*(3–4), 352–381.

Cerrone, M., and Mäekivi, N. 2021. A zoosemiotic approach to the transactional model of communication. *Semiotica, 242*, 39–62.

Chomsky, N. 2016. *What kind of creatures are we?* Columbia University Press.

Deacon, T. W. 1997. *The symbolic species: The co-evolution of language and the brain*. W.W. Norton.

Deacon, T. W. 2011. *Incomplete nature: How mind emerged from matter*. W.W. Norton.

Deacon, T. W. 2015. Steps to a science of biosemiotics. *Green Letters: Studies in Ecocriticism, 19*(3), 293–311.

Deacon, T. 2021. How molecules became signs. *Biosemiotics, 14*(3), 537–559.

Deely, J. 2001. Umwelt. *Semiotica, 134*(1–4), 125–135.

Deely, J. 2002. *What distinguishes human understanding?* St. Augustine's Press.

Di Paolo, E. A., Cuffari, E. C., and De Jaegher, H. 2018. *Linguistic bodies: The continuity between life and language*. The MIT Press.

Eco, U. 1999. *Kant and the platypus: Essays on language and cognition*. Houghtom Mifflin.

Édeline, F., and Klinkenberg, J.-M. 2024. From the meaning to the world: The catasemiosis. *Sign Systems Studies, 52*(3–4), 485–517.

Eldredge, N., and Gould, S. J. 1972. Punctuated equilibria: An alternative to phyletic gradualism. In T. J. M. Schopf (Ed.), *Models in paleobiology* (pp. 82–115). Freeman, Cooper.

Emmeche, C. 2015. Semiotic scaffolding of the social self in reflexivity and friendship. *Biosemiotics, 8*(2), 275–289.

Farina, A., and Belgrano, A. 2006. The eco-field hypothesis: Toward a cognitive landscape. *Landscape Ecology, 21*(1), 5–17.

Favareau, D. 2010. *Essential readings in biosemiotics: Anthology and commentary*. Springer.

Favareau, D. 2015. Creation of the relevant next: How living systems capture the power of the adjacent possible through sign use. *Progress in Biophysics and Molecular Biology, 119*(3), 588–601.

Favareau, D., and Kull, K. 2024. Pathways to the understanding of signs and meanings in the biosphere: Historical and contemporary perspectives. In A. A. Sharov and G. E. Mikhailovsky (Eds.), *Pathways to the origin and evolution of meanings in the universe* (pp. 27–53). Wiley-Scrivener.

Florkin, M. 1974. Concepts of molecular biosemiotics and of molecular evolution. *Comprehensive Biochemistry, 29A*, 1–124.

Frisch, K. von 1957. *Erinnerungen eines biologen*. Springer-Verlag.

Fuster, J. 2003. *Cortex and mind: Unifying cognition*. Oxford University Press.

Gibson, J. J. 1979. *The ecological approach to visual perception*. Houghton Mifflin.

Goodall, J. 1990. *Through a window: 30 years observing Gombe chimpanzees*. Houghton Mifflin.

Goodwin, B. C., Sibatani, A., and Webster, G. C. (Eds.) 1989. *Dynamic Structures in Biology*. Edinburgh University Press.

Goodwin, C. 2018. *Co-operative action*. Cambridge University Press.

Gould, S. J., and Vrba, E. S. 1982. Exaptation—A missing term in the science of form. *Paleobiology, 8*(1), 4–15.

Griffin, D. 1962. *Animal structure and function*. Holt, Rinehart and Winston.

Hediger, H. 1974. Communication between man and animal. *Image Roche, 62*, 27–40.

Hemerik, L., and Reydon, T. A. C. 2005. The history of *Acta Biotheoretica*: The world's oldest theoretical biology journal. *European Society for Mathematical and Theoretical Biology Communications, 7*, 13–17.

Henning, B. G., and Scarfe, A. C. (Eds.). 2013. *Beyond mechanism: Putting life back into biology*. Lexington Books.

Herring, E., and Radick, G. 2019. Emergence in biology: From organicism to systems biology. In S. Gibb, R. F. Hendry, and T. Lancaster (Eds.), *The Routledge handbook of emergence* (pp. 352–362). Routledge.

Hoffmeyer, J. 1996. *Signs of meaning in the universe*. Indiana University Press.

Hoffmeyer, J. 2008a. *Biosemiotics: An examination into the signs of life and the life of signs*. Scranton University Press.

Hoffmeyer, J. 2008b. The semiotic niche. *Journal of Mediterranean Ecology, 9*, 5–30.

Hoffmeyer, J. 2015a. Introduction: Semiotic scaffolding. *Biosemiotics, 8*(2), 153–158.

Hoffmeyer, J. 2015b. Semiotic scaffolding: A unitary principle gluing life and culture together. *Green Letters: Studies in Ecocriticism, 19*(3), 243–254.

Houser, N. 1996. Semiotic as cognitive science. *Cruzeiro Semiótico, 22*(25), 139–150.

Jablonka, E., and Ginsburg, S. 2022. Learning and the evolution of conscious agents. *Biosemiotics, 15*(3), 401–437.

Jablonka, E., and Lamb, M. J. 2005. *Evolution in four dimensions: Genetic, epigenetic, behavioral, and symbolic variation in the history of life*. MIT Press.

Jhally, S. 1987. *The codes of advertising*. Routledge.

Krampen, M. 1997. Models of semiosis. In R. Posner, K. Robering, and T. A. Sebeok (Eds.), *Semiotics: A handbook on the sign-theoretic foundations of nature and culture* (Vol. 1) (pp. 247–287). Walter de Gruyter.

Krause, B. 2012. *The great animal orchestra: Finding the origins of music in the world's wild places*. Little Brown, Hachette.

Kull, K. 1993. Semiotic paradigm in theoretical biology. In K. Kull and T. Tiivel (Eds.), *Lectures in theoretical biology: The second stage* (pp. 52–62). Estonian Academy of Sciences.

Kull, K. 2012. Scaffolding. In: D. Favareau, P. Cobley, and K. Kull (Eds.), *A more developed sign: Interpreting the work of Jesper Hoffmeyer* (pp.228–229). Tartu University Press.

Kull, K. 2023. Further considerations on semiosis in evolution: Arbitrarity plus semiotic fitting, and/or mutability plus natural selection. *Sign Systems Studies, 51*(1), 171–194.

Kull, K, Deacon, T., Emmeche, C., Hoffmeyer, J., and Stjernfelt, F. 2009. Theses on biosemiotics: Prolegomena to a theoretical biology. *Biological Theory, 4*(2), 167–173.

Kull, K., and Favareau, D. 2022. Semiotics in general biology. In J. Pelkey and S. W. Matthews (Eds.), *Bloomsbury semiotics, vol. 2: Semiotics in the natural and technical sciences* (pp. 35–56). Bloomsbury Academic.

Kull, K., and Favareau, D. 2023. Neurosemiotics: A brief history of its development and key concerns. In A. M. García and A. Ibáñez (Eds.), *The Routledge handbook of semiosis and the brain* (pp. 13–29). Routledge.

Laubichler, M. D. 2007. The specter of the past: What the history of theoretical biology means today. *Biological Theory, 2*(2), 131–133.

Locke, J. 1690. *An essay concerning humane understanding*. Eliz. Holt.

Longo, G., and Montévil, M. 2014. *Perspectives on organisms: Biological time, symmetries and singularities*. Springer.

Lorenz, K. 1970. *Studies in animal and human behavior*. Harvard University Press.

Lotman, J. 2005. On the semiosphere. *Sign Systems Studies, 33*(1), 215–239. (Original work published 1984)

Machtyl, K. 2024. Umwelt and time: Extending the humanistic view on temporality through umwelt theory. *Sign Systems Studies, 52*(3–4), 344–370.

Mäekivi, N. 2024. Organisms as agents in zoosemiotic perspective: The case of umwelt reversion. In J. Švorcová (Ed.), *Organismal agency: Biological concepts and their philosophical foundations* (pp. 245–261). Springer.

Magnus, R., and Mäekivi, N. 2023. Ecosemiotic analysis of species reintroduction: The case of European Mink (*Mustela lutreola*) in Estonia. *Biosemiotics, 16*(2), 239–258.

Magnus, R., Mäekivi, N., Päll, L., Maran, T., Remm, T., Kull, K. 2026. Umwelt-based analysis of multispecies places: Guidelines for application. *Biosemiotics*, 19 (in print).

Maran, T. 2021. The ecosemiosphere is a grounded semiosphere: A Lotmanian conceptualization of cultural-ecological systems. *Biosemiotics, 14*(2), 519–530.

Maran, T. 2023. Umwelt collapse: The loss of umwelt-ecosystem integration. *Biosemiotics, 16*(3), 479–487.

Markoš, A. 2002. *Readers of the book of life: Contextualizing developmental evolutionary biology*. Oxford University Press.

Marler, P. 1961. The logical analysis of animal communication. *Journal of Theoretical Biology, 1*, 295–317.

Masing, V. 1981. Consortia as elements of the functional structure of biocenoses. In L. Laasimer (Ed.), *Anthropogenous changes in the plant cover of Estonia* (pp. 64–76). Institute of Zoology and Botany.

Mivart, St. G. 1871. *On the genesis of species*. MacMillan and Co.

Morris, C. 1925. *Symbolism and reality: A study in the nature of mind*. University of Chicago.

Morris, C. 1937. *Logical positivism, pragmatism and scientific empiricism*. Hermann et Cie.

Nicholson, D. J., and Gawne, R. 2015. Neither logical empiricism nor vitalism, but organicism: What the philosophy of biology was. *History and Philosophy of the Life Sciences, 37*(4), 345–381.

Noble, D. 2021. The illusions of the Modern Synthesis. *Biosemiotics, 14*(1), 5–24.

Pagni, E., and Simanke, R. T. (Eds.). 2021. *Biosemiotics and evolution: The natural foundations of meaning and symbolism*. Springer.

Paterson, H. E. H. 1993. *Evolution and the recognition concept of species*. The J. Hopkins University Press.

Pattee, H. 1969. How does a molecule become a message? *Developmental Biology Supplement, 3*, 1–16.

Paula Gruithuisen, F. von 1823. *Ueber naturforschung*. J. Wolffische Buchhandlung.

Peirce, C. S. 1932. Ground. Object and interpretant. In C. S. Peirce, C. Hartshorne, and P. Weiss (Eds.), *The collected papers of Charles Sanders Peirce, vol. 2: Elements of logic*. Harvard University Press. (Original work published 1897)

Peirce, C. S. 1992. New Eelements (kaina stoicheia). In The Peirce Edition Project (Eds.), *The essential Peirce, vol 2: Selected philosophical writings (1893–1913)* (pp. 300–325). Indiana University Press. (Original work published 1904)

Pelkey, J., and Walsh Matthews, S. (Eds.). *Bloomsbury semiotics, vol. 2: Semiotics in the natural and technical sciences*. Bloomsbury Academic.

Peterson, E. L. 2016. *The life organic: The theoretical biology club and the roots of epigenetics*. University of Pittsburgh Press.

Pigliucci, M., and Müller, G. B. (Eds.) 2010. *Evolution: The extended synthesis*. The MIT Press.

Popper, K. 1999. *All life is problem solving*. Routledge. (Original work published 1994)

Popper, K. 2014. A world without natural selection but with problem solving. In H.-J. Niemann (Ed.), *Karl Popper and the two new secrets of life: Including Karl Popper's Medawar lecture 1986 and three related texts* (pp. 132–133). Mohr Siebeck.

Prodi, G. 2021. *The material bases of meaning*. University of Tartu Press. (Original work published 1977)

Rashevsky, N. 1938. *Mathematical biophysics: Physico-mathematical foundations of biology*. The University of Chicago Press.

Reinke, J. 1901. *Einleitung in die theoretische Biologie*. Verlag von Gebrüder Paetel.

Reinke, J. 1911. *Einleitung in die theoretische Biologie*. 2te Aufl. Verlag von Gebrüder Paetel.

Reydon, T. A. C., Dullemeijer, P., and Hemerik, L. 2005. The history of Acta Biotheoretica and the nature of theoretical biology. In T. A. C. Reydon and L. Hemerik (Eds.), *Current themes in theoretical biology: A Dutch perspective* (pp. 1–8). Springer.

Ribó, I. 2019. Semiotic alignment: Towards a dialogical model of interspecific communication. *Semiotica, 230*, 247–274.

Rocha, L. 2001. Evolution with material symbol systems. *Biosystems, 60*, 95–121.

Rosen, R. 1988. Constraints and the origin of life. In K. Kull and T. Tiivel (Eds.), *Lectures in theoretical biology* (pp. 22–26). Valgus.

Rosen, R. 1991. *Life itself: A comprehensive inquiry into the nature, origin, and fabrication of life*. Columbia University Press.

Rosen, R. 1999. *Essays on life itself*. Columbia University Press.

Rosslenbroich, B. 2023. *Properties of life: Toward a theory of organismic biology*. The MIT Press.

Rothschild, F. S. 1968. Concepts and methods of biosemiotic. *Scripta Hierosolymitana, 20*, 163–194.

Sebeok, T. A. 1963. Communication among social bees; porpoises and sonar; man and dolphin. *Language, 39*, 448–466.

Sebeok, T. A. 1998. The Estonian connection. *Sign Systems Studies, 26*, 20–41.

Sebeok, T. A. 2001. Biosemiotics: Its roots, proliferation, and prospects. *Semiotica, 134*(1), 18.

Sebeok, T. A., and Danesi, M. 2000. *The forms of meaning: Modeling systems theory and semiotic analysis*. De Gruyter Mouton.

Sebeok, T. A., and Umiker-Sebeok, J. (Eds.). 1992. *Biosemiotics: The semiotic web 1991*. Mouton de Gruyter.

Semple, S., Ferrer-i-Cancho, R., and Gustison, M. L. 2022. Linguistic laws in biology. *Trends in Ecology and Evolution, 37*(1), 53–66.

Sercarz, E., Celada, F., Mitchison, N. A., and Tada, T. (Eds.). 1988. *The semiotics of cellular communication in the immune system*. Springer.

Sériot, P. 2014. *Structure and the whole: East, west and non-Darwinian biology in the origins of structural linguistics*. De Gruyter Mouton.

Shaw, A. K. 2022. Diverse perspectives from diverse scholars are vital for theoretical biology. *Theoretical Ecology, 15*, 143–146.

Sharov, A. A., and Kull, K. 2022. Evolution and semiosis. In J. Pelkey (Ed.), *Bloomsbury semiotics, vol. 1: History and semiosis* (pp. 149–168). Bloomsbury Academic.

Sharov, A., and Tønnessen, M. 2021. *Semiotic agency: Science beyond mechanism*. Dordrecht.

Sinha, C. 2006. Epigenetics, semiotics, and the mysteries of the organism. *Biological Theory, 1*(2), 112–115.

Sonea, S. 1992. Half of the living world was unable to communicate for about one billion years. In T. A. Sebeok and J. Umiker-Sebeok (Eds.), *Biosemiotics: The semiotic web 1991* (pp. 375–392). Mouton de Gruyter.

Stepanov, Y. 1971. *Semiotika.* Nauka.

Thom, R. 1975. From animal to man: Thought and language. In R. Thom, *Structural stability and morphogenesis: An outline of a general theory of models* (pp. 297–327). W. A. Benjamin.

Tinbergen, N. 1972. *The animal in its world.* Harvard University Press.

Tommasi, L., Peterson, M. A., and Nadel, L. (Eds.). 2009. *Cognitive biology: Evolutionary and developmental perspectives on mind, brain, and behavior.* The MIT Press.

Tønnessen, M. 2009. Umwelt transitions: Uexküll and environmental change. *Biosemiotics, 2*(1), 47–64.

Tønnessen, M. 2015. Umwelt and language. In E. Velmezova, K. Kull, and S. Cowley (Eds.), *Biosemiotic perspectives on language and linguistics* (pp. 77–96). Springer.

Tønnessen, M., Magnus, R., and Brentari, C. 2016. The biosemiotic glossary project: Umwelt. *Biosemiotics, 9*(1), 129–149.

Uexküll, J. von 1928. *Theoretische biologie.* 2te Aufl. Verlag von Julius Springer.

Uexküll, J. von 1932. Die Umwelt des Hundes. *Zeitschrift für Hundeforschung, 2*(5–6), 157–170.

Uexküll, J. von 1940. *Bedeutungslehre.* Verlag von J. A. Barth. [Translation: von Uexküll, Jakob 1982. The theory of meaning. *Semiotica, 42*(1), 25–82.]

Waddington, C. H. (Ed.). 1968. *Towards a theoretical biology. 1. Prolegomena.* Aldine.

Waddington, C. H. (Ed.). 1969. *Towards a theoretical biology. 2. Sketches.* Edinburgh University Press.

Waddington, C. H. (Ed.). 1970. *Towards a theoretical biology. 3. Drafts.* Edinburgh University Press.

Waddington, C. H. (Ed.). 1972. *Towards a theoretical biology. 4. Essays.* Edinburgh University Press.

Wallace, A. R. 1891. *Natural selection and tropical nature: Essays on descriptive and theoretical biology.* MacMillan and Co.

Walsh, D. M. 2015. *Organisms, agency, and evolution.* Cambridge University Press.

West-Eberhard, M. J. 2003. *Developmental plasticity and evolution.* Oxford University Press.

Wissemann, V. 2006. Johannes Reinke (1839–1931) and his "Dominanten" theory—An early concept of gene regulation and morphogenesis. *Theory in Biosciences, 124*, 397–400.

Wright, C. 1871. *Darwinism: Being an examination of Mr. St. George Mivart's 'Genesis of Species'.* John Murray.

Yong, E. 2022. *An immense world.* Random House.

2 "The Pattern That Connects [Corrects]": Bateson, Waddington, the IUBS, and the Antireductionist Heart of Theoretical Biology

Erik L. Peterson

2.1 Prelude: A Mystery—the First IUBS Meeting at Bellagio, Italy (1966) and What Followed

This paper begins with a mystery that, on its face, seems unrelated to the rest of the essay—I will attempt to unwind it in a "Coda" at the end of the article.

Why did one prominent biologist, Ernst Mayr, marginalize the work of another prominent biologist, C. H. Waddington, when the two shared so much in common regarding the objectives of biological theory? Our mystery deepens when we learn that Conrad Hal Waddington (1905–1975; known to his scientific friends as "Wad") and Ernst Mayr (1904–2005) *partnered* to launch the International Union of Biological Sciences (IUBS) meetings at Bellagio, Italy, that serve as one of the inspirations for the book of which this essay is a chapter. It is my contention that the solution to this Mayr-Waddington mystery also serves to illuminate other fault lines in the history of biology, including the uneven adoption of the research program of biosemiotics.

The origins of this mystery originated in a few nights in the spring of 1965. Sitting at his desk, Waddington jotted down several puzzles in "biological philosophy" or "the philosophy of biology," as he deemed it. With a background in Cambridge University geology, prestigious Cambridge fellowships in philosophy, and a founding membership in the Theoretical Biology Club (TBC)—and as the president of the International Union of Biological Sciences (IUBS) and the academic chair of a practically minded genetics institute at Edinburgh—Wad had been working in this interstitial space for decades. And he sensed there was something missing, some set of questions that no one was attempting to answer. He realized he should organize a symposium or workshop to better define these bio-theoretical puzzles and to investigate whether any persuasive answers to those questions had been generated recently. So, he reached out to an old friend and sparring partner, Harvard ornithologist Ernst Mayr, already becoming well known as a defender of the neo-Darwinian Synthesis.[1]

This IUBS workshop, Wad explained, should be freewheeling and interdisciplinary, very much like his old Cambridge TBC meetings with the Needhams, "Sage" Bernal, Dorothy Wrinch, J. H. Woodger, J. B. S. Haldane, and Karl Popper.[2] They restricted those meetings—held back in the 1930s when Wad worked in the wake of Spemann's "organiser" project—to a particular kind of scholar. One needed biological skill, sure. But also, philosophical heft.

There was a real danger to these sorts of pairings, of course. Biologists, Waddington complained, tossed out "usual trivialities"; philosophers scrawled "imposing para-mathematical symbols." Little progress emerged from these sorts of meetings, unless attendees were managed carefully.[3]

Mayr responded to Waddington's letter enthusiastically from his New Hampshire farm. "Physicists" and "positivists" had dominated philosophy of biology for too long with their "semantic free-for-alls." The key to a good conference—and indeed a proper understanding of the life sciences—started by understanding the exceptional subject matter: Organisms were "stored historically tested programmed information." The existence of "highly complex historically evolved systems," like organisms, insisted Mayr, added "such a dimension of complexity to the simple systems normally studied by physicists and chemists" that philosophers and biologists alike should promote a unique "philosophy of biology" free from "that horrible tendency to Indulge [*sic*] in a 'me too' philosophy in which we tried to demonstrate that indeed life can be explained in terms of physics and chemistry and therefore that biologists are perfectly respectable scientists." The conference should be, Mayr emphasized, free from physics-drenched scientists and, frankly, most varieties of philosophers.[4]

They populated an initial list of invitees with eleven Americans; Wad added six from the UK and three from continental Europe. Wad expressed the IUBS's desire to keep the location "international" and, thus, locate the conference outside of the US or Europe.[5] But conference travel funding was tight. Wad called in a favor from the Rockefeller Institute and secured the Villa Serbelloni in Bellagio, Italy, right on Lake Como, for late August through early September 1966. The location was perfect and accommodated precisely twenty people. Significant others could not join. Also absent would be—Wad admitted with just a hint of satisfaction—"a lot of molecular biologists, physicists and one of the two philosophers."[6] And with that, the IUBS symposia that resulted in major papers and discussions collected as *Towards a Theoretical Biology* was off and running.

For Waddington, the IUBS symposia at Villa Serbelloni represented a second "golden age" of his career. The first had been his epigenetics work with the Needhams and Jean Brachett in the 1930s and 1940s culminating in the groundbreaking paper "The Epigenotype" at the advent of World War II.[7] Over the years of four IUBS symposia in the late 1960s, Waddington stitched together a large and fruitful network of biologists, mathematicians, physicists, and even philosophers.

Mayr, by contrast, left the first Villa Serbelloni disappointed and never returned to the project he helped birth. "The Bellagio meeting did not develop in the direction of my own interests," he complained to Wad months after he had returned to the US.[8] He and Wad never again collaborated. After Wad's untimely death in 1975 at the age of sixty-nine, Mayr still viewed Waddington's work with skepticism. His major conferences on the Modern Synthesis in the 1970s, and the subsequent book published from these conferences in 1980, excluded mentions of Waddington—which participant Viktor Hamburger, the renown developmental biologist, tacitly scolded Mayr for so doing. Hamburger labeled Waddington's work a key part of the mysteriously "Missing Chapter" of Mayr's evolutionary theory.[9] We will return to address this mystery again at the "Coda" to this essay.

2.2 Introduction

The present chapter resides in a book commemorating *Towards a Theoretical Biology* (1968–72) and the IUBS conferences that gave rise to the papers therein. It is my contention that the key to understanding the outcome of those IUBS meetings—including the transition toward biosemiotics derived from those papers—can be found not only in Waddington's own proclivities but also in those of his extensive intellectual network. In other words, by looking *beyond* Waddington to the work of one of his closest friends, Gregory Bateson, we can uncover not only the mystery of the dispute between Mayr and Waddington but also a major fault line in theoretical biology itself, centering around what Waddington deemed the "Conventional Wisdom of the Dominant Group" or "COWDUNG." The first part of this essay describes the relationship between Bateson and Waddington, often mediated by anthropologist Margaret Mead, that amplified Waddington's own antireductionist stance. The second part classifies the contributions to the IUBS that eventually became *Towards a Theoretical Biology* in light of that relationship. My sources include not only the published works of Waddington, Mayr, Bateson, and the IUBS meetings, but also the private correspondence that allows us to see otherwise unacknowledged intentions, motivating values, and hidden networks of influence that shape not only the publications themselves, but also who gets to attend conferences and what of their contributions becomes preserved "science."

2.3 Part 1: An Antireductionist Network

2.3.1 Gregory Bateson (1904–1980) and Margaret Mead (1901–1978): The Pattern That Connects

One of Margaret Mead's treasured memories was observing Wad argue back and forth with his old friend (and her new beau) Gregory Bateson on the Irish coast during a 1934 holiday. Both Oxbridge educated, both expansive minds interested in science, philosophy, abstract painting, communication, psychology, and everything in between, Waddington and Bateson seemed to finish each other's sentences: "One minute from embryology, the next from geology, the next from anthropology, back and forth, very freely, so that the illustrations from one spot illuminated, corrected, and expanded the one from another."[10]

What they wrangled over—what all three of them, Mead included, mulled over continually in the 1930s and beyond—was *patternedness*. Waddington saw it in the whorl of ammonite shells; he was a geologist first. Bateson and Mead saw it in their anthropological work in New Guinea. Language and behavior followed patterns, too. Even though each would drift apart into separate interests after World War II, Mead, Bateson, and Waddington stayed in contact and focused for the rest of their careers on patterns of patterns, metapatterns, or "the pattern that connects," as Bateson used to say. Even while Waddington lay in a hospital bed after suffering another heart attack in the early 1970s, he and Mead wrote back and forth about patterns in personalities, and how those patterns are themselves embedded within the patterns of whole cultures.[11]

Gregory Bateson grew up the third, and last surviving, son of William and C. Beatrice Durham Bateson. William Bateson was Britain's original champion of Mendelism at the turn of the twentieth century and partnered with Wilhelm Johannsen to define "genes," and then coined the term "genetics" in 1906. William and Beatrice named their son Gregory after Gregor Mendel.

Like his father, Gregory cut his scientific teeth at St. John's College, Cambridge University. After officially affiliating with the Cambridge Zoology Laboratory as a first-year Natural Sciences Tripos student in 1922, Gregory worked long laborious hours at taxonomic classification under J. Stanley Gardiner. For diversion, he joined an impromptu Biological Tea Club with fellow undergraduates G. Evelyn Hutchinson and Grace Pickford (both later prominent Yale University biologists). The group became instrumental in connecting Bateson with Waddington. Bateson's skills as a zoologist progressed rapidly enough to encourage other Cambridge life scientists to offer him more prestigious research opportunities. After completing the difficult Tripos and graduating in 1924, Gregory now stood, officially, as a biologist of recognizable expertise. Satisfied at his son's progress, William Bateson initiated collaborative work with Gregory on genetics in 1924.[12]

Their joint project before William Bateson died appeared in 1926 as "On Certain Aberrations of the Red-Legged Partridge *Alectoris rufa* and *saxatilis*." The Batesons detailed strange sequences of feather color banding observed in several partridges gathered from across Britain and the continent. These variant partridges resembled, the Batesons felt sure, patterns of coloration exhibited by the stripes-to-solid patterns of quagga (a recently extinct relative of the zebra)—and, tellingly, in precisely the same way that the quagga differed from zebras. Though the Batesons largely refused to speculate on the nature of the underlying genotypic mechanics leading to the unusual and precise variations, they were certain that it could not be explained in the typical American Morgan-school fashion of point mutations in genes. Instead, they were led to the work of P. W. Whiting that attributed "pigment forming metabolic activity" in moths and felines to actual "waves" propagated during development.[13] This explanation bore all the marks of William Bateson's holistic "vibratory hypothesis" of heredity first formulated in 1891. Though it feels just as out of place in a mid-1920s genetics journal as it did when first proposed three decades earlier, the Bateson vibratory hypothesis was, in retrospect, scientifically prescient.[14]

For Gregory, it was a first serious grappling with the problems of biological fields and of patternedness in an organic system. Though he rapidly moved out of genetics (at the very moment he was convincing Waddington to leave geology *for* genetics), Gregory Bateson retained twin obsessions, shared with his father: first, that finding the secret behind organized patternedness was the secret to the life sciences, and, second, that the mid-century neo-Darwinian Synthesis provided only the most anemic answers to the evolution of patternedness.

Through at times arduous ethnographic fieldwork in the 1920s, Mead and Bateson both gained a deep appreciation for patterns of meaning. Each started this work separately—Mead with American anthropological scion, Franz Boas, and Bateson through A. C. Haddon at Cambridge and in contact with Malinowski and Radcliffe Brown. Both Bateson and Mead ended up in New Guinea, working among the Sepik peoples of the northeast New Guinean coast. Mead and Bateson married soon after—he was her third and final husband, and she the first of his three wives—and continued their ethnographic work in the Southeast Pacific right through the end of the 1930s. This only intensified their search for patterns and metapatterns among and between people groups in the Southeastern Pacific archipelagos.

The coming of war drove Bateson and Mead back to the US. Mead began commuting from New York to Washington, DC, in 1942 to work alongside her not-so-secret love interest and academic mentor, Ruth Benedict, in the National Resource Council. In 1943, Bateson joined the Office of Strategic Services, the predecessor of the CIA, and found himself shipped off to Southeast Asia and India to counteract Japanese propaganda, sometimes aboard naval warships. For almost two years, Bateson attempted to inject his knowledge of communications patterns gleaned from ethnographic work and published as *Naven* (1936) into American intelligence work. But to no avail. While Mead flourished in DC, Bateson floundered in South Asia. He returned discouraged with what he saw as a kind of cavalier, know-nothing militarism mapped onto intransigent American racism perpetrated against non-white allies.[15]

Before formally joining the American war effort in 1942, Bateson and Mead conceived of the Institute for Intercultural Studies, and developmental psychologist Lawrence K. Frank joined them. Frank had moved from a position as grant officer at the Rockefeller Foundation to director of the Josiah Macy Jr. Foundation in New York, and he soon became a hub for numerous scholarly groups comprised of international luminaries, including those escaping the Nazi menace. In 1942, Frank, Mead, Bateson, and a handful of others birthed the Macy conferences that, by 1946, germinated groundbreaking work in closed, self-corrective circuits, or "feedback loops"—soon dubbed "cybernetics" by American neurophysiologist Norbert Wiener.

At the Macy cybernetics meetings, in the Institute for Intercultural Studies, and in their frequent interactions with intellectuals in the orbit of Lawrence Frank—some of whom were also in Waddington's UK and US circles—Mead and Bateson returned again and again to patterns and patterns of patterns. Only now what they traced moving in and through these patterns was not vague "culture" but precise "information." Bateson could look back to his work on partridges with his father and see how "information" could orient patterns and metapatterns across a bio-field—a developing plant or animal, for instance. Cybernetics, from Bateson's point of view in the late 1940s and early 1950s, really could be the key to a unified theory of organic life.

2.3.2 Waddington: Developing a Whiteheadian Evolutionist

After his conversations with Gregory Bateson in the 1920s, Waddington jumped feet first into the world of genetics. Yet he maintained a strong interest in the philosophy of biology—before that was even a recognized field—which rendered him an outlier among all but a handful of his peers. Wad's closest two scientific confidants in the late 1920s and through the 1930s were Joseph Needham and Joseph Woodger, both bench scientists who likewise delved into the philosophical. All three imbibed Alfred North Whitehead's philosophy as embodied in the *Principia Mathematica* (1910, 1912, 1913), *The Concept of Nature* (1920), *Science and the Modern World* (1925), *Symbolism: Its Meaning and Effect* (1927), and, especially, *Process and Reality* (1929)—which nudged Woodger into the conceptual wrangling with Needham that eventually resulted in Cambridge's Theoretical Biology Club (TBC).[16] While he spent the late 1930s in T. H. Morgan's "Fly Room" alongside Cal Bridges, Albert Sturtevant, and Theodosius Dobzhansky, Waddington never lost his dedication to Whiteheadian process philosophy.[17]

Within Whitehead's work, members of the TBC distilled three important concepts: the "chreod," "homeorhesis," and the "morphogenetic field." None were original to Whitehead.

Even the terminology was developed by others—Waddington himself coined "chreod" (originally spelled "creode") in his 1957 milestone *The Strategy of the Genes*. Nevertheless, the heart of Whiteheadian philosophy of science pulsed around this conceptual triad:

(i) *Chreod*: there are no "things" but events with relative stability on trajectories or paths.

(ii) *Homeorhesis*: deviations from these paths are not chaotic but tend to resolve themselves into new paths, even occasionally with the same developmental trajectory.

(iii) *Morphogenetic fields*: a distributed whole ordered in such a way that the position of a constituent in one region of the whole bears a definite relation to the position of a constituent in other regions of the whole.[18]

Whitehead situated this conceptual triad within a broader epistemological system that resembled other intellectuals of that time. For instance, C. S. Peirce's triadic semiotics. Whitehead interacted with Charlie Dunbar (C. D.) Broad, an English philosopher with interests in Peircian semiotics. Whitehead's epistemology could be glimpsed in, for instance, this dense quote from *The Concept of Nature* (1920):

> Our knowledge of the peculiar characters of different events depends upon our power of comparison. I call the exercise of this factor in our knowledge "recognition," and the requisite sense-awareness of the comparable characters I call "sense-recognition." Recognition and abstraction essentially involve each other. Each of them exhibits an entity for knowledge which is less than the concrete fact, but is a real factor in that fact. The most concrete fact capable of separate discrimination is the event. We cannot abstract without recognition, and we cannot recognise without abstraction. Perception involves apprehension of the event and recognition of the factors of its character.[19]

Here "recognition" and "abstraction" appear to map onto Peirce's "object" and "sign" categories, respectively. Elsewhere, Whitehead describes how this abstraction is rendered intelligible by an "expositor" to a "recipient" employing demonstrations plus descriptive phrases into "propositions."[20]

After he returned in 1939 to the Cambridge Strangeways Laboratory from T. H. Morgan's "Fly Room" at California Polytechnic (CalTech), Waddington blended this Whiteheadian philosophy of science, his 1930s work on Spemann's "organiser," conducted alongside Joseph and Dorothy Needham, and Wad's new American-style work on *Drosophila* genetics. Among many other publications and talks, the resulting period from 1939 to 1942 included a groundbreaking book heavy in theory, *Organisers and Genes* (1940), and an ahead-of-its-time article, "The Epigenotype" (1942).

Organisers and Genes introduced the "epigenetic landscape" metaphor for which Waddington is still best known (though Joseph Needham first depicted it in *Order and Life* four years earlier in 1936). Segments of a developing organism travel different paths from genotype to phenotype. Those segments can be nudged onto a new path if environmental conditions are persistent enough. Note that the potential paths themselves are not *created* by the environment—that would be a Lamarckian account. Genes still fashion the chreods, or valleys in the epigenetic landscape, the possible paths of development—he also called them "canals" and depicted them as the branching distributaries of a river delta or as tracks in a railroad switchyard. But environmental pressure "determines" which path / distributary / track a particular portion of the developing organism will travel toward eventual adulthood.

"The Epigenotype" proved to be Waddington's second major contribution in the early 1940s. Today, scientists often hold up the article as the foundational work of epigenetics.[21] At the time, however, it attracted less attention than his halfway house between Lamarckism and Darwinism he titled "Canalization of Development and the Inheritance of Acquired Characters." "Epigenotype" truly was Waddington's attempt to capture the late-1930s work leading up to *Organisers and Genes* in the language of nonspecialists. By the term "epigenotype," Waddington meant the process by which the genotype becomes the phenotype—development, in other words. To be clear, however, Waddington did *not* mean merely that development followed a genetic "program," as later molecular biologists would.

Instead, Waddington called back to the Whiteheadian foundations laid in the TBC. Instead of individual particles, like genes, biologists must regard "concatenations of processes linked together in a network, so that a disturbance at an early stage may gradually cause more and more far reaching abnormalities."[22] He stressed complexity and multicausality—a pursuit that should engage biologists trained in multiple subfields. Perhaps now "the analysis of the effects of genes" could be blended with experimental embryology: "The two methods of analysis whose *rapprochement* has for so long been no more than a pious hope can now actually and in practice come together."[23]

But, of course, Wad's hopes would not be realized in the 1940s, nor for decades afterward.[24] Some of it was the sheer complexity of these ideas—so different than Morgan's comparatively simple American fruit fly genetics. But some of it was just the War and Waddington's role in it. It pulled him away from the Oxbridge-London intellectual hub and into "practical" genetics. He worked as one of "Churchill's scientists" during World War II. Then Waddington, with help from Joseph Needham and Julian Huxley, among others, fortuitously obtained the chair in animal genetics at the University of Edinburgh vacated by F. A. E. Crew. By the late 1950s/early 1960s, Wad reached a new height in his career, publishing dozens of books and articles and delivering talks internationally on a wide range of subjects. *The Strategy of the Genes* (1957), *The Nature of Life* (1961), and *New Patterns in Genetics and Development* (1962) each stand as milestones in the creation of subfields like systems biology and epigenetics.

Still, colleagues in the upstart field of molecular biology—Francis Crick, for instance—seemed not to pay much attention.[25] Unfortunately, it was that audience Wad most wanted to reach. His desire to correct their ignorance of process, field, and pattern drove him, over those nights in the spring of 1965, to take the first decisive steps toward creating a group of good-faith interlocutors that resulted in the IUBS Theoretical Biology Symposia in Bellagio, Italy.

2.3.3 Bateson: The Pattern Behind Biosemiotics

After the War, Bateson cast about for various projects in and around Mead's New York City circle. But even before their divorce in 1950, the frayed edges around their marriage had begun to show. Mead maintained a romantic relationship with Ruth Benedict, who died from a heart attack at age sixty-one in 1948. Bateson had a child with new wife, Betty Sumner, and moved to the University of California San Francisco Medical School in 1948. He brought his dedication to cybernetic theory into a psychiatric facility, where it hybridized with other concepts floating around about epistemology and communication. Though it is unclear whether he studied C. S. Peirce's semiotics directly or encountered it secondhand, Bateson mirrored Peirce's classic work on the triadic nature of signs: the sign (or representamen), the object being referred to, and the interpretant (the meaning or effect produced by the sign). Bateson applied this triadic model to various aspects of communication and cognition in living systems, emphasizing the role of signs and symbols in mediating interactions between organisms and their environments. Furthermore, Bateson embraced Peirce's view of semiotics as a broad and interdisciplinary perspective that encompasses not only human language but also the ways in which signs operate in nature.

During the late 1950s, Bateson took a seemingly hard turn into what we would today call biosemiotics and away from his earlier work. But understanding his background, his work on patterns in partridges with his father, and his interactions with Waddington, we should see more continuity than difference in this next phase. He conducted experiments with marine animals, particularly dolphins, at multiple oceanographic institutes, starting with John C. Lilly, a pioneer in dolphin communication research. But quickly Bateson moved even farther west, landing, by the mid 1960s, in Hawaii to observe dolphin training at marine theme parks.

Bateson found far more in common in the communication patterns of marine animals and humans. Body postures, interactions with their environment and other animals, and vocalizations (at least in the case of dolphins) showed that the gap between human patterns, cetacean patterns, and cephalopod patterns of behavior was not as wide as imagined, suggesting that communication patterns can be indicators of common descent.

For instance, though the two organisms are only very distantly related, communication of both dolphin and octopus is highly contextual, with the interpretation of signals influenced by factors such as social relationships, environmental conditions, and past experiences. When observing dolphin training for performances in front of audiences, Bateson noted that trainers and the performance animal had to establish a proper context before the signs (usually whistles) made by the trainer could be properly interpreted by the dolphins. One set of whistles might mean "do a task" (e.g., jump, bounce a ball, swim quickly). But if the trainer was trying to explain to an audience the task of teaching itself, the dolphin needed it to be "explained" by certain other signals including timing. If the trainer did not set the appropriate context, the dolphin would grow irritated, sometimes smacking its fluke (tail). The performance animal had to learn that it was an *occasion for learning*, which demanded a whole new interpretive lens placed over behaviors. The dolphin might do the same actions but for entirely different reasons. Or the animal might be asked to do new actions in order achieve the old result. Understanding context mattered.

Bateson's recognition of the crucial element of communications' context he designated a second, higher communications level; for instance, "Learning 2" or "deutero-learning"—learning to learn, or metalearning—versus "Learning 1" or knowledge acquisition.[26] Above this would be Learning 3, learning to learn to learn, or meta-metalearning. And this kind of cognition is not limited to learning, per se. "Play" was another multilevel category that Bateson studied. There could be Play 1, Play 2, or even Play 3. This process of identification Bateson termed "logical typing," and it was an insight he drew from Whitehead's work, much as Waddington did when investigating epigenetics. It highlighted the importance of networks of context in the Peircean triadic nature of signs.

Mead acted as a sounding board and a go-between for both Waddington and Bateson through the rest of their lives, hosting Wad's trips to the US and Bateson's to the East Coast.

Mead also ensured that the connection between Bateson and Mead's daughter, Cathy Mead-Bateson, and her father remained firm, resulting in the partnership behind several of Gregory Bateson's touchstone works. Mead even shepherded Waddington's attempt in 1970 to entice Bateson back to the UK to teach at the University of Edinburgh.[27] The academic appointment Wad attempted to convince Bateson to take in 1970 grew into the "strong programme" of the Sociology of Scientific Knowledge typically associated with David Bloor, Barry Barnes, and their colleagues at Edinburgh's Science Studies Unit. But Bateson expressed little desire to leave almost complete intellectual freedom of working with marine animals beachside in Hawaii to be tied down to traditional academic structures in grey Edinburgh.

2.3.4 Minds and COWDUNG

Perhaps not surprisingly, both Bateson and Waddington ended their careers engaged in discussions over the nature of minds and communication between minds.

Conversations with Mead, and a Mead-prompted re-reading of Bateson's *Naven* blossomed into Waddington's popular *The Ethical Animal* in the early 1960s.[28] As part of the Gifford Lectures, for two years in the early 1970s, Waddington engaged in a four-way conversation entitled the *Phenomenon of Mind* with Oxford philosopher of religion Tony Kenny (now Sir Anthony John Patrick Kenny), Oxford philosopher of mathematics John R. Lucas, and early Artificial Intelligence researcher Christopher Longuet-Higgins (who was also involved in the IUBS meetings).

Gregory Bateson penned *Mind and Nature: A Necessary Unity* (1979) and his essay collection, *Steps Toward an Ecology of Mind* (1972), while studying human mental disorders and animal communication. Bateson's contributions to these fields are appreciated today perhaps above any of his others.[29]

Throughout their careers, both men reached similar, though not necessarily identical, conclusions about the nature of minds and communication between minds. To better understand how they positioned biological explanations—and to better elucidate the IUBS meetings—I turn to a helpful categorization by Kenneth Schaffner. Schaffner divided approaches to the life sciences into stronger and weaker versions of ontological and methodological commitments. He fit many well-known biologists and philosophers of biology squarely within his four-by-two grid. To that grid, reproduced below, I have added individuals germane to this chapter (table 2.1).

As this essay has stressed throughout, both Bateson and Waddington relied upon a long appreciation for multilevel patternedness—patterns of patterns. To study organisms, we must interact with multilevel patterned systems in which behavior and communication depends on levels above and below. Especially when it comes to the kind of thinking required to communicate, the appropriate methodological approach to studying living things is not reductionistic. They made no appeal to vitalism. But they also grew increasingly uncomfortable with glib reductionist metaphors about DNA writing "the book of life" and so on. Animal minds were certainly biological products. Yet the typical methodologies used when describing minds and the information flowing between minds proved too restrictive because they were too tied to reductionist philosophies.

To be sure, the Whitehead-inspired work of Waddington and Bateson did not become mainstream in the life sciences. By the end of his life, Waddington warned of an unthinking,

Table 2.1
Adapted from Schaffner's (1993, p. 412) schema of life science approaches

Type of approach	Reductionistic	Anti-reductionistic
Ontological, Strong	*Traditional mechanism*: Jacques Loeb	*Neo-vitalism*: Hans Driesch, *Henri Bergson*
Ontological, Weak	Ernst Nagel	Emergentism-in-principle: Claude Bernard, Paul Weiss, Michael Polanyi, Walter Elsasser, *C. S. Peirce, Alfred North Whitehead, Arthur Koestler*
Methodological, Strong	Sound biological generalizations are molecular: *Francis Crick [Wad's "Conventional Wisdom of the Dominant Group"]*	Biological explanations must involve higher levels: Paul Weiss, G. G. Simpson, Ernst Mayr, Marjorie Grene, *Tony Kenny, Gregory Bateson, Margaret Mead(?)*
Methodological, Weak	For the present, we must appeal to interlevel/interfield features: [Schaffner includes a number of philosophers.] *J. B. S. Haldane, Theodosius Dobzhansky, Julian Huxley(?), Christopher Longuet-Higgins*	For the present and future we must appeal to related but not-fully-connected higher levels: Jerry Fodor, David Hull, Alex Rosenberg, *C. H. Waddington, other members of the Cambridge Theoretical Biology Club (1930s), H. H. Pattee.*

Schaffner's original is normal type; my additions are italicized.

supposedly atheoretical "Conventional Wisdom" that he sensed in the "Dominant Group" of his biological and biomedical colleagues. "COWDUNG," he styled it.[30]

Nobel laureate Francis Crick became one of Waddington's preeminent examples of this Conventional Wisdom of the Dominant Group. In a series of 1966 lectures published as *Of Molecules and Men*, Francis Crick declared that progress in the twentieth century life sciences depended upon the removal of vitalism.[31] Waddington pointed out in his review of *Of Molecules and Men* that Crick debated straw men.[32] While privately Crick admitted that his critique lacked precision, he believed it did not matter: the reductionistic "genes-for" language Crick deployed was the way molecular geneticists described the organic world.[33] If anyone resisted the perspective of molecular biologists that all explanations in biology were ultimately reducible to microscopic physico-chemical entities, they would be discredited. The molecular approach, as William T. Astbury once reminded Waddington in an exchange published in *Nature*, "is the coming biology."[34] This was the Conventional Wisdom, and Crick was in the Dominant Group.

These sorts of revolutionary claims about the reducibility of whole organisms to microscopic entities had been made earlier. But by the 1960s and 1970s, the situation had changed, which began to worry Waddington: after World War II, the professional ethos of the life sciences slowly began to shift in accordance with economic principles favoring short-term, increasingly commercial results.[35] Waddington sensed the incongruity of this kind of pragmatism with the true aims of biology: "[R]eductionism is a recipe for action."[36] When what is called for is making a "quick (scientific) buck by discovering some useful practical information," the reductionistic approach made sense. And as the years went on, he worried that this reductionist approach of Crick and Astbury, among others, had blossomed into an "epidemic intellectual disease, which causes people to argue that the reality of anything is proportional to the precision with which it can be defined in molecular or atomic terms."[37] The slower, anti-reductionist Whiteheadian approach, he protested, was the "method for making major advances in human comprehension," like those of "Darwin, Freud, Einstein [and] the quantum physicists." However, reductionism promised predictability and control and—let's be honest here—funding.

Thus, by the end of his life, this approach appeared to take its place as the Conventional Wisdom of the Dominant Group—"COWDUNG."[38] Referencing table 2.1, this is a strong methodological reductionism. COWDUNG *could be* ontological reductionism. But—as Wad noted and Bateson, too, frequently pointed out—there travels with this epistemology a disincentive to make philosophical considerations at all, a kind of theoretical "know-nothingism." Two brief examples of what I mean here:

Waddington's contemporary, Nobel laureate Peter Medawar, demonstrated this antitheoretical attitude to historian June Goodfield's question: "I made the suggestion . . . that the attitude with which a man approached biology might help him to focus on certain problems that otherwise might be ignored." But Medawar's response was "a flat negative . . . it made no difference at all. A man was a good scientist, or he was not."[39] Philosophical reflection threatened to slow research.

And an example from our own era. The late E. O. Wilson proclaimed his distaste for philosophy: "Most of philosophy—which is a declining and highly endangered academic species, incidentally—consists of failed models of how the brain works. So, students going into philosophy have to learn what Descartes thought and then after a long while why that's wrong . . . and what Kant might of [*sic*] thought or did think. But they cannot go on from that position and historical examination of the nature of humanity to what it really is and how we might define it."[40]

In a 2005 television interview alongside James Watson, Wilson extended this critique: "Psychologists . . . and the vast majority of social sciences, who really determine a lot of the intellect that goes into our policies, you know, and our philosophies . . . have no connection at all to their true nature of humanity or what human nature is at a biological level."[41] Co-interviewee Watson nodded vigorously and added, "This century will be the coming together of psychology and biology." For Wilson and Watson, slower, more nuanced analysis of thought influenced by the humanities and social sciences always lost out to quicker experimentation and off-the-cuff theorizing—that was the Conventional Wisdom, and they were the Dominant Group.

These examples from prominent biologists in different eras do not prove that thinking theoretically in biology is a "highly endangered academic species," nor that COWDUNG reductionism is entirely atheoretical. But these examples do demonstrate how widely shared is the opinion that careful theoretical examinations of biological propositions can be glibly dismissed without professional penalty. Indeed, examining biological theory *at all* seems to set one apart in biology. What that gives us is another set of clues toward the composition and research pursuits of those involved in Wad's IUBS meetings in the late 1960s. To be interested in theory at all was also, often, to lean against reductionism, to signal one's resistance to the Conventional Wisdom implicitly espoused by the Dominant Group.

2.4 Part 2: Antireductionism in *Towards a Theoretical Biology*

His long interaction with fellow Whitehead-influenced pattern-seeker Gregory Bateson convinced Wad that a large enough group of scholars still existed who opposed blithe reductionism. As he wrote to Mayr one night in the spring of 1965, it might be interesting to discuss this the problems of reductionism and the benefits of thinking about patterns and patterns of patterns at a meeting of such scholars.[42] The Bateson–Waddington relationship, in other words, helps elucidate a major point of this chapter: why antireductionist views comprised a disproportionate number of the IUBS meetings and the resulting *Towards a Theoretical Biology*.[43] This is so both because those drawn to lengthy interactions about theoretical biology would be fewer in number and because Waddington experienced that marginalization and developed a network of those who wanted to talk theory. Thus, we should not be surprised to see what an analysis of *Towards a Theoretical Biology* shows us: a disproportionate number of talks stress the prominence of "networks of meaning" and metapatterns in biology as a theme. I take the exploration of "networks of meaning" as a root preoccupation of biosemiotics.

Below, I will highlight several chapters in Waddington's *Towards a Theoretical Biology* relevant to the development of biosemiotics that bear Bateson's intellectual fingerprints.

2.4.1 Vol. 1, Chapter 5: "The Concept of Organism" (Waddington)

In this paper, which truly set the tone for much of what followed, Waddington explores the organism as an integrated, autonomous entity that interprets and responds to its environment. He explains how organisms process various forms of information, such as chemical signals, physical stimuli, and social cues, to maintain homeostasis and adapt to changes. This interpretative process is central to biosemiotics, which studies how living beings generate and understand signs. Waddington employs Peirce's triadic model of the sign—representamen, object, and interpretant—showing how organisms continuously generate meaning from environmental cues and internal states. He underscores that understanding life requires considering the entire system of interactions and relationships, rather than merely breaking it down into individual parts. Bateson's concept of "the pattern that connects" aligns with this systemic approach, reinforcing the view that organisms should be seen as networks of interacting signs.

Waddington's ideas in this chapter will develop further in "The Concepts and Theories of Growth, Development, Differentiation, and Morphogenesis" (vol. 3). Both chapters highlight the importance of understanding organisms as integrated systems that process information in complex, recursive, nonreductive ways.

2.4.2 Vol. 2, Chapter 4: "Bohm's Metaphysics and Biology" (Marjorie Grene)

In the most explicit analysis of reductionism versus antireductionism in the entire four volumes, Grene discusses David Bohm's concepts of order and wholeness and their relevance to biology. Bohm's notion of implicate and explicate orders resonates with longstanding biosemiotic principles, where the implicate order represents the underlying, hidden structures of reality, and the explicate order manifests as observable phenomena. Grene draws comparisons with Alfred North Whitehead and with Peirce's idea of the dynamic and immediate objects in semiosis, highlighting how biological processes continuously unfold these orders, with meaning and interpretation intrinsic to life. Grene emphasizes the importance of context and the recursive nature of communication in living systems. She inadvertently echoes Bateson's view that "information is a difference that makes a difference"—again focusing on the emergent properties and complex interactions within biological systems.

2.4.3 Vol. 2, Chapter 5: The Practical Consequences of Metaphysical Beliefs (Waddington)

After Grene's characterization of him as a slightly heterodox neo-Darwinian, rather than the iconoclast he believed he was, Waddington began a deep dive into his own intellectual journey, examining how philosophical beliefs shape scientific practices and the interpretation of biological phenomena. Here, he argues that biologists' underlying assumptions about the nature of life absolutely influence their research methods and conclusions. Once again, he nods toward biosemiotics by emphasizing that interpreting biological data involves sign processes and meaning-making. Waddington echoes Peirce's pragmatic maxim, suggesting that the meaning of a concept lies in its practical effects, and explores how metaphysical beliefs impact scientific understanding and practice. (Not surprisingly, Gregory Bateson's concept of epistemology, particularly his ideas on how our mental models influence our perceptions and interpretations, reinforced Waddington here.)

Waddington's autobiographical remarks here would resonate for the rest of his life. His attempt at an autobiographical collection of papers was titled *The Evolution of an Evolutionist*—trying to navigate between the Scylla of Grene's comments that he was too conventional and the Charybdis of Mayr's later marginalizing of Waddington based on Mayr's belief (we will encounter in the coda) that Waddington was not a conventional evolutionist at all.

2.4.4 Vol. 3, Chapter 10: The Problem of Biological Hierarchy (Pattee)

In this underappreciated chapter, Pattee pulls together the various threads presented at two previous workshops (volume 1, chapter 4 and volume 2, chapter 19) to expose the fundamental incompatibility of two popular approaches: that of Crick ("complete reductionism") and Michael Polanyi ("irreducible to [physicochemical] laws").[44] Though Pattee does not subdivide their positions as I did above, he nonetheless affirms (a) just how representative their positions were within the broader intellectual space of the life sciences, (b) the degree of talking past one another that was contributing to this debate, and (c) how intractable the methodological problem of patternedness really was. (To be fair, Pattee did not say patternedness—that is my term—he spoke of constraints, though he meant the same thing: organized behavior of cells to accomplish some function).

For Pattee, the puzzle to solve in (c) is not merely biological *function*. In principle, all biological functions can be explained by reference to biochemicals, which are subject to laws of physics. Michael Polanyi and other methodological antireductionists would agree. The problem is explaining the *origins* of differences between the behavior of different levels. Chemicals do not spontaneously pass on specific instructions the way that the DNA + RNA + ribosomal system does, for instance. And all examples of this passing on of instructions that we know of happen with cells. Appeals to evolution, such as those made by reductionists like Crick, conveniently ignore that all evolutionary processes occur in the context of cellular life and only in the context of cellular life. Nonliving crystals cannot replicate across generations like cells can, Pattee stresses, even if blasé accounts like Crick's assert crystals are proper analogs.

2.4.5 Vol. 4, Chapter 12: Laws and Constraints, Symbols and Languages (Pattee)

Pattee's conclusion to his four-essay series at the IUBS is perhaps the closest he comes to an outright antireductionist ontology. It certainly stands as a seminal paper in biosemiotics. And the antireductionism apparent in the paper (especially when read in the context of the other three essays) offers another bit of evidence why biosemiotics has continually suffered from too little acceptance in the western life sciences community.

Pattee examines the intricate relationship between physical laws, constraints, and the emergence of symbolic language in both biological and human contexts. Laws, as he has laid out in earlier papers, describe regularities in nature. Constraints, as stated in the third paper, above, restrict possible states or processes. This distinction, Pattee argues, is fundamental to understanding how complex systems, particularly living organisms, operate and evolve.

Physical laws alone—at least as we currently understand them—cannot explain the behavior of biological systems. Constraints limit the range of possible actions and interactions, creating a structured environment in which life can develop and function. By imposing boundaries, constraints enable the emergence of order and organization from the underlying physical laws.

Symbols, whether in genetic codes or human languages, must adhere to these laws while simultaneously serving as abstract representations. This duality allows symbols to bridge the gap between the two levels: the physical and a higher-order informational level. In biological systems, genetic codes exemplify this principle, as they represent information while being grounded in the physical properties of nucleic acids. Language marks a significant evolutionary step, enabling organisms to process and transmit information in novel ways. This capacity for symbolic representation enhances an organism's ability to adapt.

Pattee also addresses the role of measurement in understanding biological systems. He contends that measurement itself involves symbolic representation, as it requires the abstraction and quantification of physical phenomena. This process of measurement bridges the gap between the concrete world of physical laws and the abstract realm of symbols and information. By measuring and interpreting biological systems, scientists can uncover the underlying principles that govern life.

2.5 Conclusion

Though a small selection of the dozens of essays in *Towards a Theoretical Biology*, the above summaries help to underline a central point. The overwhelming majority—at a 3-to-1 ratio, by my count—of IUBS papers presented and collected by Waddington reject reductionism and espouse, for the most part, methodological antireductionism. The majority of the papers that make this rejection explicit (e.g., Bohm, Goodwin, Lewontin, Pattee, Waddington) do so by referencing patternedness and metapatternedness. Why does this matter?

First, it shows the degree to which Waddington remained consistent in his advocacy for antireductionism in the life sciences. That thread of antireductionism linked otherwise disparate biologists, physicists, and philosophers. Beginning with his connection with Gregory Bateson, Waddington clearly continued to build relationships and create opportunities for likeminded researchers at the 1966–1968 Villa Serbelloni meetings, especially younger workers like Brian Goodwin and Richard Lewontin. Both biologists would become well respected in their own right and go on to influence other prominent scientists (e.g., Stephen Jay Gould) and develop new subfields (e.g., systems biology). Thus, the trajectory of one segment of the life sciences was undoubtedly shaped by the relationship between Wad and Bateson through the twentieth century.

Secondly, this combination of antireductionist biologists, physicists, and philosophers opened the door for conversations about biosemiotics—how meaning-making happens in biological systems—as foregrounded in Bateson's work in the 1970s.

Thirdly, it remains fair to argue that the continuous conjunction of antireductionism and theoretical biology as exemplified at these IUBS meetings contributed to the marginalization of theoretical biology by adherents of Conventional (reductionist) Wisdom, which, after all, has long been espoused, or at least not questioned, by the Dominant Group. Biosemiotics may be guilty by association, from this point of view, given the importance of multilevel patternedness in explanations of meaning.

2.6 Coda: The Uncooperative Compatriot

Finally, let's return to the snubbing of Waddington by Mayr, that mystery I alluded to at the top. I apologize to the reader that space does not allow a full defense of my argument here. But let me see if I can offer some brief thoughts about a solution.

First, let me sharpen the mystery. Like many of the IUBS participants, Mayr, too, seemed to be firmly on the antireductionist end of the spectrum. He criticized the "selfish gene" and "sociobiology" literature that appeared at roughly the same time. Even more surprising, Mayr joined his friends Francisco Ayala and Theodosius Dobzhansky at Villa Serbelloni in the early 1970s for *another* conference specifically on the problem of reductionism in biology, and at that conference, Donald T. Campbell referenced the cybernetic scholarship of Gregory Bateson![45] Their interests aligned so well, as was clear in their 1965 exchange I quoted above, that it is difficult to see any reason for any disagreement between Mayr and Waddington as the cause for Mayr's dissatisfaction and subsequent withdrawal from the IUBS meetings, and then seeming dismissal of Waddington's contributions to the Modern Synthesis.

Personality differences certainly could have been one reason. Mayr was known to be obstinate, even authoritarian, and Waddington libertine and distracted at times.[46] However, examining Mayr's correspondence well after Wad's death militates against the tendency to blame merely personal dislike.

The evidence presented above, plus archival sources, points to three other enmeshed areas of intellectual conflict: (i) the status of mathematical and physical concepts like algorithms and coding in biology, (ii) the importance of developmental processes versus the selection of adult forms, and (iii) the role of direct environmental pressure on evolution. Put more simply, Mayr and Waddington differed on (i) biophysics, (ii) embryology, and (iii) the inheritance of acquired characteristics. Let's look at each in turn.

First, Waddington repeatedly invited IUBS contributions from mathematicians like Réne Thom and Christopher Zeeman and biophysicists like Stuart Kauffman. Mayr, by contrast, distanced himself from too much influence of the physical sciences on biology.[47] Even beyond that, Mayr never wanted to stress "experimental biology" in his histories of biology, preferring "historical biogeography, behavior, and the theory of systematics."[48] Clearly, they each focused on different "levels" even within biology. Just as clearly, the IUBS meetings were too focused on biophysics and experimental biology for the liking of Mayr.

Secondly, Waddington spent much of his early career working in embryology and, even after moving to a genetics-focused lab at Edinburgh, became one of the best-known mid-century developmental biologists integrating genetics and embryology. Mayr hewed more toward the work of population genetics produced by their mutual friend Theodosius Dobzhansky. Waddington decried the Modern Synthesis "disassociation" of development and genetics was an "entirely American phenomenon with little influence in Europe."[49] That Americanized lacuna in modern evolutionary theory, unfortunately from Waddington's perspective, infected Mayr's major contributors to the Modern Synthesis as well: "One shouldn't let the fact that Dobzhansky has got hold of a nice case of a species which goes in for integration at the genetic level bamboozle one into thinking that that is the crux of the matter."[50] From Mayr's perspective, however, embryologists (he certainly implied Waddington as well) often fell victim to his favorite bogey, "typological thinking."[51]

These first two factors bled into the third—the most egregious, from Mayr's perspective. In his 1930s–1950s embryological work, Wad focused on the inheritance of acquired characteristics, better known as "Lamarckism." Wad himself stressed repeatedly that he was *not* advocating for Lamarck's biology, calling his theory "genetic assimilation." Yet as the Lysenko controversy in the USSR heated up, Wad did not sufficiently detach himself from this version of evolution, even traveling to Lysenko's lab inside the Soviet Union. Mayr, H. J. Muller, and several others, distanced themselves from Waddington for even this perceived lukewarmness toward Lysenkoism—even if Waddington did not express that himself.[52] "Somehow," Mayr lamented, "Waddington always wanted to find something intermediate between Darwinism and Lamarckism. It is curious how negative toward selection almost all embryologists have been."[53] Here, again, Mayr undoubtedly felt out of place at the first Villa Serbelloni IUBS meeting: the papers showed no special deference toward explaining all facets of biology in terms of their immediate adaptive advantage.

Why does this small story of conflict between Mayr and Waddington matter for biosemiotics? Waddington and his network, especially Gregory Bateson, supported studying the production and interpretation of signs and signals in living systems. While Mayr decried reductionism at least as much as Waddington (and so should have been a natural ally of biosemiotics), he supported the study of animal behavior known as "ethology" instead of biosemiotics. Mayr continued to support ethology even as some of its chief promotors (e.g., Konrad Lorenz) slipped into the more reductionist sociobiology camp.[54] To be fair, he was rarely explicit about it, but Mayr's prejudices against Waddington's work, above, may have overlapped with his lack of interest in biosemiotics.

Once again, we might witness through this historical episode lasting division between allied areas in the life sciences caused by two titans of their respective areas having slightly different training, values, institutions, interpersonal conflicts, and networks.

Notes

1. Waddington and Mayr knew each other through mutual friend Theodosius Dobzhansky, with whom each had worked beginning in the 1930s. Waddington participated in Mayr's 1959 Darwin Centenary at the University of

Chicago. They argued repeatedly from the 1940s through the 1960s over the so-called Baldwin Effect and the impact of the environment upon the genotype.

2. Peterson, 2017, chapter 6.

3. C. H. Waddington [hereafter CHW] to Ernst Mayr, 4 May 1965. Mayr-HU.

4. Ernst Mayr to CHW, 10 May 1965. HUG(FP) 74.7, b.12, f. 874, Mayr-HU.

5. CHW to Mayr, 25 Nov. 1965. HUG(FP) 74.7, b.12, f. 919, Mayr-HU.

6. CHW to Mayr, 4 Feb. 1966. HUG(FP) 74.7, Box 12, Folder 919, Mayr-HU.

7. Waddington, 2012 [1942].

8. Mayr to CHW, 6 Nov. 1966. HUG(FP) 74.7, Box 12, Folder 919, Mayr-HU

9. Hamburger, 1980.

10. Mead quoted in Brand, 1976, p. 38.

11. CHW to Mead, 17 Oct 1973, and Mead to CHW, 19 Oct 1973; Margaret Mead papers/South Pacific ethnography archive, Box 17, Folder 2, Library of Congress, Washington, DC, USA.

12. Though not actually hidden, his contemporary admirers have downplayed the details of Gregory Bateson's extensive biological training, as Gregory did himself, preferring to emphasize his specializations in anthropology and psychology. Yet, when challenged, Bateson did draw upon this Cambridge zoology pedigree. See, for instance, his 1945/6 application for John Simon Guggenheim funding: "Contributions to Psycho-Cultural Theory." Box O5, Folder 6, South Pacific Ethnography/Margaret Mead papers, Library of Congress, Washington, DC, USA.

13. Bateson and Bateson, 1926.

14. Paradoxically, this vibratory theory has become presently an exciting and fruitful area of research in developmental genetics. And it appears that the Batesons were largely correct in their interpretations, though the instrumentation for actually observing the process would not be developed for at least a half century. See, for instance, Newman, 2007; Palmeirim et al., 1997; Pourquié, 2003; Baker et al., 2006; Aulehla and Pourquié, 2008.

15. Lipset, 1982, pp. 174–175.

16. Peterson, 2017, pp. 62–69.

17. Peterson, 2011.

18. Haraway, 2004 [1974], pp. 35–36.

19. Whitehead, 1920, pp. 189.

20. Whitehead, 1920, p. 6; see also Stjernfelt, 2014.

21. For example, Haig, 2012.

22. Waddington, 2012 [1942], quote on p. 10.

23. Waddington, 2012 [1942], p. 13.

24. Erik L. Peterson "'Here I Go Again'—Will C. H. Waddington's Hopes Finally Be Fulfilled? Part III," Extended Evolutionary Synthesis, April 4, 2018, https://extendedevolutionarysynthesis.com/why-didnt-waddingtons-attempts-fix-the-division/.

25. Crick and Waddington corresponded throughout the 1960s and into the 1970s. During Waddington's lifetime, Crick never took Wad's epigenetics seriously, though he did attend one IUBS meeting at Wad's invitation. See Peterson (2023).

26. Pryor et al., 1967.

27. CHW to Gregory Bateson, 18 May 1970; Margaret Mead papers/South Pacific ethnography archive, B17 f.1, Library of Congress, Washington, DC, USA.

28. CHW to Mead, 17 Jan. 1960; CH Waddington papers, MS 3060.1, University of Edinburgh Library, Edinburgh, Scotland, UK.

29. Harries-Jones, 2016; Hoffmeyer, 2010; Wheeler, 2010.

30. Waddington, 1977, pp. 24–25.

31. Crick, 1966.

32. Waddington, 1967. Other reviewers concurred with Wad that Crick was conflating organicism with vitalism to make a sloppy point about biological progress (e.g., Livingston, 1968).

33. Crick to CHW, 9 Nov. 1967, and CHW to Crick, 27 Dec. 1967, Francis Harry Compton Crick papers, Wellcome Library for the History and Understanding of Medicine, London, PP/CRI/I/2/6/5.

34. Astbury, 1961.

35. Callon, 2002.

36. Waddington, 1977, p. 23.

37. Waddington, 1975, p. 21.

38. Waddington, 1977, pp. 24–25.

39. Goodfield, 1974.

40. Quote from, Junod, Tom. "E. O. Wilson: What I've Learned." *Esquire*, January 5, 2009, http://www.esquire.com/news-politics/interviews/a5375/eo-wilson-quotes-0109/.

41. "Watson and Wilson on Charles Darwin," *Charlie Rose* (air date 12/14/2005, rebroadcast 08/22/2006), https://charlierose.com/videos/18174, at 48:52 to 50:12.

42. Unfortunately, I cannot locate explicit evidence for why Bateson himself did not attend these meetings, nor even if Waddington invited him. By the time of Waddington's Serbelloni meetings, Bateson had moved to Sea Life Park, in Waimanalo, Hawaii, with new wife Lois Cammack; even Mead had trouble keeping up correspondence with him.

43. I must acknowledge that these ideas were "in the air" at the time. Aside from Waddington's IUBS meetings in Italy, three other conferences on nearly the same topics appeared at nearly the same time. Tellingly, two others appeared *in the same conference space* in Bellagio: Arthur Koestler's Alpbach Symposium (Koestler and Smythies, 1969); also the September 1972 and June 1974 conferences—"Reductionism in Biology" and "Mind in Nature"—both held at Serbelloni (Ayala and Dobzhansky, 1974; Cobb and Griffin, 1977).

44. Pattee, 1970, p. 121.

45. Campbell, 1975.

46. Robertson, 1977, and Smocovitis, 1994.

47. Milam, 2010.

48. Mayr to CHW, 15 Jul. 1975. Mayr papers, HUG(FP) 74.7, Box 24 (folder 1243), Pusey Library, Harvard University, Boston, Mass., USA. Mayr 1997 and Mayr 2004.

49. CHW to Mayr, 24 Sep. 1969. Waddington papers, MS 3041.4, University of Edinburgh, Edinburgh, UK.

50. CHW to Mayr, 15 Dec. 1952. Mayr papers, HUG(FP) 74.7, Box 24 (folder 550), Pusey Library, Harvard University, Boston, Mass., USA.

51. Witteveen, 2015.

52. Peterson, 2017, pp. 172–188.

53. Mayr to Daniel Polikoff, 18 Jul. 1981. Mayr papers, HUG(FP) 74.7, Box 29 (folder 1339), Pusey Library, Harvard University, Boston, Mass., USA.

54. See their correspondence, Mayr papers, HUG(FP) 14.17, Pusey Library, Harvard University, Boston, Mass., USA.

References

Archival Sources

Francis Harry Compton Crick papers, Wellcome Library for the History and Understanding of Medicine, London, UK

Ernst Mayr papers, Harvard University archives, Pusey Library, Cambridge, MA, USA

Margaret Mead papers/South Pacific ethnography archive, Library of Congress, Washington, DC, USA

Conrad H. Waddington papers, University of Edinburgh Library, Edinburgh, Scotland, UK

Published Sources

Astbury, W. T. 1961. Molecular biology or ultrastructural biology? *Nature*, *190*, 1124.

Aulehla, A., and Pourquié, O. 2008. Oscillating signaling pathways during embryonic development. *Current Opinions in Cell Biology*, *20*(6), 632–637.

Ayala, F. J., and Dobzhansky, T. (Eds.). 1974. *Studies in the philosophy of biology: Reduction and related problems*. University of California Press.

Baker, R. E., Schnell, S., and Maini, P. K. 2006. A clock and wavefront mechanism for somite formation. *Developmental Biology*, *293*(1), 116–126.

Bateson, W., and Bateson, G. 1926. On certain aberrations of the red-legged partridge *Alectoris rufa* and *saxatilis*. *Journal of Genetics*, *16*(1), 101–123.

Brand, S. 1976. "For God's sake, Margaret." *CoEvolutionary Quarterly*, *10*(21), 32–44.

Callon, M. 2002. From science as an economic activity to socioeconomics of scientific research: The dynamics of emergent and consolidated techno-economic networks. In P. Mirowski and E.-M. Sent (Eds.), *Science bought and sold: Essays in the economics of science* (pp. 277–317). University of Chicago Press.

Campbell, D. T. 1975. "Downward causation" in hierarchically organized biological systems. In F. Ayala and T. Dobzhansky (Eds.), *Reduction and related problems; Proceedings of a conference in Bellagio, Italy, September 1972* (pp. 179–186). Studies in the Philosophy of Biology. University of California Press.

Cobb, J. B., and Griffin, D. R. 1977. *Mind in nature: Essays on the interface of science and philosophy*. University Press of America.

Crick, F. H. C. 1966. *Of molecules and men*. University of Washington Press.

Goodfield, J. 1974. Postscript in the light of discussions at Serbelloni. In F. J. Ayala and T. Dobzhansky (Eds.), *Studies in the philosophy of biology: Reduction and related problems* (pp. 85–88). University of California Press.

Haig, D. 2012. The epidemiology of epigenetics. *International Journal of Epidemiology*, *41*, 16–20.

Hamburger, V. 1980. Embryology and the modern synthesis in evolutionary theory. In E. Mayr and W. B. Provine (Eds.), *The evolutionary synthesis: Perspectives on the unification of biology* (pp. 97–112). Harvard University Press.

Haraway, D. J. 2004 [1974]. *Crystals, fabrics, and fields: Metaphors that shape embryos*. North Atlantic Books.

Harries-Jones, P. 2016. *Upside-down gods: Gregory Bateson's world of difference*. Fordham University Press.

Hoffmeyer, J. (Ed.). 2010. *A legacy for living systems: Gregory Bateson as precursor to biosemiotics*. Springer.

Koestler, A., and Smythies, J. R. 1969. *Beyond reductionism: New perspectives in the life sciences. The Alpbach Symposium 1968*. Macmillan.

Lipset, D. 1982. *Gregory Bateson: The legacy of a scientist*. Beacon Press.

Livingston, L. 1968. Review: *Of Molecules and Men* by Francis Crick. *Yale Journal of Biology and Medicine*, *40*, 328.

Milam, E. L. 2010. The equally wonderful field: Ernst Mayr and organismic biology. *Historical Studies in the Natural Sciences*, *40*(3), 279–317.

Mayr, E. 1997. *This is biology: The science of the living world*. Belknap Press of Harvard University Press.

Mayr, E. 2004. *What makes biology unique? Considerations on the autonomy of a scientific discipline*. Cambridge University Press.

Newman, S. A. 2007. William Bateson's physicalist ideas. In M. D. Laubichler and J. Maienschein (Eds.), *From embryology to evo-devo: A history of developmental evolution*. MIT Press.

Palmeirim, I., Henrique, D., Ish-Horowicz, D., and Pourquié, O. 1997. Avian hairy gene expression identifies a molecular clock linked to vertebrate segmentation and somitogenesis. *Cell*, *91*(5), 639–648.

Pattee, H. H. 1970. The problem of biological hierarchy. In C. H. Waddington (Ed.), *Towards a theoretical biology v.3: Drafts* (pp. 117–136). Aldine Publishing Co.

Peterson, E. L. 2011. The excluded philosophy of evo-devo? Revisiting Waddington's failed attempt to embed Alfred North Whitehead's "organicism" in evolutionary biology. *History & Philosophy of the Life Sciences*, *33*, 301–320.

Peterson, E. L. 2017. *The life organic: The theoretical biology club and the roots of epigenetics*. Pittsburgh University Press.

Peterson, E. L. 2023. A 'fourth wave' of vitalism in the mid-20th century? In C. Donohue and C. T. Wolfe (Eds.), *Vitalism & its legacies in the twentieth century* (pp. 173–192). Springer Nature.

Pourquié, O. 2003. The segmentation clock: Converting embryonic time into spatial pattern. *Science*, *301*(5631), 328–330.

Pryor, K., Haag, R., and O'Reilly, J. 1967. *Deutero-learning in a Roughtooth Porpoise (*Steno bredanensis*)*. U.S. Naval Ordinance Test Station, NOTS TP 4270.

Robertson, A. 1977. Conrad Hal Waddington (1905–1975). *Biographical Memoirs of Fellows of the Royal Society*, *23*, 575–622.

Smocovitis, V. B. 1994. Organizing evolution: Founding the Society for the Study of Evolution (1939–1950). *Journal of the History of Biology*, *27*(2), 241–309.

Stjernfelt, F. 2014. *Natural propositions: The actuality of Peirce's doctrine of dicisigns*. Docent Press.

Waddington, C. H. 1967. No vitalism for Crick; Review of *Of Molecules and Men* by Francis Crick. *Nature*, *216*, 202–203.

Waddington, C. H. 1975. Fifty years on. *Nature*, *258*, 20–21.

Waddington, C. H. 1977. *Tools for thought: How to understand and apply the latest scientific techniques of problem solving*. Basic Books.

Waddington, C. H. 2012. The epigenotype. *International Journal of Epidemiology 41*(1), 10–13. (Original work published in *Endeavour*, 1942)

Wheeler, W. 2010. Gregory Bateson and biosemiotics: Transcendence and animism in the 21st century. *Green Letters*, *13*(2), 35–54.

Whitehead, A. N. 1920. *The concept of nature*. Cambridge University Press.

Witteveen, J. 2015. "A temporary oversimplification": Mayr, Simpson, Dobzhansky, and the origins of the typology/population dichotomy (part 1 of 2). *Studies in History and Philosophy of Science Part C: Studies in History and Philosophy of Biological and Biomedical Sciences*, *54*, 20–33.

3 Organisms as Meaning-Makers

Charles Wolfe

Meaning is the guiding star that biology must follow.
—Uexküll (1982, p. 43)

3.1 Introduction

Organisms, like zombies, or more gently, like a character in a play who keeps exiting and then somehow returning to the stage, keep returning or having a "revival" in bio-philosophical thought. Most of the time, they return as key elements of proud empirical claims to "overturn mechanism"—that is, truth claims (organisms *are* x, are *defined* by *properties* y and z, and so on). I have offered some friendly criticism of such bluntly empirical organicism in the past. But there is another way in which organisms can and do return: as *meaning-makers*, following a line of inquiry associated with Jakob von Uexküll's umwelt research and more recently framed as biosemiotics. It is a very different approach to organisms, to their existence, and to "what makes them tick" to say they are defined by metabolism, or organizational closure, and to say that they are defined by the production of and responsiveness to meaning.

What does it mean for an organism to be a "meaning-maker"? One of the most important answers to this question was given, programmatically, by Uexküll, who shall be my main focus, although I additionally examine the reception and transformation of this Uexküllian idea in Kurt Goldstein and Georges Canguilhem. For Uexküll, meaning is the guiding star for biology; as nicely glossed by Di Paolo, "Snowprints left by a prey are meaningful even though metabolism is not feeding on the prints themselves" (Di Paolo, 2005, p. 442). Tellingly, in a 1951 paper that appeared in French translation (by Georges and Simone Canguilhem), a kind of position paper on biology and its "epistemological status," Kurt Goldstein presents the core problem of biology as "where do order and meaning come from, in the vast chaos of isolated, contradictory facts?" And he adds that this is the "characteristic problem of the behavior of the organism" (Goldstein, 1971[1951], p. 439). This is, of course, quite an idiosyncratic vision of biology!—one that most biologists would not recognize today. But it was certainly one of Uexküll's central questions or *Leitfaden*: meaning-making inscribed in a biological frame.

The above question overlaps with, without being identical with, another: What does it mean for an organism to be a subject? And by extension, for an organism to be a meaning-maker, does it have to be primarily and originarily a subject, in a foundationalist phenomenological sense? The notion of organic or organismic unity (what Merleau-Ponty called its *forme d'unité*) seems to necessarily imply a notion of a subject—certainly in phenomenologically motivated theories, which are in esse theories of subjectivity, discourses starting from a foundational subject. Now, we are often told that this changed radically with the turn toward embodied phenomenology, but Merleau-Ponty's clear statement on what "body" means is quite revealing: "My existence as subjectivity is one with my existence as a body."[1] Bodily existence and subjective existence are one and the same—an intuition extended by Francisco Varela and in a more anti-materialist way by Evan Thompson.

Should an organism's "uniqueness" be derived from a core subjectivity? If it cannot be understood apart from a meaningful world that it both *inhabits* and *creates* (part of the core umwelt concept, to which Goldstein will add a more existential dimension involving what it really means to create a meaningful world), is this a realist ontology, or a kind of biologized or naturalized existentialism? Some thinkers, including Goldstein, would de facto reject this choice and assert that it is a quasi-objective feature of organisms that *they exist*, and qua "existents," that they experience and have an intentional relation to the world. As I have observed elsewhere (Wolfe, 2014), there is a peculiar tension in this existential account of organisms because it can be specified both in more *ontologized* terms, as noted above, and in more *constructivist* terms. Instead of claiming ontological specificity for organisms in terms of empirical features (e.g., organismic unity, homeostasis, metabolism, or organizational closure), the constructivist view treats organisms as creators of meaningful environments, as *meaning-makers*. As we will see with Uexküll, the latter idea is given a strongly Kantian formulation, albeit one which some readers find unfaithful to Kant (although others view it as a welcome naturalization of the same thinker: compare Stjernfelt, 2007, to Clarizio, 2024). A Kantian biology as presented by Uexküll explicitly, and by Goldstein and Canguilhem more implicitly, is a subject-centered biology. Which takes us back to the questions raised at the outset.

3.2 Uexküll

Uexküll's basic intuition is that *umwelt* is distinguished from *welt*, and more precisely from *Umgebung*. An organism does not *have* an umwelt but *is* an umwelt, or is in a dynamic relation like that between organism and environment as specified by Lewontin and later, Developmental Systems Theory: "The environment is not an autonomous process but a reflection of the biology of the species. Just as there is no organism without an environment, so there is no environment without an organism" (Lewontin, 1985, p. 99). The umwelt is distinct from the physical environment and does not have any independent existence in the absence of the organism; it reflects the subjective perspective of the animal on the environment, while the *Umgebung* reflects the external situation of an organism. Canguilhem in his important essay "Le vivant et son milieu" simply describes the *Umgebung* as "the banal geographic environment."[2] The paradoxes inherent in the notion of umwelt are neatly encapsulated in the linguistic fact that it can be understood as "meaningful environment" but is also rendered as "subjective universe" (Uexküll, 1982, p. 3). Umwelt connotes the subjective, perceptual, or phenomenal world of the organism (only later on did it acquire its present standard meaning in German of "the environment" in the ecological sense).

Uexküll famously gives the example of the tick and her umwelt comprising mammalian heat and light, but also, of the difference between the human sun and the mosquito sun—that is, the sun in two different umwelten: "Among the animals, with their smaller Umwelt horizons, the celestial bodies are essentially different. When mosquitoes dance in the sunset they do not see our big human sun, setting six kilometres away, but small mosquito suns that set about half a meter away. The moon and stars are absent from the sky of the mosquito" (Uexküll, 2001, p. 108).

How are these different worlds, or rather umwelten, created? By means of what Uexküll called functional circles. These create an animal's world by integrating entities into its environment: They are "the mechanism that creates (phenomenal) objects, and thus *umwelten*" (Kull, 2015, p. 620). The functional circle is literally is a circle or cycle (*Kreis*), inasmuch as it involves a circular process of recognition and action between the inside and the outside of an organism. Indeed, there are a number of functional circles involved in the animal's experience; Uexküll emphasizes that this experience is *non-mechanical*—that is, not reducible to input and output processes, stimulus and reaction. It is not that there is no reaction, but that there is a process of interpretation (of production of meaning) involved: "The living being's reaction is, even in the simplest organisms, the selection of a response habit appropriate to the circumstance" (Uexküll, 1926, p. 129). In the *Umwelt und Innenwelt der Tiere* (1909, revised 1921), Uexküll distinguishes between mechanical or mechanistic (*maschinell*) properties and supra-mechanical (*übermaschinell*) properties (Uexküll 2014, p. 27).

Now, this analysis of umwelt and functional circles has a number of implications and allows of a number of interpretations. One sees the functional circle as a kind of "organizational closure."[3] Another sees a kind of anti-subjectivist, and almost anti-specist, intuition: "There is a panorama of perceptions and ways of approaching the world that are closed to us humans but continuously lived by other life forms" (Parikka, 2010, p. 66); in Sagan's vivid description, Uexküll, "as a scientist, shamanized himself into what it must feel like to be other organisms" (Sagan, 2024, p. 105). This is ironic given Uexküll's implicit anthropocentrism.

It may seem strange to think of Uexküll in these terms, given his insistence on the plurality of umwelten. But there is at least a tacit anthropocentrism in his thought, or at least a sense that there is a crucial divide between the multiplicity of animal umwelten and the more "open" human umwelt. It is not clear if Uexküll prefers a view of "gradations" or, as in the *Bedeutungslehre* of 1940, to view all animal *Umwelten* as equally closed in contrast to a specific human openness. Stjernfelt sees Uexküll as maintaining a sharp distinction between animal *Umwelten* and the human *Umwelt* that must be given up (Stjernfelt, 2007, p. 236), whereas Pagan and Dal Pozzolo see Uexküll's thought as much more open on the issue, but "employed against his intentions to draw a clear distinction between the environmental constraint of animals and the openness to the world of humans," notably in his reception in philosophical anthropology (Scheler, Plessner, and Gehlen), which contrasts the "closed nature of animal environments" to "the human capacity to open up to the world" (Pagan and Dal Pozzolo, 2024, p. 4), but also of course in Heidegger's enthusiastic reception of Uexküll in a lazily anthropocentric direction.[4] A more popular view of Uexküll in both the biosemiotics and environmental humanities receptions of his work, is to see him, again, as an anti-specist who proposes a "walk," "stroll," or "foray" into "unknown worlds" (Uexküll, 2010 p. 41).

More relevantly, the idea of functional circle has come in for some important counterarguments.

A criticism that was made in different ways by Plessner, Buytendijk, and Goldstein and more recently by Miquel (2023) focuses on the deterministic character of this process, its rigidity (Pagan and Dal Pozzolo, 2024). To borrow an example from Di Paolo, in a rigid model of umwelt and functional circle interrelation, receiving a blow or being stung by a wasp would be equivalent examples of behaviors to punching or stinging (Di Paolo, 2005, p. 442). And indeed, Uexküll sometimes described the "fit" between organism and environment as being like that between "pegs and sockets" (Uexküll, 1926, p. 317). Yet in other texts, Uexküll allows for more "leeway" or "wiggle room" in the animal's capacity to interpret the given situation in terms (or rather by means of) different functional circles. This less rigid view of living beings precisely treats them as *meaning-carriers* and perhaps even *meaning-makers*, rather than the more mechanistic view which is, as Thure von Uexküll wrote commenting on his father's ideas, "blind to meaning (*bedeutungsblind*)."[5]

Meaning is central here: animals are not "blind to meaning" because they *select and interpret stimuli* rather than being passive receptors threreof: "Every animal is a subject, which, in virtue of the structure peculiar to it, selects stimuli from the general influence of the outer world, and to these it responds in a certain way. These responses, in their turn, consist of certain effects on the outer world, and these again influence the stimuli" (Esposito, 2020, p. 41; Uexküll, 1926, p. 126). But such a statement is actually ambiguous, as it lends itself both to a more "active" reading and to a more "passive" reading, which is not so far removed from Pavlovian and otherwise psychophysical stimuli and responses.

The more active reading fits with the way umwelt resembles an ecological (and cognitive) niche, given that a core idea of niche construction is that organisms do not merely adapt to an external world—ticks, echidnas, and orcas have different "cognitive agendas"—but rather actively construct their environmental niches (Odling-Smee et al., 2003); Uexküll speaks of the organism's *Spielraum*.[6] Like scaffolding, affordances and ecological/cognitive niches, the umwelt is like the mirror image of Bernard's *milieu intérieur* in the sense that it is a "customized" environment, an environment organized by the organism (Goldstein, 1995, p. 85). But the ecological niche is a more objective concept than the umwelt, which could be rendered "subjective ecological niche" or, better, "semiotic niche," "ecological niche as apprehended by the organism" (Hoffmeyer, 2008, p. 171). Uexküll's version of ecological theory is a theory of the subject!

To treat an organism as a subject, as somehow engaged in the reception and perhaps even production of meaning, implicitly (or explicitly!) means defending a dimension of *agency*, since a contrario, organisms understood mechanistically lack agency. But what is an agent? Di Paolo's definition is useful and emphasizes engagement with the world: "A self-constructed unity that engages the world by actively regulating its exchanges with it for adaptive purposes that are meant to serve its continued viability" (Di Paolo, 2005, p. 443). What does it mean to investigate organisms non-mechanistically?

For one thing, in a rather Gibsonian sense, it means *not* treating the sensory organs like so many cameras and other mechanical devices. Uexküll commented favorably on Driesch's vitalist embryological ideas as striking a blow against the "machine view" of life, in an early paper on experimental biology, the later *Theoretische Biologie* (1920, 2nd ed. 1928) and finally the *Bedeutungslehre*.[7] In essence, the umwelt concept opposes a mechanistic account of animal behavior focusing on reflexes and determined by its surroundings. In the revised version of *Umwelt und Innenwelt der Tiere* (1921), Uexküll replaced the chapter on reflexes with a chapter on the functional cycle. In the *Bedeutungslehre*, Uexküll notably discusses Loeb's mechanistic theory, which reduces the organism–environment relation to one of tropisms—that is, the forced orientation of organisms (classically, plants) toward external energy sources (Uexküll, 1982, pp. 45, 47).

But the umwelt theory was not meant to be a naïvely vitalistic assertion of an inner drive, an inner force, or vitality that existed apart from external constraints. On the contrary, Uexküll sketches out a (rather dense and much-debated) model we might call interactionist (in the sense of thinkers like Lewontin and Oyama), given that it seeks to articulate a dynamic relation between the organism's *Bauplan*, the possible umwelten in which it is captured (like the spiderweb, which is an extension of the spider's Bauplan but also "incorporates" or involves the life cycle of the fly), and the somewhat metaphysical-sounding *Planmäßigkeit* or "conformity to a plan" (Uexküll, 1982, p. 30). The importance of biological design (plan, order) is considered so fundamental by Uexküll that in the introduction to his *Theoretische Biologie*, he stated aphoristically that "every design is from design" (Uexküll, 1926, p. xi).[8] *Planmäßigkeit* "was and is the problem of biology."[9] And this force—for he calls *Planmäßigkeit* a force—is described at times in terms quite similar to Driesch's entelechy, as a "supra-mechanical" (*übermaschinelle*) force (Uexküll, 1926, p. 121, and 132 on Driesch).

This version of the Bauplan idea, and the organism's conformity to such a plan, has something paradoxical about it. Namely, it is both quite reminiscent of purposiveness (the application of a fixed design) and is intended by Uexküll as somehow nonidentical with classical ideas of teleology and purposiveness, including because of the Bauplan's adaptive nature. As Stjernfelt puts it, "the telos does not consist in any goal outside the functional circle, but is simply defined by it" (Stjernfelt, 2007, p. 226), and as Cassirer noted, "There was no need for [Uexküll] to introduce any special purposive 'forces': it was enough to prove that the living world, in its totality and in its details, has a stable teleological *structure.*"[10] When Uexküll criticizes von Baer's overly teleological vision of biological development (Uexküll, 1926, pp. 316–317), his point is that the telos is not the adult organism but a congruity between organism and umwelt (Ostachuk, 2020, p. 164).

Much like the ambiguity noted above, between a more "active and interpretive" vision of organism and a more "stimulus-response" vision, the Bauplan concept in Uexküll's theoretical biology is ambiguous. On the one hand, Uexküll's model seems "to open itself to a dynamic conception of the organism/environment relationship, leaving room for the creativity of the living individual," but on the other hand, "the idea of a 'conformity to a plan'" seems instead "to fix the organism to its a priori endowment, perfectly harmonized with the specific environment" (Pagan and Dal Pozzolo, 2024, p. 5). Thus, Di Paolo (2020) thinks the idea of a perfect correspondence between organism and environment has to be challenged if there is to be a "future for Uexküll." It is unclear whether or not this ambiguity was ever fully resolved in Uexküll's work, and Goldstein and Canguilhem will, as we shall see, nudge this in a less "fixed" direction.

Now, we have not yet reached the more unique and bold part of the umwelt theory, for Bauplan-talk has something purely formal about it. What Etxeberria writes about Varela and Canguilhem applies equally to Uexküll: "As was emphasized by Varela and by the enactive perspective, for which Canguilhem's work was an important inspiration (Varela 1992; Di Paolo 2005), the relationship of the organism with the environment cannot be a fixed input-output correlation because it enacts or creates meaning related to what agents value" (Etxeberria, 2020, p. 308).

To return to the basic difference between *Umwelt*, *Umgebung*, and *Welt*, what truly distinguishes the first is the dimension of meaning, reflecting the organism's engagement with the world (in Di Paolo's phrase), the enaction or creating of meaning (in Etxeberria's phrase). Umwelt here is, as has been noticed before, much like a Gibsonian affordance, "neither an objective property nor a subjective property; or it is both if you like . . . [it] points both ways, to the environment and to the observer" (Gibson, 1979, p. 121). Indeed, much of the impact of Uexküllian theoretical biology has been in the direction of overturning the subject/object distinction (one which was famously challenged from many sides in philosophy, but less noticeably in biology). Common to biosemiotics, developmental systems theory and autopoietic theory is the strong claim that organism–environment relations yield a kind of "inner exterior and external interiority" (Hoffmeyer, 1998, pp. 33, 40). But these relations are crucially "colored" and conditioned by the production of *meaning*.

Toward the end of his life, Uexküll placed more and more emphasis on "meaning." An umwelt is crucially a construction of signs, of meaning (butyric acid, heat, and light for the tick; other signs, among the hundreds emanating from mammal bodies, are meaningless for it. Only these three become "feature carriers" for the tick: Uexküll, 2010, p. 51). The *Bedeutungslehre* opens with the claim "that life can only be understood when one has acknowledged the importance of meaning (*Bedeutung*)" (Uexküll, 1982, p. 26); "*the question of meaning is, therefore, the crucial one to all living beings*" (Uexküll, 1982, 37). In one of his many vivid examples, Uexküll speaks of the interaction (one might say the dialectical relation) between an octopus and seawater, and notes that the water becomes a "carrier of meaning" (*Bedeutungsträger*) for the animal, which uses it for its movements (Uexküll, 1982, 53). Every perceptual action "imprints its meaning on the meaningless object and thereby makes it into a subject-related meaning-carrier in the respective umwelt" (Uexküll, 1982, 31)—a flower, a bud, a drop of sap can be or cannot be a meaning-carrier depending on the umwelt. This is what I mean by "organisms as meaning makers."

This vision of biological agents as crucially bearers but also producers of meaning is, once again, opposed to a more mechanistic perspective, which Uexküll says is "blind to meaning."[11] Living beings live in universes of meaning, not as "detectors" of impersonal, objective features (the mammal's blood temperature for the tick), even if some can read the Uexküllian story as one of information processing, however much biosemioticians have tried to distinguish themselves from an information-theoretic vision of biology. But in the Uexküllian vein, organisms do not just "*detect* objective signals that have certain effects on their organism but *experience* aspects of their environment and their own body in that environment" (Emmeche, 2004, p. 212).

It was on this basis that Uexküll was presented, first by Sebeok and later Kull as a pioneer of biosemiotics (Kull et al., 2011). The core biosemiotic idea, as notably argued by Hoffmeyer (1996, 1998, 2008) is that biology has to integrate the fundamentally semiotic nature of life—that is, an understanding of organism as an entity which "has to master a set of signs, of a visual, acoustic, olfactory, tactile, and chemical nature, by means of which it can control its survival in the semiosphere. . . . The semiotic demands made on an organism are vital to its success" (Hoffmeyer, 1996, p. 59). In case the tacit reference to Uexküll was not obvious, Hoffmeyer adds that "one can never hope to understand the dynamic of the ecosystem without allowing for some form of *Umwelt* theory" (Hoffmeyer, 1996, p. 59). Differently put, the biosemiotic approach treats life as a "functional interpretation of signs in self-organized material code-systems making their own Umwelten" (Emmeche, 1998, p. 11).

Now, it is significant that the discussions over the meaning of umwelt in the secondary literature (particularly the biosemiotics literature) get tangled up regarding the extent to which it means "subjective," as Uexküll often speaks of the "subjective space" (*der subjektive Raum*) of the organism. For some, Uexküll's antireductionism leads him to present "a subjectivist epistemology based on biology" (Rüting, 2004, p. 49). Others emphasize the Kantian dimension that Uexküll himself seemed to take very seriously. As he wrote in the late 1930s, "Kant had already shaken the complacent position of the universe by exposing it as being merely a human form of perception. From there on it was a short step to reinstall the umwelt space of the individual human being in its proper position" (Uexküll, 2001, p. 109). Surprisingly to us who are accustomed to the popularity of references to Kant's third *Critique* in biological theory, the constant Kantian motif in Uexküll,[12] which is a naturalized Kantianism, does not build on the third *Critique*'s theory of organisms as organized wholes, but instead on the first *Critique* (Uexküll, 1907, p. 659). Uexküll holds that the Kantian constitutive spheres of space and time were not so much absolutes but rather special conditions of variation found in all animals and entities that sense. Indeed, in his early studies on tonus (e.g., Uexküll, 1905), he describes his position in classic Kantian epistemological terms: "to observe things as ends of Nature falls under the regulative use of reason"; and he notes that the "mistake of the vitalists" was to make constitutive use of a regulative function (Uexküll, 1905, p. 376).

To be sure, there are different kinds of naturalized Kantianism. Some, like Blumenbach in early nineteenth-century German biology, pursued a project of naturalizing teleology. In contrast, what it means to be a Kantian for Uexküll, is to bring *the subject* back into biology—hence his strong claim that "all reality is subjective appearance," in the first pages of the essay on theoretical biology: "*All reality is subjective appearance*. This has to be a great, fundamental understanding also of biology. . . . The task of biology is to widen the results of Kant's research in two directions: 1. to take into account the role of our body, especially of our sense-organs and our central nervous system and 2. to explore the relations of other subjects (the animals) to the objects" (Uexküll, 1926, p. xv; Uexküll, 1982, p. 9).

And toward the end of the *Streifzüge*, he writes, "Each and every subject lives in a world in which there are only subjective realities. . . . Whoever denies the existence of subjective realities has not recognized the foundations of his or her own environment" (Uexküll, 2010, p. 125).

Some scholars, who are more concerned with the precise features of Kant's thought, find that Uexküll simplified, vulgarized, or failed to grasp Kant's full intention (Clarizio, 2024). Others hold that Uexküll aimed to extend Kant's transcendental philosophy "to the entire living realm"—however much Kant would have to reject this attempt, as organisms are expressed through regulative, not empirical, concepts—and that both "shared the same solution: subjective spontaneity" (Esposito, 2020, p. 38)—that is, a subject-centred ontology (and biology for Uexküll). The subject here is treated "as the center of initiatives and not as a recording black box" (Esposito, 2020, p. 38). Others still, who defend a more "naturalized

Kantian" reading of Uexküll, treat the umwelt not as a "transcendental structure" but as a "pragmatic arrangement": "The *Umwelt* becomes the result of a pragmatic negotiation with the "outside" of the individual and is not reducible to a purely constituted structure in terms of an idealistic-transcendental perspective" (Ostachuk, 2020, p. 161; Pagan and Dal Pozzolo, 2024)—definitely not a very a prioristic version of the Kantian program.

As we can see, Uexküll's Kantianism is primarily a theory of subjective constitution, biologized, whether or not this fits with the "real Kant." Interestingly, Plessner criticized Uexküll not for having a wrong reading of Kant but for being *too Kantian*. For Uexküll, on Plessner's view, "the positional field would actually belong to the organism, just as Kant's forms of intuition of space and time belong to the subject and only the subject. The organism would then move in its positional field like the monad in the world, like a solipsist; in a surrounding field, but not in the actual world independent of it: in absolute immanence" (Plessner, 2019, p. 188).

Indeed, there is a lot of monadic language in Uexküll; speaking of drops of dew in a meadow, he writes that "each of these myriads of drops mirrors all the world with the sun, the mountains, the forests and the shrubs" (Uexküll, 1938a, p. 47), insisting precisely that he is not thereby a solipsist.[13] And more recent work on biological individuality would add to the umwelt concept that of nested individuality and also the dimension of multiple lives, multiple individualities in one organism, and conversely, the concept of colony individuals, superorganisms.

But the subject here is not a Kantian mind (nor, Uexküll would undoubtedly add, a materialistically understood "body," given his hostility to materialism); it is the organism.[14] And conversely, the organism thus understood is a "knower," a knowing subject (an original feature of Uexküll's umwelt concept is that it transcends the *Leib/Körper* distinction, or renders it invalid). And one should not downplay the idealist features in Uexküll's thought! Anne Harrington quotes an early paper of his on biology as *anschaulich*, which is oddly reminiscent of Canguilhem's defense of vitalism as a "lack of determinations" (Canguilhem, 1965, p. 86): "When I for example lay out the *Bauplan* of a worm, and in so doing use any convenient physical schema, it doesn't occur to me at all to touch upon a physical problem. One may always think of any other force as at work in the same object. I am not concerned with that. I seek only for a fitting expression in order to make the plan of the animal *anschaulich*."[15]

For Uexküll, the thinker of "subjective appearance" and "Anschauung," "structure is not a material thing" and "biology treats of the immaterial relationships of material parts united in a body" (Uexküll, 1930, p. 9, in Cassirer, 1969, p. 200). His *philosophie biologique*, unlike Goldstein's or Canguilhem's, includes explicit anti-materialist statements: "The constancy of matter, about which the materialists boast, is not a solid base for a comprehensive view of the world. The constancy of subjects is substantiated far better than the constancy of objects. The materialists will, however, interject that subjects also consist of matter. That is correct; but body-matter that subjects have must be built up anew from one generation to the next" (Uexküll, 1982, p. 71).

This strident anti-materialism is not present in Goldstein or Canguilhem. The new biology of umwelt ends up treading on very old metaphysical ground: Must a theory of the subject (of the organism as subject) be anti-materialist? I will leave this question aside and return to the question of the subjective point of view with Canguilhem, who presents his own post-

Kantian version of a theory of life and living beings in which these are crucially understood as knowing subjects, and before that, in Goldstein's version of the organism-as-subject and organism-as-meaning-maker theory.

3.3 Goldstein and Canguilhem

In his study of damaged brain–mind–world relations, Goldstein sketched out a new approach that was holistic and organismic *while operating at a primarily heuristic, non-ontological level*, and rejecting what he called "transcendental and vitalistic factors" (Goldstein, 1995, p. 28). In fact, Goldstein's original methodological sketch moves back and forth between a prudent *epistemological* stance (reminiscent of the Kantian regulative ideal of the third *Critique*) for which organisms are real and special *by virtue of the way we cognitively constitute them*, and a strong *ontological* stance for which organisms are real by virtue of fundamental, intrinsic characteristics.[16] I shall not go into the details of Goldstein's account, which at times sounds positively Heideggerian ("the organism is a being enduring in time," and even "in eternal time, for it does not commence with procreation, certainly not with birth, and does not end with death")[17]—although to be fair, these "existential" motifs crop up in Canguilhem too, and a number of "agential" organismic theories in recent decades have stressed temporality.[18]

A key claim in Goldstein's organicism is that biology "deals with individuals who exist and tend to exist, that is to say, to realize their capacity as best they can in a given environment. The performance of the organism in life is only understandable in terms of its relation to this fundamental tendency, i.e., only as an expression of the organism's process of self-actualization."[19] Sounding as if he were clarifying Uexküll's more murky metaphysical prose, Goldstein states, "The environment of an organism is by no means something definite and static but is continuously forming commensurably with the development of the organism and its activity," and crucially, "the environment emerges from the world through the being or *actualization* of the organism" (Goldstein, 1995, p. 85, emphasis mine). The organism's activity as self-actualization is an important addition to the theory under discussion, which was not so explicitly present in Uexküll.

Yet Goldstein does not assert a kind of Romantic subjectivity-in-nature, a vision of mind and life opposed to a "merely materialist" standpoint, as Uexküll but also many contemporary enactivists would. Rather, he primarily adopts a bottom-up approach, giving the example of learning how to ride a bicycle.[20] It is striking that Goldstein's theory of organism dispenses with ontological claims like "Living structures are thus ontologically emergent with respect to mere physical structures. They constitute a new order of nature that is qualitatively distinct from the merely physical order" (Thompson, 2007, p. 75), which almost perfectly restates Uexküll on the ontological status of structure: "Structure is not a material thing: it is the unity of immaterial relationships among the parts of an animal body. . . . Biology treats of the immaterial relationships of material parts united in a body so as to reconstitute the structure in imagination" (Uexküll, 1930, p. 9, in Cassirer, 1969, p. 200).[21] But Goldstein's bottom-up approach has, of course, a distinctly existential focus, which can be framed in a Kantian way, as Canguilhem will (differently from Uexküll's "naturalized Kantianism"). What I am calling a bottom-up (and methodological) approach, Clarizio calls

Goldstein's "pragmatist stance," which "finally breaks down the boundary between the empirical and the transcendental, asserting the normative primacy of the organism and its cognitive activity over any a priori categorical synthesis" (Clarizio, 2024).

Rather than presenting organisms as ontologically unique because of some "laundry list" of properties of life (a definition-focused strategy criticized in more detail in Wolfe, 2023b, and Wolfe, 2010), Goldstein's organicism significantly does *not* invoke definitional properties of life (or barely). He suggests that rather than state what is unique about the biological, we look to the *observer*: To be an organism is to have a *point of view* on organisms (something with echoes in second-order cybernetics[22])—one which produces intelligibility, which reveals organisms as meaning-producing beings. Goldstein credits Uexküll with some of this view: "Accurate investigations have shown that the individual organism is always fitted into a very specific environment, and that its existence, in spite of all variability, hinges ultimately on an environment that is adequate for it. Uexküll's research is basic to this point and is so generally valid that it no longer meets with much opposition. In cases of brain injuries, our experience has everywhere shown the equivalent results."[23] And he restates the umwelt theory in his own terms, while remaining quite close to Uexküll: "For each organism, not everything that occurs in the external world belongs to its environment [*Milieu* in German—CW]. Its existence and its 'normal' performance are tied to the fact that there can be such an adaptation between its structure and the environmental processes leading to the formation of an adequate milieu. . . . Every organism has its milieu, as Uexküll has so clearly shown."[24] However, Goldstein also revises Uexküll's theory, noting that the environment of an organism "is by no means something definite and static but is continuously forming commensurably with the development of the organism and its activity" (Goldstein, 1995 p. 85). And "the situation is not like this"—that is, is not so definite—for "each organism lives in a world that by no means contains only such stimuli as are adequate for it. It lives not merely in its 'own environment' [*Milieu*] but in a world in which all possible sorts of stimuli are present and act on it. The organism must cope with this 'quasi-negative' environment [*negative Umwelt*]" (Goldstein, 1995, p. 85). And this was a familiar point of Uexküll, who never denied that organisms respond to stimuli but insisted that these stimuli depend on a prior selection-and-interpretation process occurring in and through the organism's perceptual dynamics (this is his core biosemiotic claim, which is also his "vulgar Kantian" claim). As Clarizio nicely observes, "while for Kant the subjective character of the organism hinders the possibility of constituting it as an object of knowledge, for Goldstein this subjective character is, on the contrary, precisely what is to be known and what allows access to the very idea of organism" (Clarizio, 2024).

A version of this idea, which is absent in Uexküll but prominent in Goldstein (and especially Canguilhem), stresses the role of dysfunction and error as at the heart of vital normativity. Canguilhem added a celebrated chapter to the revised edition of *The Normal and the Pathological* entitled "A New Concept in Pathology: Error" (Canguilhem, 1972, p. 207f), which Foucault perfectly encapsulated in the formula "life is what is capable of error" (Foucault, 1989, p. 22; Talcott, 2013). Only living systems can make errors; as Kull puts it, "Meaning may appear only in systems in which something sometimes goes wrong" (Kull, 2015, p. 617).

In essays and lectures starting in the late 1940s, Canguilhem associates Uexküll and Goldstein, articulating a kind of "organic self-realization" theory that he sees in Uexküll dealing with animals and in Goldstein dealing with persons in contexts of illness and dysfunction (Canguilhem, 1965, p. 96). The draft version of his important essay on the topic, "The Living and Its Milieu" (in Canguilhem, 1965), includes notes on Uexküll's *Streifzüge*,[25] but his thoughts on umwelt, Uexküll, and Goldstein occur across a number of publications, including his essay "Aspects of Vitalism" (in Canguilhem, 1965). His general thought is that with Uexküll and Goldstein, "biologists came to understand that the very essence of life (*le propre du vivant*) is to compose its own environment," and that "it is characteristic of the living being that it makes its environment (*milieu*, translating Uexküll's *Umwelt*) for itself, that it composes its environment."[26]

Canguilhem is a very favorable reader of Uexküll and endorses the latter's critique of the overly deterministic concept of *milieu*. But Canguilhem, too, can be enlisted in a critique, not of the concept of umwelt *per se*, but of that of functional circle. In a recent essay appraising Canguilhem for theoretical biology, Miquel notes that Canguilhem criticizes the closed aspect of the functional circle in his essay on the milieu (Canguilhem, 1965, p. 144) because he understands the organism as fundamentally heterogeneous: "The individuality of the living does not stop at its ectodermal borders, not more than it begins with the cell" (Canguilhem, 1965, p. 144) There is never full physiological closure (Miquel, 2023 p. 197). That is, "on one hand, the environment is nothing but a part of the living organism. But on the other hand it is more. The role played by the functional biosemiotics circle is already an extension of the role played by the internal milieu, so that internal milieu and functional biosemiotics circle are already *entangled*. The organism cannot be defined only by its internal milieu, as asserted by Claude Bernard" (Miquel, 2023, p. 196). On this view, Canguilhemian vital normativity is clearly distinct from Uexküllian functional circles, notably since it depends on other, heterogeneous levels of organization.

Miquel's observation closely resembles Plessner's critique of Uexküll. For Plessner, "the functional circle is not configured as a closed monad-like order even at the level of lower animals" (Pagan and Dal Pozzolo, 2024, n.p.). Like Di Paolo, for whom an umwelt-and-functional-circle system cannot differentiate between punching and being punched, they note the absence of a reflexive dimension. For Plessner, the organism–environment relation "is not an absolutely rigid bond . . . , but rather a bond within certain boundaries that provide the living being with leeway [*Spielraum*]" (Plessner, 2019, p. 228). By implication, to summarize both Plessner's and Miquel's insights, in a rigid relation between *Umwelt* and *Bauplan*, there is no learning, no plasticity. Canguilhemian vital normativity points instead to the organism's capacity to reshape and recompose its functional circle. As Canguilhem writes when discussing regulations in the organism and in society, "an organism is an entirely exceptional mode of being, because there is no real difference, properly speaking, between its existence and the rule or norm of its existence. From the time an organism exists, is alive, that organism is possible, i.e., it fulfils the ideal of an organism; the norm or rule of its being [*existence*] is given by its existence itself" (Canguilhem, 2002a, p. 106–107). Canguilhem attributes this capacity for reshaping, this plasticity, to both Uexküll and Goldstein, but it is actually much more developed in Goldstein (and depending on how deterministically one reads Uexküll, could be more of a "wedge" between Goldstein and Canguilhem on the one hand, and Uexküll on the other hand): It is the idea of the organism as an epistemic agent, of life and knowledge as interwoven and mutually dependent processes.

Beyond Canguilhem's explicit references to Goldstein (who he had had translated into French in 1951, with Merleau-Ponty's help), Goldsteinian themes are visible in key aspects of his work: the idea of vital normativity but, more importantly here, the approach to organisms understood as interpretive entities, as *creators of meaning*. The Goldstein-Canguilhem approach to organisms holds that to be alive, to act, means, in addition to other metabolic processes, to be engaged in processes of meaning-production. In Weber and Varela's terms, which read like a literal paraphrase of Canguilhem channeling Goldstein, "organisms are subjects having purposes according to values encountered in the making of their living" (Weber and Varela, 2002, p. 102). It is not just that organisms are "meaning-carriers," in Uexküll's phrase, because there is an additional *projective* dimension—that is, what I have referred to in earlier work on Canguilhem as an "attitudinal vitalism."[27] A vitalism of projection, attitude, and meaning is implied in Canguilhem's idea that the vitalist *considers* life in one way rather than another and, moreover, emphasizes the capacity of living beings *to consider each other* in such terms rather than theorizing "vital force" or "stuff" as some neutral, objective entity. As Foucault phrased it well in his commentary on Canguilhem, "The biologist must grasp what makes life a specific object of knowledge and thereby, what makes it such that in the world of living beings, and *because* they are living beings, there are beings with the capacity of knowledge, ultimately the capacity of knowing life itself" (Foucault, 1989, p. 20, trans. modified).

One of the unusual features of Goldstein's analysis reappears in Canguilhem's distinctive approach to the singularity of organisms: how he oscillates between a "Kantian" *epistemological* stance in which organisms are real and special *by virtue of the way we cognitively constitute them*, and a strong *ontological* stance in which organisms are real by virtue of fundamental, intrinsic characteristics that happen to be present. Yet—and here we return to the vexed question of the "subject point of view in biology"—this convenient distinction between the epistemological (projective, constitutive) view of biological entities and the ontological view of the same entities (the latter resembling a form of strong vitalism) becomes somewhat blurred when Canguilhem introduces a vitalist complication: that an objective ("ontological") characteristic of living beings may lie in the fact that these entities *are "interpreting" sensu* Goldstein and Uexküll, and more specifically that these entities *need to view* other entities as organismic, purpose-driven, vital entities ("Aspects of Vitalism," in Canguilhem, 1965/2008a).

And Canguilhem's account of living beings is distinctly existential, which complicates the otherwise straightforward ontological dimension mentioned above, particularly when he describes this interpretive position as belonging to a kind of fundamental existential attitude: "The life of a living being . . . only recognises the categories of health and illness on the level of experience . . . not on the level of science. Science explains experience but does not thereby cancel it out" (Canguilhem, 1972, p. 131). Experience cannot be wholly "cancelled out" by science: This sounds irreducibly subjective. Canguilhem, like Goldstein and Uexküll, does not think the specificity of organisms is grasped via neutral, third-person descriptions of their properties (like homeostasis). Yet he does not opt for a flat-out subjectivism either.

On the one hand, Canguilhem's notion of organism is Uexküllian: it is a "relational" notion, "because the organism establishes a connection between physiological regulation and vital needs and ecological aspects enacting meaning and values" (Etxeberria, 2020, p. 325). On the other hand, his notion of organism is Goldsteinian: "The purpose of the organism is internal to the organism, and consequently, the ideal we must restore is the organism itself."[28] What unites these three authors is an "organism-centred" conception of organism–environment interaction, organism as an autonomous subject, the bearer of its own normativity that generates meaning and values. But Canguilhem naturalizes Uexküll further.[29]

3.4 Conclusion

I have examined Uexküll's, Goldstein's, and Canguilhem's answers to the question of how and why organisms are subjects, and meaning-makers. The story could be told in strictly cumulative, linear terms, in which each thinker provides specific elements, cumulatively (perhaps with some overlap, as with certain Uexküllian insights that are shared and reworked by Goldstein, and then taken over by Canguilhem). Or one could treat them as full, self-contained monadic contextual-umwelten, so to speak: Uexküllian cognitive ethology, Goldsteinian neuropsychological theory, Canguilhemian philosophy of medicine. And these are not just "merely contextual" aspects: it is striking that the environment plays a greater role in Uexküll, while the "self-actualization" features of organisms are more emphasized by Goldstein and Canguilhem (including in his celebrated idea of vital normativity).

But it is also important to note differences and tensions in the ways the core theme is addressed.

For instance, only some organicist arguments invoke a deeper interiority or self, as in Thompson, echoing the Varelian project of a "first-person science" (Varela and Shear, 1999): "Life realizes a kind of interiority, the interiority of selfhood and sense-making."[30] The extent to which Canguilhem would subscribe to such foundationalism of subjectivity either "in the body" or as an irreducible characteristic "of the body" depends on his conception of the organism as subject, as "knower," and as locus of self-actualization. This particular commitment is not present in the Goldsteinian-Uexküllian account in which the environment emerges from the world "through the being or actualization of the organism" (Goldstein, 1995, p. 85). Yet this account does not dispense with the subject. Similarly, Uexküll's idea that "all reality is subjective appearance" brings with it a certain amount of baggage (Kantian, subjectivist), but on some readings, it can easily be naturalized, including as a biosemiotic theory.[31] This works quite easily with the umwelt theory taken on its own, but if we take Uexküll and Goldstein together, it is harder to see how such a result could emerge.

Uexküll, Goldstein, and Canguilhem were, I think, on to something when they insisted that rather than say what is unique about the biological, we look to the *observer*: To be an organism is to have a *point of view* on organisms—one which produces intelligibility, which reveals organisms as meaning-producing beings. Put in more biology-friendly language, close to Uexküll's, living beings are "*generators of meaning*," which "mine their environment (including their own bodies) for things that have meaning for them: moisture, nutrients, warmth" (Ball, 2023, p. 6); "it is precisely by virtue of being a thing . . . that can autonomously attribute meaning, that an entity can be said to be alive" (Ball, 2023, p. 37). For some, the organism-dependence of meaning necessarily implies the existence of minds. But as Ball continues in this accidentally Uexküllian vein, "For some feature of an organism's environment to acquire meaning, the organism doesn't have to be 'aware' of it . . . life can be considered to be a *meaning generator*" (Ball, 2023, p. 37).

Unlike Goldstein's focus on mind, Uexküll and Canguilhem would endorse this more biologistic approach to meaning, in which the construction and production of meaning "is enabled by the plastic attunement and re-organization of the web of adaptive processes in an experience-dependent manner thus producing through the organism's ontogeny new parameters to observe and new responses to their changes" (Di Paolo, 2005, pp. 441–442). Notice that the more fixed, more static aspects of *Bauplan* and *Planmäßigkeit* have dropped out of view here; perhaps they were not part of the "future of Uexküll." Conversely, the self-actualization dimension, presented as crucial for understanding organisms by Goldstein and Canguilhem, is less present in Uexküll.

I have sketched out both an account of organisms as meaning-makers and some challenges (and internal instabilities) facing this account. Is it necessarily a theory of organisms as subjects? How important is the Kantian motif in these three thinkers? And is even that aspect fully internally consistent (e.g., Canguilhem's Kantianism is more constructivist, Uexküll's seems to hang on a theory of subjective cognitive constitution, and Goldstein's is somewhere in between)? In the end, Uexküll and his reception did not yield a clear, sharp, coherent theory of meaning. As many attempts already exist to present Uexküll as a concrete source of models for biosemiotics (and its future), I did not try to repeat that exercise, but rather focused on the philosophical power and originality of this idea of organisms as meaning-makers, first presented by Uexküll, further fine-tuned by Goldstein and Canguilhem with sometimes different emphases and valences. Its ultimate "lack of determination," to borrow from Canguilhem, is not a virtue but a vice in my view.

Acknowledgments

Versions of this paper have been presented in 2024 at the Séminaire sur le vivant (ENS Paris), the meeting on Logic and Life: Signs of Vitalism (Ghent), and the conference The Organic and the Normative (Ljubljana). Thanks to Snait Gissis, Dries Josten, Paul-Antoine Miquel, and Gertrudis Van de Vijver for helpful comments and suggestions.

Notes

1. Merleau-Ponty, 1945/2012, p. 470. Thompson states rather peremptorily that Husserl's thought is not egological or subjectivity-centered, adding that the unpublished works show this (Thompson, 2007). I find this really less than convincing. See, for example, Schaeffer, 2007, p. 118, but also Canguilhem's trenchant critique of Merleau-Pontyian subjectivity in his late lecture on health (Canguilhem, 2002b; Wolfe, 2023a).

2. "*Umwelt*, désigne le milieu de comportement propre à tel organisme; *Umgebung*, c'est l'environnement géographique banal et *Welt*, c'est l'univers de la science" (Canguilhem, 1965, p. 144).

3. As also noted by Plessner, who speaks in this regard of "the closed form, the organizational idea of the animal." On organizational closure, see Mossio and Moreno, 2010.

4. It is known that Uexküll was not only favorably taken up by Heidegger (and subsequently others in what one might charitably call non-naturalistic philosophies of life) Wolfe, 2018, but Uexküll himself wrote approvingly about the similarities between their views: Uexküll, 1937, p. 199, cit. Harrington, 1996, p. 54n83, and for more on Uexküll and Heidegger, Buchanan, 2008. I acknowledge that my references to Plessner are very partial here.

5. Uexküll, 1980, p. 42–43, cit. in Rüting, 2004, p. 67.

6. On the history of the idea of organism–environment interaction, see Pearce, 2014.

7. For a short but incisive comparison of Driesch and Uexküll, see Cassirer, 1969, p. 199f.

8. "An der Satz: *Omnis cellula e cellula* darf man den Satz hinzufügen: Alles Planmässige aus Planmässigem."

9. "Die Planmäßigkeit der Organismen war und ist das Problem der Biologie" (Uexküll, 1908, p. 79, and 1913, p. 26, cit. Kull, 2001, p. 5); Uexküll, 1938a, p. 87, in Pagan and Dal Pozzolo, 2024, p. 7. On *Planmäßigkeit*, see Hoffmeyer, 2008, p. 173f.

10. Cassirer, 1969, p. 200; for more on Cassirer's reading of Uexküll, see Stjernfelt, 2011.

11. Uexküll, 1980, pp. 42–43, cit. in Rüting, 2004, p. 67.

12. From early articles to his better-known monographs, including *Umwelt und Innenwelt der Tiere*, *Theoretische Biologie*, and the *Bedeutungslehre*; Buchanan, 2008, p. 12f; Samain, 2021; Stjernfelt, 2007.

13. Stella and Kleisner refer to Friedrich Brock, a long-term student and collaborator of Uexküll's and member of his Institut für Umweltforschung, as introducing, in a 1934 discussion of Uexküll, the concept of an *organismus-umwelt-monade*, which he presented as crucial in the development of a new discipline, idealistic biology (Stella and Kleisner, 2010, p. 41). A number of authors in the early generations of Uexküll-reception speak of his ideas in "monadic" terms, including Plessner, who describes Uexküll's thought as a "biological monadology" (Plessner, 2019, p. 192). Uexküll refers explicitly to Leibniz and monads in his 1940 novel *The Stone of Werder* (*Der Stein von Werder*: Cheung, 2006, p. 237).

14. Further, Uexküll's rough-and-ready Kantian subjectivism does not yield a "perspectivism" in our sense because his quasi-monadological metaphysics (he happily describes his theory as a metaphysics: Uexküll, 1982, p. 42) insists that there is nevertheless a "world," even if it is composed of an infinity of umwelten (small mosquito suns and larger human suns) united by a set of Bauplans and indeed by a deeper *Planmäßigkeit*: "Planmäßigkeit includes not only the 'fitting in' of a living entity into its Umwelt, but also the 'fitting in' of all Umwelten into the whole of nature" (Schnödl and Sprenger, 2022, p. 189n18); Samain suggests that *Planmäßigkeit* might be a constitutive notion in Kantian terms (Samain, 2021, p. 122n30). Even if Uexküll, taking note of Einstein's discoveries, insisted that there is no absolute space and time, so that indeed the "world" is a multiplicity of subjective spaces.

15. Uexküll, 1905 p. 377, trans. in Jennings, 1909, p. 333, and Harrington, 1996, p. 40. Uexküll speaks of *Anschauung* even earlier, in a 1902 paper (Schnödl and Sprenger, 2022, p. 123n43). See the lengthy (and partly critical) discussion of Anschauung and Anschaulichkeit in Uexküll in Jennings, 1909, p. 229f.

16. Recall that the Kantian concept of organism is explicitly built around the concept of a regulative ideal (*Critique of Judgment* §§ 67, 73). Organism according to Kant is a "reflexive" construction rather than a "constitutive" feature of reality, and reflective judgments are incapable of justifying objective assertions. It might seem obvious to us that the methodological and heuristic (and thus non-ontological, or ontologically neutral) dimension of Goldstein's theory is a positive feature, but this was by no means self-evident to all of his readers: Jonas, who defended an anti-naturalistic vision of the organism aiming at a positive ontology of Life, considers that Goldstein's methodological reading of his own work is "too modest" (Jonas, 1959, p. 162).

17. Goldstein, 1995, p. 387. Interestingly, in a late paper, Goldstein pointed to the differences between his point of view and that of existential psychiatry (Goldstein, 1959, pp. 11–12).

18. On temporality in Uexküll (in addition to the more prominent "spatial" features of *Planmäßigkeit*), see Kull, 200, p. 6.

19. Goldstein, 1971, p. 440; cf. also Goldstein, 1995, pp. 101, 290, 308, 325, 391

20. Goldstein, 1995, p. 307, a very "Merleau-Pontyian" example

21. This more bottom-up approach in Goldstein I think also explains why Canguilhem was much more interested in Goldstein than in Merleau-Ponty (Wolfe, 2023a).

22. See the early work of Gertrudis Van de Vijver.

23. Goldstein 1995, p. 84, with a footnote indicating the 1926 translation of *Theoretische Biologie*. There are actually two parts to Goldstein's Uexküll-reception and discussion, the better known being the positive but critical discussion of umwelt, but also, a wholly positive discussion of Uexküll and Bethe's earlier work on excitation and reflexes, in the chapter of *The Structure of the Organism* dealing with reflex action (Harrington, 1996, p. 42; Rüting, 2004, p. 40).

24. Goldstein, 1995, pp. 105–106. See also Goldstein, 1995, pp. 73, 84, 88–90, 106, 343.

25. Canguilhem, G. 1946–1947. ms. GC.12.1.8, "Le vivant et son milieu," feuillets 95–96, Fonds Canguilhem, Centre documentaire CAPHES, ENS, Paris.

26. Canguilhem, 1947b in Canguilhem, 2015, p. 300; Canguilhem, 2008a, p. 111 (translation modified). We need to be careful regarding the specific meanings, histories, and contexts of terms like umwelt, "milieu" and environment, which are not synonymous; but both Goldstein and Canguilhem casually slip from one to another, as noted, sometimes using *milieu* in German, sometimes translating *umwelt* as *milieu* as Canguilhem does.

27. Wolfe, 2011; for more on Canguilhem as an "attitudinal vitalist," see Van de Vijver and Haeck, 2023. I had used attitude-and-projection language in earlier work but happily note that Di Paolo speaks (in his case referring to Varela) of a "projective teleology" of sense-making in a very similar way (Di Paolo, 2005, p. 433).

28. Canguilhem, "Le problème des régulations dans l'organisme et dans la société," in Canguilhem, 2002a, p. 108.

29. Canguilhem uses Uexküll and Goldstein against a deterministic concept of milieu, for both scientific and ethico-political reasons (Feuerhahn, 2009, p. 438). *Le normal et le pathologique* is an explicit attack on a conservative umwelt worldview in which "The world in which the baker does 'baker things' and Mr. Schulz perceives 'Schulz things' is to stay the way it is, precisely because it cannot be otherwise" (Schnödl and Sprenger, 2022, p. 17, referring to Uexküll, 1938b, p. 64). Schnödl and Sprenger's book, along with Stella and Kleisner's important article, is a rare attempt to integrate the worst sides of Uexküll's work (notably his *Staatsbiologie*) into an overall analysis of his thought, rather than passing it under silence. Ironically, Canguilhem, a very favorable reader of Uexküll, is one of the most articulate and lucid critics of illegitimate "conceptual transfers" from the biological to the social, in his discussions of milieu and regulation.

30. Thompson, 2007, p. 238; on "sense-making" as a distinctive feature of enaction, in autopoietic systems (p. 139).

31. Not everyone is fully convinced by the biosemiotic reading of Uexküll, though: for Samain, we must acknowledge that "Uexküll's semiotics is more like a theory of affordances translated into Goethean formulas than a theory of meaning" (Samain, 2021, p. 130).

References

Ball, P. 2023. *How life works. A user's guide to the new biology*. University of Chicago Press.

Bianco, G., Wolfe, C., and Van de Vijver, G. 2023. Introduction. In G. Bianco, C. Wolfe, and G. Van de Vijver (Eds.), *Canguilhem and continental philosophy of biology* (pp. 2–8). Springer.

Buchanan, B. 2008. *Onto-ethologies: The animal environments of Uexküll, Heidegger, Merleau-Ponty, and Deleuze*. SUNY Press.

Canguilhem, G. 1947a. Note sur la situation faite en France à la philosophie biologique. *Revue de Métaphysique et de Morale*, *52*(3–4), 322–332.

Canguilhem, G. 1947b. Milieu et normes de l'homme au travail. *Cahiers internationaux de sociologie*, *2*(3), 120–136.

Canguilhem, G. 1965. *La connaissance de la vie* (Revised and expanded ed.). Vrin. (Original work published 1952)

Canguilhem, G. 1972. *Le normal et le pathologique* (3rd revised ed). PUF. (Original work published 1943, second much expanded edition 1966)

Canguilhem, G. 2002a. *Écrits sur la médecine*. Seuil.

Canguilhem, G. 2002b [1988]. La santé, concept vulgaire et question philosophique. In G. Canguilhem (Ed.), *Écrits sur la médecine* (pp. 49–68). Seuil.

Canguilhem, G. 2008a. *Knowledge of life* (S. Geroulanos and D. Ginsburg, Trans.). Fordham University Press. (A translation of Canguilhem, 1965)

Canguilhem, G. 2008b. Health: Crude concept and philosophical question. *Public Culture*, *20*(3), 467–477. (Translation of Canguilhem, 2002b, by T. Meyers and S. Geroulanos)

Canguilhem, G. 2015. *Œuvres complètes, tome IV: Résistance, philosophie biologique et histoire des sciences (1940–1965)* (C. Limoges, Ed.). Vrin.

Cassirer, E. 1969. *The problem of knowledge*. Yale University Press.

Cheung, T. 2006. Cobweb stories: Jakob von Uexküll and the Stone of Werder. *Place and Location: Studies in Environmental Aesthetics and Semiotics*, *5*, 237–241.

Clarizio, E. 2024. Le concept d'organisme de Kant à Goldstein: de la philosophie transcendantale au pragmatisme. *Philosophia scientiae*, *28*(3), 37–54.

Di Paolo, E. 2005. Autopoiesis, adaptivity, teleology, agency. *Phenomenology and the Cognitive Sciences*, *4*(4), 429–452.

Di Paolo, E. 2020. Afterword. A future for Jakob von Uexküll. In F. Michelini and K. Köchy (Eds.), *Jakob von Uexküll and philosophy* (pp. 252–256). Routledge.

Emmeche, C. 2004. Organicism and qualitative aspects of self-organization. *Revue internationale de philosophie*, *228*(2), 205–217.

Esposito, M. 2020. Kantian ticks, Uexküllian melodies, and the transformation of transcendental philosophy. In F. Michelini and K. Köchy (Eds.), *Jakob von Uexküll and philosophy* (pp. 36–51). London: Routledge.

Etxeberria, A. 2020. Regulation, milieu, and norms: Canguilhem's individual organisms as relations. In P.-O. Méthot and J. Sholl (Eds.), *Vital norms. Canguilhem's* The normal and the pathological *in the twenty-first century* (pp. 295–332). Hermann.

Feuerhahn, W. 2009. Du milieu à l'Umwelt: Enjeux d'un changement terminologique. *Revue philosophique de la France et de l'étranger, 134*(4), 419–438.

Foucault, M. 1989. Introduction to G. Canguilhem. In *The normal and the pathological* (C. Fawcett, Trans.) (pp. 7–24). Zone Books.

Gibson, J. J. 1979. *The ecological approach to visual perception.* Houghton Mifflin.

Goldstein, K. 1934. *Der aufbau des organismus: Einführung in die biologie unter besonderer berücksichtigung der erfahrungen am kranken menschen.* Martinus Nijhoff.

Goldstein, K. 1995. *The organism: A holistic approach to biology derived from pathological data in man.* Zone Books. (A translation of Goldstein 1934)

Goldstein, K. 1959. Notes on the development of my concepts. *Journal of Individual Psychology, 15*, 5–14.

Goldstein, K. 1971. Remarques sur le problème épistémologique de la biologie (1951). In K. Goldstein (Ed.), *Selected papers/ausgewählte schriften* (A. Gurwitsch, E. M. Goldstein Haudek, and W. E. Haudek, Eds.) (pp. 439–442). Martinus Nijhoff.

Harrington, A. 1996. *Reenchanted science: Holism in German culture from Wilhelm II to Hitler.* Princeton University Press

Hoffmeyer, J. 1996. *Signs of meaning in the universe.* Indiana University Press.

Hoffmeyer, J. 1998. Surfaces inside surfaces. *Cybernetics and Human Knowing, 5*(1), 33–42.

Hoffmeyer, J. 2008. *Biosemiotics: An examination into the signs of life and the life of signs.* University of Scranton Press.

Jennings, H. S. 1909. The work of Uexküll on the physiology of movements and behavior. *Journal of Comparative Neurology and Physiology, 19*, 313–336.

Jonas, H. 1959. Kurt Goldstein and philosophy. *American Journal of Psychoanalysis, 19*, 161–164.

Kull, K. 2001. Jakob von Uexküll: An introduction. *Semiotica, 134*(1–4), 1–59.

Kull, K. 2015. Semiosis stems from logical incompatibility in organic nature: Why biophysics does not see meaning, while biosemiotics does. *Progress in Biophysics and Molecular Biology, 119*, 616–621.

Kull, K., Emmeche, C., and Favareau, D. 2011. Biosemiotic research questions. In C. Emmeche and K. Kull (Eds.), *Towards a semiotic biology* (pp. 67–90). Imperial College Press.

Lewontin, R. 1985 [1983]. The organism as the subject and object of evolution. In R. Levins and R. C. Lewontin, *The dialectical biologist* (pp. 85–106). Harvard University Press.

Loeb, J. 1916. *The organism as a whole, from a physiological viewpoint.* Knickerbocker Press.

Merleau-Ponty, M. 2012 [1945]. *Phénoménologie de la perception.* Gallimard.

Miquel, P.-A. 2023. What is biological normativity? In G. Bianco, C. Wolfe, and G. Van de Vijver (Eds.), *Canguilhem and continental philosophy of biology* (pp. 195–208). Springer.

Mossio, M., and Moreno, A. 2010. Organisational closure in biological organisms. *History and Philosophy of the Life Sciences, 32*(2–3), 269–288.

Ostachuk, A. 2020. The organism and its umwelt: A counterpoint between the theories of Uexküll, Goldstein and Canguilhem. In F. Michelini and K. Köchy (Eds.), *Jakob von Uexküll and philosophy* (pp. 158–171). Routledge.

Odling-Smee, J., Lala(nd), K., and Feldman, M. 2003. *Niche construction: The neglected process in evolution.* Princeton University Press.

Pagan, M., and Dal Pozzolo, M. 2024. From the harmony to the tension: Plessner and Goldstein's readings of Jakob von Uexküll. *History and Philosophy of the Life Sciences, 46*(9), 1–23.

Parikka J. 2010. *Insect media: An archaeology of animals and technology.* University of Minnesota Press.

Pearce, T. 2014. The origins and development of the idea of organism-environment interaction. In G. Barker, E. Desjardins, and T. Pearce (Eds.), *Entangled life* (pp. 13–32). Springer.

Plessner, H. 2019. *Levels of organic life and the human: An introduction to philosophical anthropology* (M. Hyatt, Trans.). Fordham University Press.

Rüting, T. 2004. History and significance of Jakob von Uexküll and of his institute in Hamburg. *Sign Systems Studies, 32*, 35–72.

Sagan, D. 2024. James Lovelock, Gaia, and the remembering of biological being. In Y. Hui (Ed.), *Cybernetics for the 21st century, I: Epistemological reconstruction* (pp. 101–108). Hanart Press.

Samain, D. 2021. Le béhaviorisme sémiotique de Jakob von Uexküll. *Histoire Épistémologie Langage, 43*(2), 113–136.

Schaeffer, J.-M. 2007. *La fin de l'exception humaine.* Gallimard.

Schnödl, G., and Sprenger, F. 2022. *Uexküll's surroundings. Umwelt theory and right-wing thought.* Meson Press.

Stella, M., and Kleisner, K. 2010. Uexküllian *Umwelt* as science and as ideology: The light and the dark side of a concept. *Theory in Biosciencies, 129*, 39–51.

Stjernfelt, F. 2007. A natural symphony? Von Uexküll's *Bedeutungslehre* and its actuality. In Stjernfelt (Ed.), *Diagrammatology* (pp. 225–240). Springer.

Stjernfelt, F. 2011. Simple animals and complex biology: Von Uexküll's two-fold influence on Cassirer's philosophy. *Synthese, 179*, 169–186.

Talcott, S. 2013. Georges Canguilhem and the philosophical problem of error. *Dialogue, 52*, 649–672.

Thompson, E. 2007. *Mind in life: Biology, phenomenology, and the sciences of mind.* Harvard University Press.

Uexküll, J. 1905. Studien über den Tonus III: Die Blutegel. *Zeitschrift für Biologie, 46*, 372–402.

Uexküll, J. 1907. Die Umrisse einer kommenden Weltanschauung. *Die neue Rundschau, 18*, 641–661.

Uexküll, J. 1908. Die neuen Fragen in der experimentellen Biologie. *Rivista di Scienza 'Scientia', 4*(7), 72–86.

Uexküll, J. 1913. *Bausteine zu einer biologischen weltanschauung. Gesammelte aufsätze* (F. Gross, Ed.). F. Bruckmann.

Uexküll, J. 1926. *Theoretical biology* (D. L. Mackinnon, Trans.) Harcourt Brace & Company.

Uexküll, J. 1928. *Theoretische biologie* (2nd revised ed.). Springer.

Uexküll, J. 1930. *Die lebenslehre*. Müller & Kiepenheuer.

Uexküll, J. 1937. Die neue umweltlehre: Ein bindeglied zwischen natur- und kulturwissenschaften. *Die Erziehung: Monatschrift fur den Zusammenhang von Kultur und Erziehung im Wissenschaft und Leben, 13*, 185–199.

Uexküll, J. 1938a. *Der unsterbliche geist in der natur. Gespräche*. Wegner.

Uexküll, J. 1938b. Zum verständnis der umweltlehre. *Deutsche Rundschau, 256*(7), 64–66.

Uexküll, J. 1982. The theory of meaning. *Semiotica, 42*(1), 25–82. (Trans. of *Bedeutungslehre*, 1940)

Uexküll, J. 2001. An introduction to Umwelt (1937). *Semiotica, 134*(1–4), 111–123.

Uexküll, J. 2010. *A Foray into the worlds of animals and humans, with a theory of meaning* (J. D. O'Neil, trans. D. Sagan, intr.). University of Minnesota Press. (Translation of *Streifzüge durch die Umwelten von Tieren und Menschen*, 1934)

Uexküll, J. 2014 [1921]. *Umwelt und innenwelt der tiere* (B. Herrmann and F. Mildenberger, Eds.). (First edition 1909)

Uexküll, T. (Ed.). 1980. *Jakob von Uexküll. Kompositionslehre der natur: Biologie als undogmatische naturwissenschaft. Ausgewählte schriften.* Verlag Ullstein.

Uexküll, T. 1982. Introduction: Meaning and science in Jakob von Uexküll's concept of biology. *Semiotica, 42*(1), 1–24.

Van de Vijver, G., and Haeck, L. 2023. Canguilhem's divided subject: A Kantian perspective on the intertwinement of logic and life. In G. Bianco, C. T. Wolfe, and G. Van de Vijver (Eds.), *Canguilhem and continental philosophy of biology* (pp. 123–146). Springer.

Varela, F. J. 1992. Autopoiesis and a biology of intentionality. In B. McMullin (Ed.), *Proceedings of the workshop "Autopoiesis and Perception"* (pp. 4–14). Dublin City University.

Varela, F. J., and Shear, J. 1999. First-person methodologies: Why, when and how. *Journal of Consciousness Studies, 6*(2–3), 1–14.

Weber, A., and Varela, F. J. 2002. Life after Kant: Natural purposes and the autopoietic foundations of biological individuality. *Phenomenology and the Cognitive Sciences, 1*, 97–125.

Wolfe, C. T. 2010. Do organisms have an ontological status? *History and Philosophy of the Life Sciences, 32*(2–3), 195–232.

Wolfe, C. T. 2011. From substantival to functional vitalism and beyond: Animas, organisms and attitudes. *Eidos, 14*, 212–235.

Wolfe, C. T. 2014. The organism as ontological go-between: Hybridity, boundaries and degrees of reality in its conceptual history. *Studies in History and Philosophy of Biological and Biomedical Sciences, 48*, 151–161.

Wolfe, C. T. 2018. Review of D. Tarizzo, Life: A modern invention, trans. M. Epstein (Minneapolis: University of Minnesota Press, 2017), *Notre Dame Philosophical Reviews*

Wolfe, C. T. 2023a. Canguilhem and the promise of the flesh. In G. Bianco, C. Wolfe, and G. Van de Vijver (Eds.), *Canguilhem and continental philosophy of biology* (pp. 181–191). Springer.

Wolfe, C. T. 2023b. Varieties of organicism–a critical analysis. In M. Mossio (Ed.), *Organization in biology* (pp. 41–58). Springer.

Wolfe, C. T. 2024. A note on the situation of biological philosophy. *Revue internationale de philosophie, 307*(1), 95–110. (Special issue: Canguilhem)

II SKETCHES

4 From Stochasticity to Ecological Intelligence and the Nature of Choice

Raymond Noble and Denis Noble

4.1 Overview

Life involves both problem-solving and creativity. This is the origin of purpose. All organisms encounter uncertainty when addressing challenges or evaluating their environments. Their interactions with the surrounding world are inherently dynamic and creative. In this process, whether intentionally or unintentionally, organisms leave signals of their presence and activities These signals can be significant, depending on the species, and may influence other organisms. For instance, plants may produce signals to attract bees and other pollinators, or they may release substances to deter potentially harmful species (Bouwmeester et al., 2019; Clarke et al., 2013; Fürstenberg-Hägg et al., 2013). Similarly, animals mark their home ranges and territories to indicate their presence and identity to others (Johnson, 1973). For a variety of reasons, organisms are surrounded by intelligent data that holds cognitive significance.

4.2 Intelligence and Choice

Intelligence is communicated directly between organisms or groups of organisms. This communication includes information about mood, emotional states, and well-being, as well as intentions, their whereabouts, and how successful they are at foraging for food, as in Herring Gulls communicating over several kilometers (Frings et al., 1955). Organisms also anticipate the behavior of others, and this intelligence has played a significant role in the evolution of sensory and cognitive abilities. Moreover, organisms do not simply passively receive sensory data; they actively engage with their habitat and anticipate incoming information (Ede et al., 2018).

Survival requires continuous adaptation and the ability to change behavior in real time. Additionally, survival is influenced by identity and the acknowledgment of individuals or groups (Smith-Vidaurre et al., 2023). This recognition is essential for the emergence of creativity, and many of the factors involved are semiotic in nature.

An ongoing assessment of others, whether they are members of the same species or different ones, also contributes to self-identity (Yorzinski, 2017). This assessment forms a crucial component of living beings as conscious agents: awareness, anticipation, and readiness to act (Noble et al., 2014).

This active engagement is volitional and involves choices. Thus, a significant question is what possesses such volition and agency, and why these concepts cannot be fully understood through a reductionist approach. Organisms are functionally more than just the sum of their parts. However, throughout the twentieth and the first quarter of the twenty-first century, modern biology has emphasized genetics and the genome as the regulatory systems issuing instructions to cells and organisms, controlling them "body and mind." The organism in this view is regarded as a vehicle to maintain genes in a "gene pool," and genes are seen as both primary causal with the overriding objective of maintaining them (Dawkins, 1976). We argue that by reversing this perspective, we can better recognize and understand the agency of organisms; that is, genes are tools of the system and not its master. The entity with volition and agency is the organism and/or social groupings of organisms.

We write this article in recognition of the influence of biosemiotics on our own work on the origins of purpose and intentionality (Emmeche and Kull, 2011). We agree that "biosemiotics can no longer be seen as a marginal school of thought"; it is central to an understanding of life (Kull et al., 2011, p. 4) and lies at the core of the concept of ecological intelligence in our own book, *Understanding Living Systems* (Noble and Noble, 2023). In this article, we argue that (1) the abilities to plan or imaginatively solve problems, make choices and anticipate outcomes, whether in our own behavior or in the behavior of others, has evolved alongside other abilities, such as sensory perception and movement. (2) We argue that meaning, reason, and situational logic are key ingredients of ecological intelligence, presenting a "non-material" aspect to how organisms perceive their surroundings, influencing their motivational states and the choices they may make. (3) This integrative view provides a compatibility for the concept of free will with a materialistic system. Furthermore, (4) what has been essential in the crafting of these faculties and their creative use is the fluidity and openness of the material processes involved.

4.3 Signs Have Both Material and Immaterial Association

Signs and interpretations of signals cannot simply be reduced to the physical manifestations of those signals; signals have contextual meaning and purpose. In physiology, discussions about signaling often overlook this distinction, treating the identification of physical processes as sufficient for complete analysis. Science tends to dislike teleological explanation (Noble and Noble, 2023); reason, purpose, and motive are difficult to weigh or measure and lack the discreteness of a physical event. However, this difficulty should not blind us to the significance of purpose and intention, for they are components of this ecological language. Chemical signals affect the nervous systems through chemical and biophysical processes that follow deterministic laws. If all biological events can be described in this way, then any other explanation might be considered superfluous. In this viewpoint, "why" is considered beyond the realm of scientific enquiry or simply a deception. This sidelining of purpose arises partly from a longstanding philosophical issue regarding how nonmaterial factors—such as meaning and logic—influence physical processes. This problem became more difficult to resolve since Descartes introduced his concept of dualism, the separation of mind and body, a consequence of which was to disengage reason and logic from the material being (Descartes, 1641). In The Sixth Meditation, Descartes makes the distinction clear: defining the body as a thing that does not think, and the mind a thing that thinks and not an extended thing. One of the aims of this article is to explain why common forms of dualism in biology fail to address a critical issue: how physical processes align with nonmaterial interpretations of value and meaning.

4.4 Organisms Are Not Machines

We (organisms) are not machines; our material existence does not function like the mechanical workings of fixed cogs and wheels. While we sometimes represent the brain's processes diagrammatically in such a way, machines typically operate as rigid objects with linear functionality: A causes B, which produces an output C, leading to Z; this is a straightforward cause-and-effect, sequential relationship.

In contrast, living organisms are not regulated by such fixed mechanisms. Biological processes involve a continuous ebb and flow that conditions and predisposes potential actions. We experience these processes as emotions, desires, hopes, fears, or anxieties, which can also manifest as physiological changes, such as heart rate, blood pressure, and temperature, along with a variety of other physiological states.

Our inputs, processes, and outputs are flexible and can alter our sense of being. Even when we perform tasks, the manner in which we do them evolves in the doing, as does the potential for further actions. We do not respond mechanically to specific inputs, although many reflexes help us maintain balance while moving. Our actions influence our senses, and our interactions with others can change the things we do and how we do them, both in the moment and from our memories.

We learn to improve our abilities, use situational awareness, and make logical decisions. We may also collaborate with others to decide and act together or encourage them to do things differently or in unison. Instead of being a mere collection of conditioned reflexes, our behavior is purposefully shaped and finely tuned to achieve our goals (Ginsburg and Jablonka, 2007; Ginsburg and Jablonka, 2019).

4.5 Nested Function with No Privileged Level of Causation

What underlies this functional yet directed lability? We propose four key factors: (1) Living systems use stochasticity (variability) at all functional levels (Noble and Noble, 2018); (2) This functionality is nested; organelles operate within cells; cells function within tissues; tissues are part of organs; organs are integrated into systems; and systems exist within organisms, all interconnected through social and ecological interactions (D. Noble and R. Noble, 2023); (3) There is no single direction for causality in what is an open system (D. Noble and R. Noble, 2021; R. Noble and D. Noble, 2021). However, we argue that if there is a preferred level of causality, it lies at the more integrative and psycho-social levels (Noble et al, 2019). We also include the concepts of readiness, anticipation, and integrative awareness (consciousness). All these features are essential for intentional agency, or free will.

Living organisms harness stochasticity not at any particular point in an otherwise algorithmic decision process but as the essential ingredient of life itself—variability. Variability exists at all levels of organization, from the molecular through the cellular, tissue, organs, and whole-body function, including social arrangements and behavior. It is not just a random number generator at the bottom level of the molecular organization of living systems or at any moment of creative decision. Creative decisions are directed, not by chance alone. The latter can be subject to reason, assessment, and choice. Crucially in living systems, this is harnessed in nested integrative functions.

At the cellular level, stochasticity is harnessed to create the electrochemical gradients necessary for cells to function, secrete, contract, and signal. A mechanistic view of the behavior of organisms needs to suppose that these processes necessarily yield to thermodynamic descriptions of the global behavior of large numbers of stochastic events. Nevertheless, the evidence from neurophysiology is that this is not the case. Neurons often perform at the very edge of stochasticity. When physiologists record from single nerve cells, they listen to this crackling by transducing the electrical changes to sound waves, thus giving the experimenter an immediate perception of the extent of stochasticity involved. Thus, the postsynaptic potentials evoked by tiny packets of transmitter released from presynaptic boutons are heard as a variable waffling when recorded with an electrode, while impulses generated at threshold voltage give distinct clicks. By constraining and controlling these potentials, the networks control the impulse (signal) threshold allowing specific inputs to trigger controlled trains of impulses. While signals (impulses) arise from and against this background waffle, the continuously variable postsynaptic potentials significantly affect their generation, the frequency and pattern, and are regulated up or down in a continuous analogue process by other controlled inhibitory of excitatory influences. Thus, neuronal networks do not operate like switches on or off; they are in a continuous modulation in integrative function (Brown et al., 1987; Noble and Short, 1989). Noble and Short demonstrated this for the somatosensory system, where a preceding stimulus changes the strength of an incoming signal from the receptive field through presynaptic and postsynaptic modulation (Maxwell and Noble, 1987; Noble and Short, 1989; Noble and Riddell, 1989). In this way, both the strength of an incoming signal, altering the release of chemical transmitters, and the sensitivity of the postsynaptic transmission cells are changed in the moment; such adjustments provide greater acuity in locating the stimulus on the skin and its overall effects on ongoing perception. In this way, process is a better concept than mechanism (Dupré, 2020).

Similarly, when the heart rhythm is recorded electrically in normal physiological conditions, the frequency is found to vary stochastically from beat to beat. This variability is a critical indicator that normal physiological control processes are functioning, such as the sympathetic and parasympathetic neuronal systems, harnessing cardiac function in relation to bodily actions or emotions. Furthermore, the heart has its own sophisticated neural network that is capable of learning (Herring and Paterson, 2021; Pedroni et al., 2024). In contrast, absence of variability is a sign of abnormality (Vries et al., 1982; Zizzo et al., 2020). Thus, for example, obstetricians will monitor the strength of both fetal heartbeat and its variability to assess fetal well-being (Heart Rate Variability, 1996). Despite this variability, and because of its underlying causes, the cardiovascular system produces controlled changes in heart rate and blood pressure and flows in physiological functions. This is not a linear sequence of causality; it is modulation by factors working interactively to produce a physiologically meaningful (purposeful, system-directed) outcome.

4.6 Clues in the Functional Boundaries of Biological Systems

Consider a random event—a particle that can move spontaneously in any direction. That direction is not predictable precisely, but what is predictable is that it will move. Consider now an event precipitated by the moving particle, or more likely, the movement of thousands or millions of such particles. Let us contain them in a chamber and consider that, as the particles move, the "pressure" increases (that is definitively so). If we knew this relationship existed, we could predict that pressure would rise in the chamber if particles moved faster, for example, with change in temperature. Thus, a prediction can be reasonable even if the underlying events are random.

This physical phenomenon is Brownian motion, named after the botanist who first described it, Robert Brown, in 1827, and it led Einstein to his first Nobel Prize in 1921: the random motion of particles suspended in a liquid or a gas resulting from their collision with the fast-moving atoms or molecules in the gas or liquid. According to the theory, a substance's temperature is proportional to the average kinetic energy with which its molecules are moving or vibrating. Biological systems harness such stochasticity at all levels to generate and respond to signals and create the conditions suitable for the complex chemical reactions of life.

Cell membranes are not just simple barriers; they play a vital role in controlling cell function. They can allow certain electrolytes and materials to pass through, but not others, making them variably permeable. This function of membranes is crucial for life on Earth. Through messengers, cell membranes regulate genome expression and other cell processes (for a review, see Brown, 2002). They act as an interface between the cell and other cells and tissues, forming part of the body's regulatory systems. The conditional logic in cell function, which is like the "IF X THEN Y ELSE Z" clauses in a computer program, lies in the membranes and the proteins (channels, exchangers, receptors) that are sensitive to the internal and external membrane environment. An example of this is how channel proteins change shape, altering the permeability to ions as the voltage across the membrane changes (voltage-dependent channels). Changing shape is vital to the function of "carrier proteins" transferring other molecules across the membrane (Alberts et al., 2002). There is no "program of life" in a genome since the genes only provide faculties that the living organism uses. It is easy enough to consider these extraordinary and complex biochemical changes to be sufficient to explain what is happening in the control of the genome; but this intercellular ecology is in turn influenced by the ecological processes of organisms, providing a line for social factors controlling the expression of the genome. A mother's touch of her baby, for example, can be a key player in fine-tuning its genome (Pickles et al., 2013). If touch can do it, then why not sight, smell, sound, and well-being (Dunbar, 2010)?

Cells influence other cells by changing the ion concentrations and the presence of other key chemicals in the solution bathing them. They release chemicals that influence neighboring cells. For example, this is a key ingredient of local control of the flow of blood to the tissues for the transfer of nutrients and elimination of waste products. But they can also secrete chemicals that will influence other cells and organs at a distance. This endocrine system modifies and conditions functions in the body. The ebb and flow of such signals influence the state of functionality. It creates an informative internal environment of signals used in the priming and coordination of the cells and tissues.

The body is a buzz of electrochemical processes that at any moment can be harnessed to produce movement from the contraction of muscles, just as they are involved also in the motivation to move muscles to achieve objectives at any moment. Such electrochemical processes are involved in the function of the billions of neurons and the billions of synaptic connections in our brains from moment-to-moment shaping or modifying function in goal-directed ways.

4.7 The Bubbling Cauldron of Synaptic Function

The human brain is made up of billions of functional entities, including neurons and support networks of cells. These neurons create functional networks with billions of synapses, which are the functional connections between them. The function of these networks ebbs and flows, rather than stopping and starting abruptly. At the synapses, there is much that is stochastic, but also much that is shaped and harnessed to produce meaningful change in output or signals. This ebb and flow involve both excitatory and inhibitory networks that modulate the process of inputs and outputs. The function of the brain is not an aggregate of neuronal function. On the contrary, states of functionality exist at the network and whole brain level of organization, which can be observed when recording the electrical activity and whole brain functions. These states in turn influence function.

Much of the function of the brain is anticipatory of events rather than simply receptive. Neural pathways from the brain modulate sensory reception and perception. Synapses are not functionally fixed. Sensory stimulation or activation is needed to shape the organization of the billions of connections in the mammalian cortex. Additionally, the function of synapses changes over time and in response to experience. This facilitation process may create a form of memory of neuronal networks and synaptic pathways. It takes humans twenty years to hone these connections.

Lomo and Bliss discovered that a burst of stimuli to neurons in the brain can produce long-term potentiation of the subsequent release of neurotransmitters, a phenomenon of synaptic plasticity (Bliss and Lomo, 1973). Furthermore, environmental enrichment is associated with the activation of "plasticity genes" in the hippocampus and cortex, leading to increased numbers of dendrites and synapses. This cascade of environmental, neuroendocrine, and genetic processes restores and maintains learning and memory, both of which are important for problem-solving (Fischer et al., 2007).

4.8 Gene-Centric Dualistic View

There has been a tendency in science to equate "causality" with determinism. This is particularly true for reductionist approaches that emphasize upward causation from molecular behavior to organism behavior. The gene-centered perspective has further distorted this understanding by focusing on concepts like "codes," "blueprints," or the "book of life." As a result, the gene has been treated as a trump card, especially in the context of the "selfish gene" theory of evolution. This perspective becomes absurd when attempts are made to quantify the contributions of "genes" or "environment" to complex traits such as intelligence.

We propose an alternative view: Genes are tools that are used and controlled by the organism. Additionally, organisms leverage stochastic processes, even at the gene level, where errors can be corrected. Genes are essential in creating and maintaining the faculties to sense and act in the environment, but they are not used in the process of doing so. These faculties have evolved in relation to ecological intelligence and language (signs, their significance, and interpretation), but that intelligence and language can only be understood in relation to situational logic and awareness. Ecological intelligence and situational awareness work in harmony in the choices made by organisms.

4.9 The Self and the Symphony That Is the Forest

This leads us to a proposition: Language facilitates the sharing of perception and experiences and becomes part of creativity, but it is not necessary for abstract perception as such, even as much as it facilitates thought. Organisms can make assumptions (abstract) about situations without language. Thus, it is wrong to conclude that language is necessary (1) for consciousness (self-awareness or being alert to the active engagement of experiencing) and (2) for abstracting and using situational logic. The dominance of linguistics in human thought processes should not close our eyes to thought, which is initially nonlinguistic. Creating symbols or representations of experience might be the progenitor of language and facilitate associative learning. Categorizing events can also occur without language, as it uses the memory of past events and makes assumptions about or prediction of future events. The forest is a symphony of signs.

Monkeys tend to be noisy, chattering and shrieking, particularly to convey alarm, want, and feelings, to encourage, or to prevent behavior in others. The great apes are less noisy but will hoot and scream in numerous ways. It has primarily been considered "genetic." However, while genes might be involved in categories of behavior, genes are not involved in the precise or specific use; the sounds are molded in the moment and can uniquely identify an individual. Thus, warning other group members is situational, as is using sounds or other behavior, such as facial expression, expressing or alerting others to feelings, such as anger and other emotions, consciously (intentionally) or subconsciously (unintentionally). Indeed, a group might be made aware of the feelings of others from their behavior and other signs without those making the signs doing so with a conscious intention. However, this does not rule out such conscious intent; a relatively simple repertoire used unintentionally in some situations can also be used with intent in others, particularly after associated learning of their effect on others. We should distinguish the faculty (repertoire) from the specific use of such abilities in given situations. Chimpanzees have been observed to communicate deliberately (with reason) with an intelligence that can only be explained by situational logic. Chimpanzees will also create signs to convey specific intent. The assumption that chimpanzees cannot understand human language is precisely that—an assumption—but it is not the point. They can learn signs (Jensvold et al., 2023), which is significant because it demonstrates the abstractness and awareness attributed to humans.

4.10 Home Range, Self, Signs, and Ecological Intelligence

The concept of home range in behavioral ecology was introduced by Burt in 1943. He defined it as "that area traversed by an individual in its normal activities of food gathering, mating, and caring for young." However, Darwin had previously noted in 1861 that animals tend to restrict their movements within certain areas, and that home ranges are not just defined geographically but also by what the animals are doing. This behavior is significant in the process of natural selection or adaptation for survival. With the advent of telemetry and tracking, it became possible to assess home ranges in a more dynamic sense. Another way of considering the concept of home range is to see it as an area about which an animal has ongoing cognizance, or from which it obtains and gives information regularly. This is what we refer to as ecological intelligence in our recent book, *Understanding Living Systems* (Noble and Noble, 2023).

Ecological intelligence involves a distinction between self and others, including species, and group for identity relationships. The ability to recognize individuals or their relatedness plays an important role in mammalian social behavior. For instance, a giant panda may be considered a solitary mammal, yet it has an ongoing awareness of other pandas within or overlapping with its home range, as well as other organisms. Various cues such as sights, sounds, scents, and markings provide a dynamic interactive assessment of what is happening in the home range, which in turn influences an animal's behavior within it. This range is a kaleidoscope of signs that can evoke memories of past events and anticipation of the future. Information from a range of senses, such as visual recognition of physical features, or vocal cues, can be used for discrimination among conspecifics (Johnson, 1973).

The significance of any particular location cannot be judged by the time spent there. It would require insight into the ecological intelligence to understand its significance. The frequency of visits might reflect the anticipation of change, and what would be significant or prioritized would change with the seasons, and possibly from visit to visit over time. Additionally, new signs might be found or old signs renewed by the visitor, making the assessment of the home range an ongoing process.

4.11 Situational Logic and Biological Function

Situational logic, though not a material entity, is an integral part of an individual's perception and comprehension of a situation. For instance, if Jack were to climb up the hill along with Jill, we could presume that they have a reason for doing so, such as fetching a pail of water, or that they were in love. Even if a dog were not capable of putting it into words, it could still make assumptions based on association (Ginsburg and Jablonka, 2019), and assume that Jack and Jill were fetching water, or make any other assumptions. The crux of the matter is not whether they were fetching water, but that the possibility of assuming it exists. The process of assuming logic or reason in a perceived event is functional and has biological correlates in brain function.

Situational logic is aligned with function, in the sense that the situation influences an individual's biological state of being, encompassing perception, anticipation, emotions, desires, needs, and actions. This process of assumption does not require language, but it does require situational logic and intelligence (signs and their interpretations). The assumptions about Jack and Jill require an ability to "imagine" that things could be other than they are or might be—simply put, an ability to make such assumptions about the behavior of others. The development of language at the psycho-social level facilitates this in providing for the communication of narratives. In this sense, living organisms have an ongoing and modifiable narrative. Wittingly or unwittingly, they have a tale to tell other organisms.

Understanding situational logic as part of biological function has a long history in natural philosophy. It is similar to the "doctrine of signs" proposed by Locke in 1689, who referred to it as a third branch of science. Locke elaborates on the nature of this third category explaining it in the following terms: "Thirdly, the third branch [of sciences] may be termed . . . the doctrine of signs, the most usual whereof being words, it is aptly enough termed also . . . logic; the business whereof is to consider the nature of signs the mind makes use of for the understanding of things or conveying its knowledge to others" (Locke, 1689). Organisms in numerous ways make a continuous assessment and reassessment of their home range and, in the process, also leave significant signs for others. Such assessments impact on behavior, such as, for example when it changes with the population density.

4.12 Conclusion

In this paper, we have made the case that the concept of free will involves choices made by an organism to achieve objectives that are motivationally and rationally derived. The key components of free will include (1) signs related to ecological intelligence; (2) the ability to anticipate the significance of such signals; and (3) the capacity to use these signals to influence the behavior of others purposefully. In this context, consciousness can be understood as a state of integrative awareness that encompasses the anticipation of social, ecological, or situational logic. Signs, gestures, and language form a substrate for an immaterial interdependency—a psycho-social manifestation of living beings—not a separate entity in parallel with the material being but created by and functionally intertwined with it.

References

Alberts, B., Johnson, A., Lewis, J., et al. 2002. *Molecular biology of the cell* (4th ed.). Garland Science. (Ch. Principles of Membrane Transport)

Bliss, T. V., and Lomo, T. 1973. Long-lasting potentiation of synaptic transmission in the dentate area of the anaesthetized rabbit following stimulation of the perforant path. *The Journal of Physiology*, *232*(2), 331–356.

Bouwmeester, H., Schuurink, R. C., Bleeker, P. M., and Schiestl, F. 2019. The role of volatiles in plant communication. *The Plant Journal*, *100*(5), 892–907.

Brown, A. G., Koerber, H. R., and Noble, R. 1987. An intracellular study of spinocervical tract cell responses to natural stimuli and single hair afferent fibres in cats. *The Journal of Physiology*, *382*, 331–354.

Brown, T. A. 2002. *Genomes* (2nd ed.). Wiley-Liss, Oxford. (Ch 12. Regulation of Genome Activity)

Burt, W. H. 1943. Territoriality and home range concepts as applied to mammals. *Journal of Mammalogy*, *24*(3), 346–352.

Clarke, D., Whitney, H., Sutton, G., and Robert, D. 2013. Detection and learning of floral electric fields by bumblebees. *Science*, *340*(6128), 66–69.

Dawkins, R. 1976. *The Selfish Gene*. Oxford University Press.

Descartes, R. 1641. *Meditations on First Philosophy*. Cambridge University Press.

de Vries, J. I., Visser, G. H., and Prechtl, H. F. 1982. The emergence of fetal behaviour, I: Qualitative aspects. *Early Human Development, 7*, 301–322.

Dunbar, R. I. M. 2010. The social role of touch in humans and primates: Behavioural function and neurobiological mechanisms. *Neuroscience & Biobehavioral Reviews, 34*(2), 260–268.

Dupré, John. 2020. Life as process. *Epistemology and Philosophy of Science, 57*(2), 96–113.

Ede, F. van, Chekroud, S. R., Stokes, M. G., and Nobre, A. C. 2018. Decoding the influence of anticipatory states on visual perception in the presence of temporal distractors. *Nature Communication, 9*(1), 1449.

Emmeche, C., and Kull, K. (eds.). 2011. *Towards a semiotic biology: Life is the action of signs*. Imperial College Press.

Fischer, A., Sananbenesi, F., Wang, X., Dobbin, M., and Tsai, L.-H. 2007. Recovery of learning and memory is associated with chromatin remodelling. *Nature, 447*, 178–182.

Frings, H., Frings, M., Cox, B., and Peissner, L. 1955. Auditory and visual mechanisms in food-finding behavior of the Herring Gull. *The Wilson Bulletin, 67*(3), 155–170.

Fürstenberg-Hägg, J., Zagrobelny, M., and Bak, S. 2013. Plant defense against insect herbivores. *International Journal of Molecular Sciences, 14*(5), 10242–10297.

Ginsburg, S., and Jablonka, E. 2007. The transition to experiencing, II: The evolution of associative learning based on feelings. *Biological Theory, 2*, 231–243.

Ginsburg, S., and Jablonka, E. 2019. *The evolution of the sensitive soul: Learning and the origins of consciousness*. MIT Press.

Heart rate variability. 1996. Standards of measurement, physiological interpretation, and clinical use. Task Force of the European society of cardiology and the North American society of pacing and electrophysiology. *European Heart Journal, 17*, 354–381.

Herring, N., and Paterson, D. J. 2021. The heart's little brain: Shedding new light and CLARITY on the "black box." *Circulation Research, 128*(9), 1297–1299.

Jensvold, M. L., Dombrausky, K., and Collins, E. 2023. Sign language studies with chimpanzees in sanctuary. *Animals, 13*(22), 3486.

Johnson, R. P. 1973. Scent marking in mammals. *Animal Behaviour, 21*(3), 521–535.

Kull, K., Emmeche, C., and Hoffmeyer, J. 2011. Why biosemiotics? An introduction to our view on the biology of life itself. In C. Emmeche and K. Kull (Eds.), *Towards a semiotic biology: Life is the action of signs* (pp. 1–21). Imperial College Press.

Locke, J. 1689. *An essay concerning human understanding* (Book 4: chapter 21). Thomas Basset.

Maxwell, D. J., and Noble, R. 1987. Relationships between hair-follicle afferent terminations and glutamic acid decarboxylase-containing boutons in the cat's spinal cord. *Brain Research, 408*(1–2), 308–312.

Noble, D., and Noble, R. 2021. Rehabilitation of Karl Popper's ideas on evolutionary biology and the nature of biological science. In Z. Parusniková and D. Merritt (Eds.), *Karl Popper's science and philosophy* (pp. 193–209). Springer.

Noble, D., and Noble, R. 2023. How purposive agency became banned from evolutionary biology. In P. A. Corning, S. A. Kauffman, D. Noble, J. A. Shapiro, R. I. Vane-Wright, and A. Pross (Eds.), *Evolution "on purpose": Teleonomy in living systems* (pp. 221–235). MIT Press.

Noble, D., Noble, R., and Schwaber, J. 2014. What is it to be conscious? In J. R. Smythies, L. R. Edelstein, and V. S. Ramachandran (Eds.), *The claustrum: Structural, functional, and clinical neuroscience* (pp. 353–363). Academic Press.

Noble, R., and Noble, D. 2018. Harnessing stochasticity: How organisms make choices. *Chaos, 28*, 106309.

Noble, R., and Noble, D. 2021. Can reasons and values influence action: How might intentional agency work physiologically? *Journal for General Philosophy of Science, 52*, 277–295.

Noble R., and Noble, D. 2023. *Understanding living systems*. CUP.

Noble, R., and Riddell, J. S. 1989. Descending influences on the cutaneous receptive fields of postsynaptic dorsal column neurons in the cat. *The Journal of Physiology, 408*(1), 167–183.

Noble, R., and Short, A. D. 1989. Spatial spread of in-field afferent inhibition in the cat's spinocervical tract. *The Journal of Physiology, 413*, 107–118.

Noble, R., Tasaki, K., Noble, P. J., and Noble, D. 2019. Biological relativity requires circular causality but not symmetry of causation: So, where, what and when are the boundaries? *Frontiers in Physiology, 10*, 827.

Pedroni, A., Yilmaz, E., Del Vecchio, L., Bhattarai, P., Vidal, I. T., Dai, Y-W. E., Koutsogiannis, K., Kizil, C., and Ampatzis, K. 2024. Decoding the molecular, cellular, and functional heterogeneity of zebrafish intracardiac nervous system. *Nature Communications, 15*, 10483.

Pickles, A., Hill, J., Breen, G., Quinn, J., Abbott, K., Jones, H., et al. 2013. Evidence for interplay between genes and parenting on infant temperament in the first year of life: Monoamine oxidase a polymorphism moderates effects of maternal sensitivity on infant anger proneness. *Journal of Child Psychology and Psychiatry*, *54*(12), 1308–1317.

Smith-Vidaurre, G., Pérez-Marrufo, V., Hobson, E. A., Salinas-Melgoza, A., and Wright, T. F. 2023. Individual identity information persists in learned calls of introduced parrot populations. *PLoS Computational Biology*, *19*(7), e1011231.

de Vries, J. I., Visser, G. H., and Prechtl, H. F. 1982. The emergence of fetal behaviour, I: Qualitative aspects. *Early Human Development*, *7*, 301–322.

Yorzinski, J. L. 2017. The cognitive basis of individual recognition. *Current Opinion in Behavioural Sciences*, *16*, 53–57.

Zizzo, A. R., Kirkegaard, I., Hansen, J., Uldbjerg, N., and Mølgaard, H. 2020. Fetal heart rate variability is affected by fetal movements: A systematic review. *Frontiers in Physiology*, *11*, 578898.

5 Signs of Life: Interkingdom Communication and the Sympoietic Organism

Scott F. Gilbert

5.1 An Introduction to a New Biological Order

The concept of organism has changed rapidly during the past twenty years. Indeed, if the twentieth century has been characterized as "the century of the gene" (Keller, 2002), the twenty-first century may become characterized as the century of the holobiont relationship. Holobionts are defined as the host organism plus the symbionts living on or in it (Theis et al., 2016). Mammals, for instance, are organisms whose anatomy, physiology, immunity, development, mental health, and evolution are generated in concert with other symbiotic organisms. Rather than being seen as the readout of a single zygotic genome established at fertilization, organisms are increasingly being appreciated as holobiont collectives composed of several species of organisms, each playing a role in generating and sustaining the holobiont organism.

This essay will expound the following two principles for modeling and discussing living beings:

First, living beings are holobionts, and a holobiont is both an organism and a collection of ecological communities (Gilbert, 2019, 2023; Suarez and Stencel, 2020). These communities change as an organism develops, such that holobiont ontogeny entails both embryological development and ecological succession. The symbionts of a holobiont are each other's environment, and symbiotic relations are critical for uniting organisms into their respective environments. Symbiosis is the norm of life. "Life, we claim, is typically found at the collaborative intersections of many lineages, and we even suggest that collaboration should be seen as a central characteristic of living matter" (Dupré and O'Malley, 2009). This holobiont nature of life brings biosemiotics to the center of biological discussions since mutualistic symbiosis is an existential endeavor in communication, mediated through molecules that have context-dependent signaling functions.[1]

Second, metabolism is the mechanism by which organisms use and circumvent the general entropic pull of the Second Law of Thermodynamics. This metabolic flux constitutes the physical constraint that provides both identity and changing substance to organisms. There is a dialectic between an organism and "its" metabolism. While metabolism can be seen as a function of the organism, the organism can also be seen as function of its metabolism. Indeed, metabolic flows exist between organisms and help distinguish between the organism and its environment (Jonas, 1966; Landecker, 2013). By coupling chemical reactions that acquire energy from sunlight with energy-requiring reactions that mediate the synthesis of ordered structures, life can persist. As Hoffmeyer (2008, p. 44) reflected, "In this way life sidesteps the entropy law while it creates highly organized (low entropy) structures—living organisms—without violating the thermodynamic requirement that entropy must increase with each spontaneous process."

The confluence of these two areas—holobiont theory and metabolism—provides a new view of evolution, one in which competition and cooperation act together in context-related manners and wherein the organism is created by multiple distributed sources of agency. It is a view of evolution where semiotic relationships—signal production, transmission, reception, and interpretation—become central both within and between organisms.

5.2 Our Holobiont Heritage and the Entanglement of Sympoiesis

5.2.1 Symbiosis

If we are to model evolution, we must first understand that living beings are holobionts. Each organism is a collection of species living and acting together. Moreover, not only do organisms exist as consortia of symbionts, there are "grand symbioses" that allow life to exist throughout the planet and that are mediated through holobiont agents. These include the terrestrial and marine ecosystems made possible by the extension of plant roots by fungi and by the fixation of nitrogen by rhizobium bacteria living in conjugation with legumes and diatoms (Tschitschko et al., 2024). These ecosystems also include the coral reefs provided by photosynthetic algae residing within the coral cells and the tidal ecosystems facilitated by the symbiotes of clams and sea grass. Fungi and other detritovores allow the decomposition of plants and animals to form nutrient-rich soil (something found only on this planet), and competition (parasitic symbiosis) occurs as a fine-tuning mechanism within the structure provided by mutualistic symbioses that are mediated by holobionts (Margulis, 1999). Cooperation and competition work together.

Life is a single clade (Hermida, 2016; Mariscal and Doolittle, 2018), and this allows communication. Informational molecules are only informational in the context of the reader (Oyama, 1985). Each organism can respond to the others because it uses and recognizes the same signs, although each organism may have evolved to interpret a sign differently. Moreover, the evolution of such semiotic differences can be occasioned by symbionts. Grass has different values to herbivores and carnivores, and even the butyric acid that is necessary for reactivating the metabolism of the cattle tick is made by the microbes in the cattle's gut. Symbionts can alter the Umwelt of the organism, and organisms form consortia with other organisms because specific signs are received and interpreted. Life is symbiosis, and symbiosis is a series of semiotic encounters.

5.2.2 Sympoiesis

The body is constructed through the reciprocal exchange of information as organs develop, involving the semiotic flux of production, transmission, reception, and interpretation of signals (Gilbert and Epel, 2015). When thyroid hormone courses through the blood of a tadpole, this same hormone causes its limb cells to proliferate, its tail cells to die, and its nerve cells to alter their trajectories. The cell types respond differently due to their prior developmental history. When bone morphogenetic protein-4 is produced in the vertebrate embryo, it tells the cartilage cells to ossify, the cells between our fingers to die, and the heart cells to form muscle protein. Again, the signal may be the same, but the response differs. When a predatory fly larvae eats a *Daphnia* water flea, the chemicals from its gut inform other *Daphnia* that predators are present, and the remaining *Daphnia* alter their development to become less palatable to the larval fly. Here, evolution has turned a gut chemical of a fly larvae into a trans-species informational signal. So the body is already a semiotic system that has formed itself as a remarkably integrated communications network.

Holobionts add a new layer of semiotic complexity to development and evolution. How are these symbiotic signals made, delivered, and interpreted? One of the discoveries has been that symbioses are not just between mutually consenting adults. When one looks at the creation of symbioses, one often finds that symbiotic interactions take place during development to *generate* the adult. This set of mutualistic partnerships bringing about normal development is more accurately called "sympoiesis," a "making-with" process, rather than autopoiesis, "self-formation." Sympoiesis refers to the observation that multicellular eukaryotic organisms employ symbiotic organisms (especially microbes) to co-construct their organs (Clarke and Gilbert, 2022; Dempster, 1998; Haraway, 2016). Mammals first receive their microbiomes as they pass through the birth canal. Thus, our human bodies become seeded by our mothers' microbes as a normal part of childbirth (Funkhouser and Bordenstein, 2013; Koren et al., 2012). Moreover, immediately after birth, the infant mammal is fed mother's milk, a fluid that contains two sets of nutrients. One set of nutrients feeds the infant, while a second compartment of mother's milk feeds the species of bacteria (primarily *Bifidobacteria*) that promote the colonization of the gut by other sympoietic bacterial partners (Muinck and Trosvik, 2018; Zivkovic et al., 2011).

This "starter-set" of microbes helps complete our development, and animal (and plant) development is predicated on such sympoiesis (see Gilbert, 2023). The neurons that promote hearing and gut peristalsis, the lymphocytes that comprise the immune system, the intestinal capillaries that move nutrients from the gut to the body, and even portions of the mammalian brain are generated or matured by microbial symbionts. Indeed, the rumen that allows cattle to eat grass is constructed in response to signals from the bacteria whose progeny will dwell within it (Baldwin and Conner, 2017; Sander et al., 1959).[2] In invertebrates, mothers often transfer bacteria into their eggs or early embryos, so such sympoiesis can occur before birth. In the nematode *Brugia malayi*, *Wolbachia* bacteria are responsible for the proper orientation of the second mitotic division, and the same genus of bacteria are also needed for the maintenance of ovaries in the *Asobara* wasp. Most all animals appear to be cocreated through interactions between their zygotically derived cells and their microbes from other kingdoms of life.[3]

Symbiotic microbes are crucial partners in the development of mammalian immune and nervous systems, the major cognitive interfaces that enable us to interact with our environments (Ratsika et al., 2022; Tauber, 2013). Germ-free mice (rendered symbiont-deficient for experimental research) have significant neurological and immunological deficiencies. The immune systems of germ-free mice have fewer lymphocytes, less active intestinal macrophages, reduced vascularity, lower cytokine production, and lower titers of serum immunoglobulin than genetically identical mice with normal gut microbes. In such symbiont-free mice, the B-cells and T-cells of the gut-associated lymphoid tissue fail to form (Rhee et al., 2004). Therefore, immunity is not merely a property of the host. Rather, the immune system is a holobiont function generated through sympoiesis (Gilbert and Tauber, 2016; Pradeu, 2019; Schneider, 2021; Thomas, 1973).

The mammalian brain also develops sympoietically (Nagpal and Cryan, 2021). Compared to conventionally bred mice, germ-free mice have lower levels of the mRNAs for transcription factor Egr1 and the paracrine factor BDNF in relevant portions of their brains, while having elevated levels of the neural hormone serotonin (Diaz Heijtz et al., 2011; Clarke et al., 2013). These changes correlate with behavioral differences between the germ-free and conventionally raised mice, leading Diaz Heijtz and colleagues (2011, p. 3047) to conclude that "during evolution, the colonization of gut microbiota has become integrated into the programming of brain development, affecting motor control and anxiety-like behavior."

Although mammalian embryos are often thought to develop within a sterile amnion, about 30 percent of the small molecules carried in the maternal bloodstream of a pregnant mouse arise directly or indirectly from symbiotic microbes (McFall-Ngai et al., 2013; Nicholson et al., 2012). In this way, the microbes in the maternal gut can create compounds that circulate throughout the body, cross the placenta, and affect the developing fetus. When the gut microbes of a pregnant mouse digest plant fiber, they produce butyrate and propionate that enter the maternal bloodstream, pass into the fetal blood circulation, and reach the organs of her developing embryos. Here, these bacterial products activate particular genes in the embryonic pancreas, nervous system, and intestines. These changes in gene expression generate proteins that mature the sympathetic neurons and confer life-long obesity-resisting metabolic phenotypes to the offspring (Kimura et al., 2020). Other soluble products of the maternal gut microbiome stimulate the maturation of auditory neurons in the fetal mouse brain (Vuong et al., 2020). Remarkably, gut bacteria have evolved to send signals to the developing vertebrate brain that regulate complex outcomes for adult vertebrate phenotypes. There is critical interkingdom communication from the gut microbes to the developing mammalian brain.

Gut microbes are also critical for the development of new neurons in adult mice. Neural stem cells generate new neurons in the adult mouse hippocampus. In the adult mouse, indole, synthesized from dietary tryptophan by the mother's gut microbiome, is carried through the blood to the hippocampus. Here, it activates the aryl hydrocarbon receptor of the neural stem cells, converting the receptor into a functional transcription factor that activates the genes responsible for neuron production (Wei et al., 2021). Since these are the neurons thought to be critical for memory and learning, the bacteria are truly a part of "who" one is. As Wei and colleagues (2021, p. 2) claim, "Gut microbes are an evolving, prokaryotic component of the meta-organismal self." It should be noted that hippurate, indole, and short-chained fatty acids are byproducts of bacterial metabolism, excreted from the bacteria. However, they have become "informational molecules" in that animals have evolved receptors that recognize them. The bacteria have become affordances through "semiotic co-option" (Maran and Kleisner, 2010).

If the microbes of mammalian guts are critical for "basic neurogenerative processes such as the formation of the blood-brain barrier, myelination, neurogenesis, and microglia maturation" (Sharon et al., 2016, p. 915), then could such microbes also be critical for normal mental functioning? There is now evidence that by helping generate the brain, the microbiome may also be stimulating mammalian social behaviors (Stilling et al., 2018; Suzuki et al., 2025). Germ-free mice have autism-like behaviors, many of which can be normalized by providing the germ-free mice with normal gut bacteria (from either mice or humans) early in postnatal life (Desbonnet and colleagues, 2014; Sharon et al., 2019). The brain might be "the most complicated object in the known universe,"[4] but it can be controlled, even co-constructed, by bacteria from another biological kingdom.

Social behaviors in evolution are therefore not just generated within a species. They can be constructed between the species that constitute the organism. Since these sympoieses contribute to physical development and to the development of social behaviors, they are critical to the completion of life cycles. This has recently been shown in invertebrates. The lipopolysaccharide (LPS) in the capsule of the biofilm-forming bacterium *Cellulophaga lytica* is obviously a structural molecule for the integrity of the bacterial cell wall. Recently, however, it was also found to be an informational molecule necessary for the settlement and metamorphosis of marine organisms such as *Hydroides elegans*, the common tube worm (Freckelton et al., 2022). Moreover, the LPS of different bacterial strains appears to induce the settlement and metamorphosis of different species of marine invertebrates (Freckelton et al., 2024; Hu et al., 2024). This means that larval settlement depends upon how a larva recognizes and interprets the LPS of different bacteria. This, in turn, means that the biogeography of many marine organisms (e.g., corals, tube worms) depends on where certain bacteria are located.[5]

Sympoiesis therefore should be foregrounded as a crucial characteristic of life. First, sympoiesis means the end of our thinking of organisms as the readout of a single genome. We can no longer model life as being a set of monogenomic, self-contained, self-making individuals (Dempster, 1998; Gilbert et al., 2012; Haraway, 2016). As Donna Haraway remarked, we literally "become with others." We receive 22,000 different genes from the zygote, while we receive over 8 million different genes from our symbionts (Funkhouser and Bordenstein, 2013). And if human phenotypes cannot be considered as mere readouts of the zygotic genome, then precision genetic medicine has to include (at least) the genomes of the holobiont's microbes, whose genes encode proteins that can inhibit or augment drug treatment (Sariola and Gilbert, 2020; Weersma et al., 2020).

Second, if organisms are to be perceived as individuals, we should be perceived as individual teams. Moreover, just like teams, organisms are formed through a mixture of cooperative and competitive processes. One competes to become part of a cooperating team, and "making the team" can be intensely competitive. The *Euprymna* squid requires *Vibrio fisheri* bacteria to form its light organ. To acquire these bacteria, the newborn squid absorbs environmental microbes onto its ventral surface. Then, the squid's immune system selectively winnows away gram-positive bacteria, and then kills all the gram-negative bacteria that are not *Vibrio fisheri* (Koehler et al., 2018). As biosemiotician Jesper Hoffmeyer (2008, p. 100) reflected, "When organisms that are so fundamentally different that they belong to different species (or even to different kingdoms) are going to cooperate, they are forced to overcome a host of obstacles concerning the establishment of unambiguous reciprocal interactions at all levels, from chemistry to social behavior." Indeed, one of the critical components of evolution and development is the ability to distinguish signal from noise (Odling-Smee, 2024). Such competition between organisms may be critical for eliminating confusing or conflicting signals for symbiosis.

Therefore, for the development of most animals (and probably plants and fungi, as well), there is no centralized autonomy, and autopoiesis (wherein an organism is defined as having the ability to create itself, *sensu* Maturana and Varela, 1980) is confined to cellular self-maintenance.[6] Rather, animals are generated through sympoiesis, which is an entangled integration of embryology and ecology, cooperation and competition. The co-symbionts mutually scaffold and provide affordances for one another's other's existence (Chiu and Gilbert, 2020; Clark and Gilbert, 2022). The symbionts in the rumen allow the development and persistence of cattle; the cattle allow the persistence and propagation of their ruminal bacteria. There is no single "host" in the holobiont relationship (Haraway, 2016). Animals constitute the environment of bacteria, as bacteria constitute the environment of the animals (Formosinho et al., 2022). Moreover, if bacteria help make us into healthy and social animals, perhaps we animals are the bacteria's way of making more niches for their descendants (Gilbert, 2018). Both the animal and microbial partners benefit.

Thus, sympoiesis has important consequences for evolution; for evolutionary speciation is often brought about, as Lynn Margulis and Dorian Sagan (1992) announced, through "acquiring genomes." Indeed, evolution occurs at many levels, and in many cases, what gets selected in evolution can be the "team" of organisms rather than the individual players (Osmanovic et al., 2018; Roughgarden et al., 2017; Roughgarden, 2020). On the micro-evolutionary level, new combinations of symbionts have given rise to new selectable phenotypes. Whether a pea aphid is green or red, thermotolerant or thermolabile, or immune or susceptible to parasitoid wasp infection depends on the bacteria associated sympoietically with the developing aphid body (Dunbar et al., 2007; Oliver et al., 2009; Tsuchida et al., 2010).

What is critical is that acquiring new symbionts changes the Umwelt of an organism. In acquiring a new symbiont, an organism can acquire a reinterpretation of its world. Chiu and Gilbert (2020) have shown that the acquisition of plant-digesting and poison-degrading symbionts enabled grass to be used as a nutritional affordance. The entire physiological and ecological mode of herbivory was now possible for vertebrates. The world was literally a different place for herbivores. What had been an inert part of the environment can now be food. The acquisition of new symbionts has enabled the red turpentine beetle to infect and kill millions of Chinese pine trees (Sun et al., 2013; Taerum et al., 2013). What had been a part of the landscape has now become a place to lay one's eggs and acquire nutrients. New symbionts that metabolize insecticides have allowed the *Riptortus* bean bug to passively acquire the resistance to those poisons. This resistance spread immediately and did not need the generations that gene-mediated inheritance would need (Kikuchi et al., 2012; Kim et al., 2016).

In addition, major evolutionary innovations may have been made through new sympoietic pathways. These include multicellularity (Dayel et al., 2011), meiosis (Woznica et al., 2017), herbivory (Gilbert, 2020), and the mammalian uterus (Emera and Wagner, 2012). Each novel feature appears to have arisen through new combinations of symbiotic organisms (Gilbert, 2019). Even the animal nervous system may have evolved partly out of the advantage of being able to orchestrate the complex interactions between animals and their associated micro-organisms (Augustin et al., 2017; Klimovich and Bosch, 2018). Symbionts thereby add a new modular dimension to evolution. These symbiotic modules can be tied tightly to an animal's physiology and development (such the pea aphid's acquiring microbial symbionts that enable the aphid complete nutrition; Monnin et al., 2020) or loosely tied to an animal's physiology and development (as in the above-mentioned cases for color, thermotolerance, and parasitoid resistance in pea aphids). As Žukauskaitė (2020) points out, "sympoietic systems carry different bits of information in their components. . . . This makes sympoietic systems more flexible and adaptive, in the sense that they can easily adapt to changing environments, and also create something new, produce new forms of organization (in this regard they are allopoietic)."

What evolves, as Lewontin (1983) claimed, is the relationship. These can be the relationships between transcription factor proteins and their DNA enhancers (which create evolutionary novelties such as bat wings, snake ribs, and insect limbs), the relationships between symbionts within a body, or the relationships between the holobiont organism and its environment (Chiu and Gilbert, 2020; Gilbert and Epel, 2015). These relationships are each accomplished through sign-based interactions, thus confirming the importance of semiosis for development and evolution, as suggested by Kull (2015).

But what constitutes the boundary of the holobiont? There is neither closure nor a self-defined boundary to the symbiotic team. Originally, Sagan and Margulis (1991) had opined, following Maturana and Varela (1980), that the organism ended at the cell membrane. However, the cell is not self-contained, and closure and autopoiesis can no longer characterize holobionts. Coming from semiotics, Hoffmeyer (2007, p. 241) saw the immune system as having generated a mechanism for self-identity by converting the negative quality of "nonself" into a positive property, "an index of foreignness." Coming from immunology and developmental biology, Gilbert and Tauber (2016) extended that view, proposing that the boundary of the holobiont is the immune system, a system created through the interactions of host and symbionts. This system defines "self" in a dynamic manner, changing constantly as new microbes and new meals are encountered.

Whatever these interactions are between microbe and host, the holobiont is being continuously constructed. Harmony is not something given, but rather something that requires interactive agency throughout the lifespan of the organism.

5.3 Metabolism

5.3.1 Intracellular Metabolism

Life manifests itself in the world as metabolism. As Hans Jonas (1966) realized, "The exchange of matter with the environment is not a peripheral activity engaged by a persistent core: it is the total mode of continuity (self-continuation) of the subject of life itself."

The living state is, for Jonas, the set of relationships wherein one retains individual identity by changing one's parts. This is the meaning of metabolism, and it distinguishes the biotic from the abiotic. An organism "is never the same materially and yet persists as its same self, *by* not remaining the same matter" (Jonas, 1966, p. 76; see Landecker, 2017). Thus, the living organism is characterized by what Jonas (1966, p. 80) has called "a needful freedom." Metabolism constitutes both the identity of the organism and, simultaneously, its dependence on its environment. As Anne Sophie Meincke (2022) has expounded, this dialectic allows one to perceive an organism as an identity-maintaining individual unit as well as part of a non-individualistic world of metabolic flux. This dialectic is critical for perceiving holobionts as co-metabolizing wholes.

This contemporary view of metabolism, wherein the components of an organism are in a constant state of chemical renewal, was formulated by biochemist Rudolf Schoenheimer. He was the first investigator to tag nutrients with radioactive atoms and watch their transformation as they became different molecules and different body parts (Guggenheim, 1991; Landecker, 2013; Schoenheimer, 1942; Schoenheimer and Rittenberg, 1935). These insights were formulated into metabolic pathways by Schoenheimer and several other professors, many of whom had fled Hitler to come to Great Britain or the US (Kennedy, 2001). Hans Krebs, Otto Meyerhoff, and Fritz Lipmann, for instance, helped delineate the pathway whereby oxygen, the byproduct of photosynthesis, served as the receptor for electrons from dietary nutrients, thereby allowing cells to synthesize the energy-storing ATP molecules that allowed life to persist against entropic gradients. The idea of coupled reactions, whereby the ATP generated by a dietary degradation was used as an energy source for synthetic reactions, allowed life to play "jiujitsu" with the second law of thermodynamics, using the energy produced by degradative reactions to catalyze synthesis.[7] As Schoenheimer (1942, p. 65) wrote, "The new results suggest that all constituents of living matter, whether functional or structural, of simple or of complex constitution, are in a steady state of rapid flux. The finding of the rapid molecular regeneration . . . suggests that the biological system represents one great cycle of closely linked chemical reactions."

This view that the body was a temporary form of chemicals in flux had a history prior to biochemistry (Gilbert, 1982). Thomas Huxley (1868), for instance, celebrated that remarkable "transubstantiation" by which lobster could be turned into Huxley, and if shipwrecked, Huxley could be reciprocally transformed into lobster. Embryologist E. B. Wilson (1896, p. 431) concluded that "inheritance is the recurrence, in successive generations, of like forms of metabolism."

Metabolism sustains living beings by allowing "some new principle of refreshment," for it is able "to preserve order amid change and to preserve change amid order" (see Whitehead, 1978 [1929], p. 399). The goal of life is not a thermodynamic end state (which is death), but rather the maintenance of form far from equilibrium through metabolic processes that are, themselves, mediated by thermodynamics (see Kauffman, 2002).

5.3.2 Holobiont Co-metabolism

Whereas metabolism was originally defined as a cellular phenomenon, metabolism is now seen as a communal enterprise (Dupré and O'Malley, 2009; Kelty and Landecker, 2019). Moreover, the set of symbiotic organisms constructing the holobiont (including the "host") participate in each other's metabolisms. The woodrats (*Neotoma lepida*) of the Mojave desert feed almost exclusively on creosote bushes, ingesting quantities of creosote that are poisonous to all other herbivores. This allows them to survive and flourish without much competition. What makes this possible is their gut microbiome, which contains bacteria that detoxify creosote (Kohl et al., 2014). What is poison to one species is food to another. Symbionts can make that interpretation possible.

In the mealy bug *Planococcus*, the production of the amino acid phenylalanine begins with enzymes encoded by the bug's symbionts, *Tremblaya* bacteria. The metabolites generated by these enzymes then travel into another bacterial species, *Moranella*, which are symbionts within the *Tremblaya* bacteria. The newly made metabolites from *Moranella* then return to the *Tremblaya* bacteria, wherein they are converted into a product that can be metabolized into phenylalanine by enzymes encoded by the *Planococcus* genome (McCutcheon and Dohlen, 2011). In mammals, bacterial products, especially the short-chain fatty acids, course from the gut throughout the body, where they sustain digestion, blood circulation, and other physiological functions (Kelty and Landecker, 2019; McFall-Ngai et al., 2013). In 2013, Smith and colleagues coined the term "co-metabolism" to denote the pathways wherein foods become metabolized by both zygote-derived and microbe-derived enzymes, and that the products of one species are often substrates for the other. We have an entangled metabolism.

5.3.3 Holobiont Co-metabolism: Gaia

Hannah Landecker (2013) has pointed out that metabolism can be seen as processes of an organism that simultaneously tie it to and separate it from its external environment(s). Metabolism produces "the two-ness of organism and environment," while at the same time integrating them together.

Metabolism is a planetary process whereby organism and environment are co-constructed, and Gaia is characterized (Margulis and Sagan, 2000, p. 78) by its "incessant self-organizing metabolism." As Bruno Latour (2017, p. 142) remarked, "Gaia is only the name proposed for all the intermingled and unpredictable consequences of the agents, each of which is pursuing its own interests by manipulating the environment." This idea that organisms, as well as cells and organisms, were enmeshed in metabolic webs was first appreciated in community ecology. G. Evelyn Hutchinson (1940, p. 268) proposed that a community of living organisms could be regarded as an organism, and "it should be possible to study the metabolism of that organism." And here we have the beginnings of the Gaia hypothesis, which arose by name from James Lovelock's consideration that the world might indeed be an entire organism. In this view, geology was not merely a "stage" for life, and Earth was not a "Goldilocks planet" where water, oxygen, and the temperature were perfect for life. Rather (in a way that resembles Niche Construction writ large), life was the agent that terraformed the planet.

First, living organisms are and have been responsible for the remarkable composition of Earth's atmosphere, lithosphere, and oceans (Lenton and Dutreuil, 2020; Lenton et al., 2020; Lovelock, 1975, 1979). Without photosynthetic organisms, the atmosphere would contain only one oxygen atom in 10^{12} rather than one atom in five. Methane is thought to be 10^{30} times more abundant on earth than it would be at an abiotic equilibrium. Archaea originally expelled the methane, while cyanobacteria made the oxygen. Second, earth has soil, not regolith. The flux of phosphorus through living systems is probably 1,000-fold over what might be expected geologically, and organisms have been critical in hydrating the earth's crust. Much of earth's minerals are oxygen-containing carbonates or silicates, made from life's excess oxygen. And third, without life, earth would probably be a pressure cooker with a temperature of 50°C. Photosynthetically derived oxygen created the ozone layer and induced a cold trap to form in the upper atmosphere, preventing water and hydrogen from escaping. All the world is not a stage. It is a co-construction, a byproduct of organisms.

The world's ecosystems are, like the symbionts on our bodies, both contained and interacting. Phosphorus, nitrogen, calcium, and atmospheric gases are being cycled between organisms in food webs of planetary proportions. In Canada's Great Bear Rainforest (the largest temperate rainforest on the planet), "eighty percent of the nitrogen in the forest's trees comes from the salmon" (Muller, 2022; Wagner and Reynolds, 2019). And nearly 100 percent of the carbon in their trunk and branches come from the carbon dioxide expelled by organisms and fossil fuels.

Lynn Margulis (1995) argued, "Lovelock would say that Earth is an organism. I disagree with this phraseology. . . . I prefer to say that Earth is an ecosystem, one continuous enormous ecosystem composed of many component ecosystems." As Bruce Clarke (2020, p. 87) has noted, one of Margulis's mantras was "Gaia is symbiosis seen from space," and that this conception of life provided the context for her discussions of any particular example of symbiosis. Interestingly, this notion of earth as a single confederation of component ecosystems was also the characteristic of another one of Margulis's conceptions, the holobiont (Margulis, 1991). For Margulis, Gaia could be seen an autotrophic (photosynthetic) holobiont whose symbiotic components are in constant co-metabolism (Matyssek and Lüttge, 2013).

Clarke (2020) has also shown that Margulis extended Lovelock's first-order (control-mediated) cybernetics by fusing it with the Maturana and Varela's concept of autopoiesis to create a second-order cybernetics based on autonomy, recursion, and closure. The autonomous living agents came together in mutualistic metabolism to form a new autopoietic whole.[8] Clarke has also pointed out (Clarke, 2020, p. 238) that "the qualities of the holobiont taken altogether recall a prior biological scenario—the comparable inversion of control as propounded decades earlier in the Gaia hypothesis."

I would amend this view by saying that Gaia's existence, like that of all holobionts, is sympoietic, not autopoietic (see Clarke and Gilbert, 2022). Gaia resembles a developing system more than a physiological system. Indeed, Margulis (Margulis, 1990, p. 866) characterized Gaia as not being in a state of physiological homeostasis, but in an embryological state of "homeorhesis." This was Waddington's (1957) term for the embryologists' observations that embryonic cells changed as they interacted with other cells, eventually differentiating into the component cells of integrated organisms (Matsushita and Kaneko, 2020).

The holobiont body, as we have seen, is an interconnected temporally changing web of ecosystems, wherein the fingers, mouth, gut regions, and reproductive orifices each support different communities of organisms. The holobiont is thus a team of organisms, a consortium expressing an integrated, enmeshed, and distributed development and metabolism. Patrice Manglier (2021, pp. 67–71), following Bruno Latour, has argued that a key feature of Gaia, perhaps its most important characteristic, is that it is the "continuity of entanglements." Once one accepts the reality of Jonas's view of metabolism and our current view of holobionts, Gaia is neither magical nor extraordinary—or perhaps no more magical or extraordinary than any co-metabolizing holobiont. In terms of modeling, Gaia can be considered as a photosynthetic holobiont, continually synthesizing and reorganizing the metabolites it creates. Indeed, Gaia would have better been named for her daughter, Rhea, the goddess of flux.

Moreover, Gaia and holobionts are communities based on interspecies communication. Both are predicated on the ability to produce, transmit, receive, and appropriately interpret signals. This is a fundamental principle of embryology. The act of fertilization is a mixture of cooperation and competition such that the sperm and egg that are the best listeners (the oocyte that can detect the smallest amount of pituitary hormones, the sperm that can detect the signals from the egg) are those that engage to make the zygote. The mammal that develops inside the uterus comes from a zygote that divided into hundreds of nearly identical cells that interact through signals that amplify small differences. These different immature cell types then influence each other as they differentiate to create the organs of the embryo. And this is made possible by the hormones of the mother influencing the embryo, as the embryo makes hormones that alter the physiology of the mother. At childbirth, maternal microbes colonize the newborn infant, and they help finish the infant's development while the infant provides these symbionts with nutrients and a place to procreate.

This holobiont then interacts with other holobionts to form communities, and the communities that interact to form biomes—it is semiosis and reciprocal maturation all the way up and down. It is all sympoiesis, which means that it is all a matter of semiosis—of producing and interpreting signals. As Donna Haraway (2008, p. 228), remarks, reciprocal induction "is the name of the game," and it is "reciprocating complexity all the way down." Indeed, she concludes (Haraway, 2008, p. 26) that "relationships are the smallest possible pattern for analysis."

And the signals that each organism produces and secretes influence the development of Gaia. This creates vulnerability. Glyphosate used in weed killers can destroy the growth pathways of certain symbiotic bacteria and reduce the communication between the symbiotic bacteria and the honeybee (Motta and Moran, 2020, p. 2024). Anthropogenic global warming disrupts algae-coral symbioses as well as the symbiosis between *Euprymna* squid and *Vibrio* bacteria. Deborah Rose Bird (2017) has called our attention to these situations where symbioses make organisms vulnerable, where "relationships unravel, mutualities falter [and] dependence becomes a peril rather than a blessing."

The atmosphere, lithosphere, and hydrosphere are each their own attractors, and they are integrated into Gaia, which is a metabolic attractor state whose component parts are in dynamic stable relationships with each other. Gaffney and Steffan (2016) have modeled Gaia as an attractor and have pointed out that there have been (and still are) other possible attractors. There is the attractor of "snowball earth" (times of global glaciation) and the attractor of "hothouse earth" (which would result from catastrophic climate change). Different ecosystems would evolve for such planets, and in neither of these attractor states would earth be habitable for *Homo sapiens* and the food we now grow. Other attractor states have also been imagined. These include "plantationocene earth" (Haraway et al., 2015), where some human beings' extractive management has diminished biodiversity solely to capitalistic Western humans and their (often coerced) companion species, and "techno-earth" (Ellul, 1964)[9] where nature has all but vanished, and humans, if existing at all, are the workers who maintain an integrated technological matrix of artificially intelligent entities.

There may also be (e)utopian attractors, where humans actually promote the health and diversity of the biosphere. Latour's (2017) "terrestrial attractor," seeking a balance between emphases on physical (local) habituation and access to the global world, is depicted as a sane alternative to other, more pathological, attractors. A "polyphonic" earth is envisioned by Robinson (2020, p. 562) and is championed by authors as diverse as Haraway (2016) and Kimmerer (2013). In this attractor basin of Gaia, economic well-being is decoupled from planetary exploitation, and its inhabitants "would not cease the layered, curious practice of becoming-with others for a habitable, flourishing world" (Haraway, 2016, p. 168). While the dystopian attractors can be realized through inaction, these eutopian states would have to be achieved by active political and social efforts. This is why imagining such possibilities is crucially important. "How can we begin to move towards ecological sustainability," writes Kimmerer (2013, p. 6), "if we cannot even imagine what the path feels like?" Indeed, "utopian thinking regarding sustainable development could result in more integrated and holistic visions of future society in climate science and policy" (Hjerpe and Linnér, 2009, p. 234). Thus, it may be more productive to envision joyful futures than to envision escapes from horrific ones.

The attractor state of Gaia has changed before in earth's history. Roughly 2.3 billion years ago, cyanobacteria evolved photosynthesis and poisoned the air with oxygen. This "Great Oxidation Event" caused the mass extinction of the anaerobic microbes that had thrived in the Gaian attractor basin that possessed a reducing (hydrogen-abundant) atmosphere (Lyons et al., 2014).[10] This massive elevation in oxygen levels created a new equilibrium, producing the planet as we know it, a planet full of metabolisms/organisms that evolved to use oxidative phosphorylation (the use of oxygen to manufacture ATP) as a means of accumulating energy. Thus, the cyanobacteria changed the attractor into the present form of Gaia.

5.4 Concluding Remarks

The formation and evolution of sympoietic holobionts creates a new domain for biology and for notions of distributed agency and semiosis. It combines Howard Pattee's (1968) idea that "life is matter controlled by symbols" with Hans Jonas's (1966) idea that life is the organization of metabolic flux, and with Lynn Margulis's (1999) idea that life is symbiotic. The biosemiotics community has been prepared for the notions of reciprocal agency and holobionts. Indeed, Günter Witzany (2008) has written on the extension of semiosis to symbiotic interspecies interactions, and this present essay extends those interactions to the *formation* of organisms. Here we find a messy mixture of "endosemiosis" and "exosemiosis" (Hoffmeyer and Stjernfelt, 2016); for the holobiont is both an organism and a biome, and cooperation on one level may be competition at another. Hoffmeyer (2007, p. 214) discusses the ways that endosymbiosis and exosymbiosis meet and interact in symbiosis. Indeed, he uses such examples to point to "the impossibility of upholding any sharp dividing line between biochemical and semiotic activity, however reasonable such a borderline may once have appeared."

Sebeok (2001) and Kull (2009) have argued that semiosis is co-extensive with life. Sensing the mixture of biochemical and semiotic signaling, semiotics may actually go beyond that, extending beyond the biosphere into what Latour (2014) called the "critical zone." Here, Gaia resides on the crust of the planet, where organisms regulate the flow of informative signals and other resources necessary for life to develop, persist, and evolve. The critical zone contains a network of co-generated, coevolving, and co-metabolizing holobionts. Both the biotic components of the critical zone and the abiotic components (e.g., tides, droughts, soil consistency) may be exchanging signals and providing the possibilities of new relationships. The evolution of holobionts in the context of Gaia provides a fascinating frontier for biosemiotic research.

Acknowledgments

I wish to thank brush, Bruno Clarke, Michael Flower, Jay Geller, Phil Gilbertson, Deborah Gordon, Donna Haraway, Deborah Heath, Rusten Hogness, Anne Sophie Meincke, Olyssa Starry, and A. I. Tauber for their constructive suggestions in revising the manuscript. Their important assistance does not imply complete agreement with the hypotheses presented.

Notes

1. See Castro and Méndez-Veras, 2023. "I contain multitudes": Biosemiotic notes on the holobiont and individuation in microbiome research. 22nd Gathering in Biosemiotics, Olomouc, Czech Republic. doi: 10.13140/RG.2.233956.69762 Also, Méndez-Veras, R. "The Holobiont as a *Biosemiotic* System: Health and Dysbiosis Signals in the Human Gut Microbiome". *Biosemiotics* (2025). https://doi.org/10.1007/s12304-025-09627-9

2. The bacterial chemical that induces the rumen to form is butyric acid. This is the same compound that Uexküll showed was necessary to awaken the tick to drop onto the cow (Uexküll, 1934). The signal provided by this small molecule is interpreted differently by different organisms.

3. *Sympoiesis* is a concept in developmental biology referring to developmental inputs from several sources. This has often been confused with the notion of *symbiogenesis*, which is a concept in evolutionary biology meaning that evolutionary events such as speciation and the formation of the first cells arose through the acquisition of new genomes (rather than by the mutation of existing ones).

4. Kaku, 2014, Behold the most complicated object in the known universe, https://www.wnyc.org/story/michio-kaku-explores-human-brain/

5. The interpretation of the LPS moiety is a critical component regulating symbiotic placement and pathogenesis. In its pathogenic mode, LPS is known as "endotoxin."

6. Although "autopoietic" means "self-generating," maintenance, not development, was the criterion that Margulis and Sagan (1986) used for being autopoietic: "Living systems are autopoietic: they maintain themselves."

7. As Hoffmeyer (1986) pointed out, these processes of photosynthesis and respiration are performed by chloroplasts and mitochondria, two elements of he cells that had formerly been bacterial symbionts.

8. Although Lynn Margulis was a major supporter of the notion of autopoiesis, she did not know of sympoiesis (a term coined by Dempster in 1998), and her views regarding the "operational closure" of autopoietic systems would presently create a problem for her theories of linked, symbiotic associations in a holobiont (Clarke, 2020; Clarke and Gilbert, 2022). Rather, for Sagan and Margulis (1991), symbiosis was a conglomeration of *adult* organisms.

9. See also Hawking and Cellan-Jones, 2014, Stephen Hawking warns artificial intelligence could end mankind, BBC News, https://www.bbc.com/news/technology-30290540

10. See also Aiyer, 2022. The great oxidation event: How cyanobacteria changed life, https://asm.org/Articles/2022/February/The-Great-Oxidation-Event-How-Cyanobacteria-Change

References

Augustin, R., Schröder, K., Murillo Rincón, A. P., Fraune, S., Anton-Erxleben, F., Herbst, E. M., et al. 2017. A secreted antibacterial neuropeptide shapes the microbiome of Hydra. *Nature Communications*, *8*, 698.

Baldwin, R. L. IV, and Conner, E. E. 2017. Rumen function and development. *Veterinary Clinics: Food Animal Practices*, *33*, 427–439.

Barbaras, R. 2008. Life, movement, and desire. *Research into Phenomenology*, *38*, 3–17.

Bird, D. R. 2017. Shimmer: When all you love is being trashed. In A. Tsing, H. Swanson, E. Gan, and N. Bubandt (Eds.), *Arts of living on a damaged planet: Ghosts of the Anthropocene* (pp. G51–G63). University of Minnesota Press.

Chiu, L., and Gilbert, S. F. 2020. Niche construction and the transition to herbivory: Phenotype switching and the origination of new nutritional modes. In H. Levine, M. Jolly, P. Kulkarni, and V. Nanjundiah (Eds.), *Phenotype switching: Implications in biology and medicine* (pp. 459–482). Elsevier.

Clarke, B. 2020. *Gaian systems: Lynn Margulis, neocybernetics, and the end of the Anthropocene*. University of Minnesota Press.

Clarke, B., and Gilbert, S. F. 2022. Margulis, autopoiesis, and sympoiesis. In C. A. Jones, N. Bell, and S. Nimrod (Eds.), *Symbionts: Contemporary artists and the biosphere* (pp. 63–78). MIT Press.

Clarke, G., Grenham, S., Scully, P., Fitzgerald, P., Moloney, R. D., Shanahan, F., Dinan, T. G., and Cryan, J. F. 2013. The microbiome-gut-brain axis during early life regulates the hippocampal serotonergic system in a sex-dependent manner. *Molecular Psychiatry, 18*(6), 666–673.

Dayel, M. J., Alegado, R. A., Fairclough, S. R., Levin, T. C., Nichols, S. A., McDonald, K., and King, N. 2011. Cell differentiation and morphogenesis in the colony-forming choanoflagellate *Salpingoeca rosetta. Developmental Biology, 357*, 73–82.

Dempster, M. B. L. 1998. *A self-organizing systems perspective on planning for sustainability* [MA thesis, environmental studies]. University of Waterloo.

de Muinck, E. J., and Trosvik, P. 2018. Individuality and convergence of the infant gut microbiota during the first year of life. *Nature Communications, 9*, 2233.

Desbonnet, L., Clarke, G., Shanahan, F., Dinan, T. G., and Cryan, J. F. 2014. Microbiota is essential for social development in the mouse. *Molecular Psychiatry, 19*, 146–148.

Diaz Heijtz, R., Wang, S., Anuar, F., Qian, Y., Björkholm, B., Samuelsson, A., Hibberd, M. L., Forssberg, H., and Pettersson, S. 2011. Normal gut microbiota modulates brain development and behavior. *Proceedings of the National Academy of Sciences of the United States of America, 10*, 3047–3052.

Dunbar, H. E., Wilson, A. C. C., Ferguson, N. R., and Moran, N. A. 2007. Aphid thermal tolerance is governed by a point mutation in bacterial symbionts. *PLoS Biology, 5*, e96.

Dupré, J., and O'Malley, M. 2009. Varieties of living things: Life at the intersection of lineage and metabolism. *Philosophy and Theory in Biology, 1*, e0003.

Ellul, J. 1964. *The technological society*. Vintage Books.

Emera, D., and Wagner, G. P. 2012. Transposable element recruitments in the mammalian placenta: Impacts and mechanisms. *Briefings in Functional Genomics, 11*, 267–276.

Formosinho, J., Bencard, A., and Whiteley, L. 2022. Environmentality in biomedicine: Microbiome research and the perspectival body. *Studies in History and Philosophy of Science, 91*, 148–158.

Freckelton, M. L., Nedved, B. T., Cai, Y.-S., Cao, S., Turano, H., Alegado, R. A., and Hadfield, M. G. 2022. Bacterial lipopolysaccharide induces settlement and metamorphosis in a marine larva. *Proceedings of the National Academy of Sciences of the United States of America, 119*, e220795119.

Freckelton, M. L., Nedved, B. T., and Hadfield, M. G. 2024. Bacterial envelope polysaccharide cues settlement and metamorphosis in the biofouling tubeworm *Hydroides elegans. Communications Biology, 7*(1), 883.

Funkhouser, L. J., and Bordenstein, S. R. 2013. Mom knows best: The universality of maternal microbial transmission. *PLoS Biology, 11*(8), e1001631.

Gaffney, O., and Steffen, W. 2017. The Anthropocene equation. *The Anthropocene Review, 4*, 53–61.

Gilbert, S. F. 1982. Intellectual traditions in the life sciences: Molecular biology and biochemistry. *Perspectives in Biology and Medicine, 26*, 151–162.

Gilbert, S. F. 2018. Perspective: Rethinking parts and wholes. In S. Gissis, E. Lamm, and A. Shavit (Eds.), *Landscapes of collectivity in the life sciences* (pp. 123–132). MIT Press.

Gilbert, S. F. 2019. Evolutionary transitions revisited: Holobiont evo-devo. *Journal of Experimental Zoology Molecular Development Biology, 332*, 307–314.

Gilbert, S. F. 2020. Developmental symbiosis facilitates the multiple origins of herbivory. *Evolution and Development, 22*, 154–164.

Gilbert, S. F. 2023. Holobiont development: Embryology and ecological succession. *Human Development, 67*, 257–272.

Gilbert, S. F., and Epel, D. 2015. *Ecological developmental biology*. Sinauer Associates.

Gilbert, S. F., Sapp, J., and Tauber, A. I. 2012. A symbiotic view of life: We have never been individuals. *Quarterly Review of Biology, 87*(4), 325–341.

Gilbert, S. F., and Tauber, A. I. 2016. Rethinking individuality: The dialectics of the holobiont. *Biology and Philosophy, 31*, 839–853.

Guggenheim, K. Y. 1991. Rudolf Schoenheimer and the concept of the dynamic state of body constituents. *Journal of Nutrition, 121*, 1701–1704.

Haraway, D. J. 2008. *When species meet*. University of Minnesota Press.

Haraway, D. J. 2016. *Staying with the trouble: Making kin in the Chthulucene*. Duke University Press.

Haraway, D., Ishikawa, N., Gilbert, S. F., Olwig, K., Tsing, A. L., and Bubandt, N. 2015. Anthropologists are talking—about the Anthopocene. *Ethnos, 81*(3), 535–564.

Hermida, M. 2016. Life on Earth is an individual. *Theory in Biosciences*, *135*, 37–44.

Hjerpe, M., and Linnér, B.-O. 2009. Utopian and dystopian thought in climate change science and policy. *Futures*, 41, 234–245.

Hoffmeyer, J. 2007. *Biosemiotics: An examination into the signs of life and the life of signs*. University of Scranton Press.

Hoffmeyer, J., and Stjernfelt, F. 2016. The great chain of semiosis: Investigating the steps in the evolution of semiotic competence. *Biosemiotics*, *9*, 7–29.

Hu, X. M., Peng, L., Wu, J., Wu, G., Liang, X., and Yang, J. L. 2024. Bacterial c-di-GMP signaling gene affects mussel larval metamorphosis through outer membrane vesicles and lipopolysaccharides. *NPJ Biofilms Microbiomes*, *10*, 38.

Hutchinson, G. E. 1940. Bio-ecology. *Ecology*, *21*(2), 267–268.

Huxley, T. H. 1868. *On the physical basis of life*. The College Courant.

Jonas, H. 1966. *The phenomenon of life: Towards a philosophical biology*. Harper and Row.

Kauffman, S. 2002. *Investigations*. Oxford University Press.

Keller, E. F. 2002. *The century of the gene*. Harvard University Press.

Kelty, C., and Landecker, H. 2019. Outside in: Microbiomes, epigenomes, visceral sensing, and metabolic ethics. In *After practice: Thinking through matter(s) and meaning relationally* (The Laboratory for the Anthropology of the Environment and Human Relations, Ed.) (pp. 53–65). Panama Verlag.

Kennedy, E. P. 2001. Hitler's gift and the era of biosynthesis. *Journal of Biological Chemistry*, *276*, 42619–42631.

Kikuchi, Y., Hayatsu, M., Hosokawa, T., Nagayama, A., Tago, K., and Fukatsu, T. 2012. Symbiont-mediated insecticide resistance. *Proceedings National Academy of Sciences USA*, *109*, 8618–8623.

Kim, J. K., Park, H. Y., and Lee, B. L. 2016. The symbiotic role of O-antigen of *Burkholderia* symbiont in association with host *Riptortus pedestris*. *Development and Comparative Immunology*, *60*, 202–208.

Kimmerer, R. W. 2013. *Braiding sweetgrass: Indigenous wisdom, scientific knowledge, and the teachings of plants*. Milkweed Editions.

Kimura, I., Miyamoto, J., Ohue-Kitano, R., Watanabe, K., Yamada, T., Onuki, M., Aoki, R., Isobe, Y., Kashihara D., et al. 2020. Maternal gut microbiota in pregnancy influences offspring metabolic phenotype in mice. *Science*, *367*, eaaw8429.

Klimovich, A. V., and Bosch, T. C. G. 2018. Rethinking the role of the nervous system: Lessons from the hydra holobiont. *BioEssays*, *40*, 1800060.

Koehler, S., Gaedeke, R., Thompson, C., Bongrand, C., Visick, K. L., Ruby, E., and McFall-Ngai, M. 2018. The model squid-vibrio symbiosis provides a window into the impact of strain- and species-level differences during the initial stages of symbiont engagement. *Environmental Microbiology*, *21*(9), 3269–3283.

Kohl, K. D., Weiss, R. B., Cox, J., Dale, C., and Dearing, M. D. 2014. Gut microbes of mammalian herbivores facilitate intake of plant toxins. *Ecology Letters*, *17*(10), 1238–1246.

Koren, O., Goodrich, J. K., Cullender, T. C., Spor, A., Laitinen, K., et al. 2012. Host remodeling of the gut microbiome and metabolic changes during pregnancy. *Cell*, *150*(3), 470–480.

Kull, K. 2009. Vegetative, animal, and cultural semiosis: The semiotic threshold zones. *Cognitive Semiotics*, *4*, 8–27.

Kull, K. 2015. Evolution, choice, and scaffolding: Semiosis is changing its own building. *Biosemiotics*, *8*(2), 223–234.

Landecker, H. 2013. The metabolism of philosophy, in three parts. In B. Malkmus and I. Cooper (Eds.), *Dialectic and paradox: Configurations of the third in modernity* (pp. 193–224). Peter Lang.

Landecker, H. 2017. Metabolism, autonomy, and individuality. In S. Gissis, E. Lamm, and A. Shavit (Eds.), *Landscapes of collectivity in the life sciences* (pp. 225–248). MIT Press.

Latour, B. 2014. Some advantages of the notion of "Critical Zone" for geopolitics. *Procedia Earth and Planetary Science*, *10*, 3–6.

Latour, B. 2017. *Facing Gaia: Eight lectures on the new climate regime*. Polity Press.

Lenton, T. and Dutreuil, S. 2020. What exactly is the role of Gaia? In B. Latour and P. Weibel (Eds.), *Critical zones: The science and politics of landing on Earth* (pp. 168–175. MIT Press.

Lenton, T. M., Dutreuil, S., and Latour, B. 2020. Life on Earth is hard to spot. *The Anthropocene Review*, 7, 248–272.

Lewontin, R. C. 1983. Gene, organism, and environment. In D. S. Bendall (Ed.), *Evolution from molecules to men* (pp. 273–285). Cambridge University Press.

Lovelock, J. E 1975. Thermodynamics and the recognition of alien biospheres. *Proceedings of the Royal Society of London*, *189*, 167–181.

Lovelock, J. E. 1979. *Gaia—A new look at life on Earth.* Oxford University Press.

Lyons, T. W., Reinhard, C.T., and Planavsky, N. J. 2014. The rise of oxygen in Earth's early ocean and atmosphere. *Nature, 506*, 307–315.

Manglier, P. 2021. *Le philosophe, la terre et le virus: Bruno latour expliqué par l'Actualité.* Les Liens Qui Libèrent.

Maran, T., and Kleisner, K. 2010. Towards an evolutionary biosemiotics: Semiotic selection and semiotic co-option. *Biosemiotics, 3*, 189–200.

Margulis, L. 1990. Kingdom Animalia: The zoological malaise from a microbial perspective. *American Zoologist, 30*, 861–875.

Margulis, L. 1991. Symbiogenesis and symbionticism. In L. Margulis and R. Fester (Eds.), *Symbiosis as a source of evolutionary innovation: speciation and morphogenesis* (pp. 1–14). MIT Press.

Margulis, L. 1995. Gaia is a tough bitch. In J. Brockman (Ed.), *The third culture: Beyond the scientific revolution* (pp. 129–146). Simon & Schuster.

Margulis, L. 1999. *Symbiotic planet.* Basic Books.

Margulis, L., and Sagan, D. 1986. *Microcosmos: Four billion years of evolution from our microbial ancestors.* University of California Press.

Margulis, L. and Sagan, D. 1992. *Acquiring genomes: a theory of the evolution of species.* Basic Books.

Margulis, L., and Sagan, D. 2000. *What is life?* University of California Press.

Mariscal, C., and Doolittle, W. F. 2018. Life and life only: A radical alternative to life definitionism. *Synthese, 197*, 2975–2989.

Matsushita, Y., and Kaneko, K. 2020. Homeorhesis in Waddington's landscape by epigenetic feedback regulation. *Physical Review Research, 2*, 023083.

Maturana, H. R., and Varela, F. J. 1980. *Autopoiesis and cognition: The realization of the living.* Reidel.

Matyssek, R., and Lüttge, U. 2013. "Gaia: The planet holobiont." *Nova Acta Leopoldina, 111*(391), 325–344.

McCutcheon, J. P., and von Dohlen, C. D. 2011. An interdependent metabolic patchwork in the nested symbiosis of mealybugs. *Current Biology, 21*, 1366–1372.

McFall-Ngai, M., Hadfield, M. G., Bosch, T. C. G., Carey, H. V., Domazet-Lošo, T., Douglas, A. E., et al. 2013. Animals in a bacterial world, a new imperative for the life sciences. *Proceedings of the National Academy of Sciences, 110*, 3229–3236.

Meincke, A. S. 2022. Biological subjectivity: Processual animalism as a unified account of personal identity. In J. Noller (Ed.), *The unity of a person: Philosophical perspectives* (pp. 100–126). Routledge.

Monnin, D., Jackson, R., Kiers, E. T., Bunker, M., Ellers, J., and Henry, L. M. 2020. Parallel evolution in the integration of a co-obligate aphid symbiosis. *Current Biology, 30*(10), 1949–1957.

Motta, E. V. S., and Moran, N.A. 2020. Impact of glyphosate on the honey bee gut microbiota: Effects of intensity, duration, and timing of exposure. *mSystems, 5*, 00268–00220.

Motta, E. V. S., and Moran, N. A. 2024. The honeybee microbiota and its impact on health and disease. *Nature Reviews Microbiology, 22*, 122–137.

Muller, B. 2022. *Blue architecture: Water, design, and environmental futures.* University of Texas Press.

Nagpal, J., and Cryan, J. F. 2021. Microbiota-brain interactions: Moving toward mechanisms in model organisms. *Neuron, 109*, 3930–3953.

Nicholson, J. K., Holmes, E., Kinross, J., Burcelin, R., Gibson, G., Jia, W., and Pettersson, S. 2012. Host-gut microbiota metabolic interactions. *Science, 336*, 1262–1267.

Odling-Smee, J. 2024. *Niche construction: How life contributes to its own evolution.* MIT Press.

Oliver, K. M., Degnan, P. H., Hunter, M. S., and Moran N. A. 2009. Bacteriophages encode factors required for protection in a symbiotic mutualism. *Science, 325*, 992–994.

Osmanovic, D., Kessler, D. A., Rabin, Y., and Soen, Y. 2018. Darwinian selection of host and bacteria supports emergence of Lamarckian-like adaptation of the system as a whole. *Biology Direct, 13*, 24.

Oyama, S. 1985. *The ontogeny of information: developmental systems and evolution.* Princeton University Press.

Pattee, H. H. 1968. The physical basis of coding a reliability in biological evolution. In C. H. Waddington (Ed.), *Towards a theoretical biology* (Vol. 1) (pp. 67–93). Edinburgh University Press.

Pradeu, T. 2019. Philosophy of biology: Immunology and individuality. *eLife, 8*, e47384.

Ratsika, A., Cruz Pereira, J. S., Lynch, C. M. K., Clarke, G., and Cryan, J. F. 2022. Microbiota-immune-brain interactions: A lifespan perspective. *Current Opinions in Neurobiology, 78*, 102652.

Rhee, K.-J., Sethupathi, P., Driks, A., Lanning, D. K., and Knight, K. L. 2004. Role of commensal bacteria in development of gut-associated lymphoid tissues and preimmune antibody repertoire. *Journal of Immunology, 172*, 1118–1124.

Robinson, K. S. 2020. *Ministry for the future: A novel.* Orbit.

Roughgarden, J. 2020. Holobiont evolution: Mathematical model with vertical vs horizontal microbiome transmission. *Philosophy, Theory and Practice in Biology, 12*, 2.

Roughgarden, J., Gilbert, S. F., Rosenberg, E., Zilber-Rosenberg, I., and Lloyd, E. A. 2017. Holobionts as units of selection and a model of their population dynamics and evolution. *Biological Theory, 13*, 44–65.

Sagan, D., and Margulis, L. 1991. Epilogue: The uncut self. In A. I. Tauber (Ed.), *Organism and the origins of self* (pp. 361–374). Kluwer.

Sander, E. G., Warner, R. G., Harrison, H. N., and Loosli, J. K. 1959. The stimulatory effect of sodium butyrate and sodium propionate on the development of the rumen mucosa in the young calf. *Journal of Dairy Science, 42*, 1600–1605.

Sariola, S., and Gilbert, S. F. 2020. Toward a symbiotic perspective on public health: Recognizing the ambivalence of microbes in the Anthropocene. *Microorganisms*, 8, 746.

Schneider, T. 2021. The holobiont self: Understanding immunity in context. *History & Philosophy of the Life Sciences, 43*, 99.

Schoenheimer, R. 1942. *The dynamic state of body constituents.* Cambridge: Harvard University Press.

Schoenheimer, R., and Rittenberg, D. 1935. Deuterium as an indicator in the study of intermediary metabolism. *Science, 82*(2120), 156–157.

Sebeok, T. A. 2001. Biosemiotics: Its roots, proliferation, and prospects. *Semiotica, 134*, 61–78.

Sharon, G., Cruz, N. J., Kang, D. W., Gandal, M. J., Wang, B. et al. 2019. Human gut microbiota from autism spectrum disorder promote behavioral symptoms in mice. *Cell, 177*, 1600–1618.

Sharon, G., Sampson, T. R., Geschwind, D. H., and Mazmanian, S. K. 2016. The central nervous system and the gut microbiome. *Cell, 167*, 915–932.

Smith, M. I., Yatsunenko, T., Manary, M. J., Trehan, I., Mkakosya, R., Cheng, J., Kau, A. L., et al. 2013. Gut microbiomes of Malawian twin pairs discordant for Kwashiorkor. *Science, 339*, 548–554.

Stilling, R. M., Moloney, G. M., Ryan, F. J., Hoban, A. E., Bastiaanssen, T. F., Shanahan, F., Clarke, G., Claesson, M. J., Dinan, T. G., and Cryan, J. F. 2018. Social interaction-induced activation of RNA splicing in the amygdala of microbiome-deficient mice. *eLife, 7*, e33070.

Suarez, J., and Stencel, A. 2020. A part-dependent account of biologic individuality: Why holobionts are individuals and ecosystems simultaneously. *Biological Reviews, 95*, 1308–1324.

Sun, J., Lu, M., Gillette, N. E., and Wingfield, M. J. 2013. Red turpentine beetle: innocuous native becomes invasive tree killer in China. *Annual Review Entomology, 58*, 293–311.

Suzuki, T. A., et al. 2025. *Selection and transmission of the gut microbiome alone shifts mammalian behavior.* bioRxiv. https://doi.org/10.1101/2025.01.21.634013.

Taerum, S. J., Duong, T. A., de Beer, Z. W., Gillette, N., Sun, J.-H., Owen, D. R., and Wingfield, M. J. 2013. Large shift in symbiont assemblage in the invasive red turpentine beetle. *PLoS One, 8*, e78126.

Tauber, A. I. 2013. Immunology's theories of cognition. *History & Philosophy of the Life Sciences, 35*(2), 239–264.

Theis, K. R., Dheilly, N. M., Klassen, J. L. et al. 2016. Getting the hologenome concept right: An eco-evolutionary framework for hosts and their microbiomes. *mSystems, 1*(2).

Thomas, L. 1971. An earnest proposal. *New England Journal of Medicine, 285*, 1132–1133.

Tschitschko, B., Esti, M., Philippi, M. et al. 2024. Rhizobia–diatom symbiosis fixes missing nitrogen in the ocean. *Nature, 630*, 899–904.

Tsuchida, T., Koga, R., Horikawa, M., Tsunoda, T., Maoka, T., Matsumoto, S., Simon, J.-C., and Fukatsu T. 2010. Symbiotic bacterium modifies aphid body color. *Science, 330*, 1102–1104.

Uexküll, J. 1934. *A stroll through the worlds of animals and men: A picture book of invisible worlds.* University of Minnesota Press.

Vuong, H. E., Pronovost, G. N., Williams, D. W., Coley, E. J. L., Siegler, E. L., Qiu, A., Kazantsev, M., Wilson, C. J., Rendon, T., and Hsiao, E. Y. 2020. The maternal microbiome modulates fetal neurodevelopment in mice. *Nature, 586*, 281–286.

Waddington, C. H. 1957. *The strategy of the genes: A discussion of some aspects of theoretical biology.* George Allen and Unwin.

Wagner, M. A., and Reynolds, J. D. 2019. Salmon increase forest bird abundance and diversity. *PLoS ONE, 14*, e0210031.

Weersma, R. K., Zhernakova, A., and Fu, J. 2020. Interaction between drugs and the gut microbiome. *Gut*, *69*, 1510–1519.

Wei, G. Z., Martin, K. A., Xing, P.Y., Agrawal, R., Whiley, L., Wood, T. K., Hejndorf, S., Ng, Y. Z., Low, J. Z. Y., Rossant, J., Nechanitzky, R., Holmes, E., Nicholson, J. K., Tan, E. K., Matthews, P. M., and Pettersson, S. 2021. Tryptophan-metabolizing gut microbes regulate adult neurogenesis via the aryl hydrocarbon receptor. *Proceedings of the National Academy of Sciences of the USA*, *118*, e2021091118.

Whitehead, A. N. 1978 [1929]. *Process and reality* (D. R. Griffin and D. W. Sherburne, Ed.;corrected ed.). Free Press.

Wilson, E. B. 1896. *The cell in development and inheritance*. Macmillan.

Witzany, G. 2008. Biocommunication of unicellular and multicellular organisms. *tripleC*, *6*, 24–53.

Woznica, A., Gerdt, J. P., Hulett, R. E., Clardy, J., and King, N. 2017. Mating in the closest living relatives of animals is induced by a bacterial chondroitinase. *Cell*, *170*, 1175–1183.

Zivkovic, A. M., German, J. B., Lebrilla, C. B., and Mills, D. A. 2011. Human milk glycobiome and its impact on the infant gastrointestinal microbiota. *Proceedings of the National Academy of Sciences of the USA*, *108*, 4653–4658.

Žukauskaitė, A. 2020. Gaia theory: Between autopoiesis and sympoiesis. *Problemos*, *98*, 141–153.

6 Semiotic Agency: Life Is Meaning-Making

Alexei A. Sharov

6.1 Limitations of Physics and Input-Output Mapping in Explaining the Nature of Life and Cognition

Over centuries of research, mathematics and physics have accumulated a large number of theoretical templates for predicting processes and optimizing system monitoring and regulation. But it remains questionable if such computational methods can help to answer theoretical questions in biology, such as reconstructing possible scenarios of the origin and evolution of life and cognition, or improving regulatory networks in living systems. Attempts to explain life and cognition on the basis of physics were stimulated by the discovery of self-organization in nonlinear thermodynamic systems, which somewhat resemble processes in living organisms. Moreover, it was proposed that phenomena of life and cognition can be explained just by such complexity itself. Stuart Kauffman (1996) has argued that self-organization processes can generate "order for free," which can be further improved by natural selection or learning. Most biologists similarly believe that the origin of life can be explained solely by the complexity of organic compounds in the primordial soup and that the biological functions of these chemicals can emerge spontaneously by self-organization. However, artificially constructed complex systems have not been shown to self-organize into self-reproducing and evolving units.[1] Nor do biological functions emerge for free; however, once emerged via adaptive evolution, their recipe can be copied almost for free from parents to progeny.

Another strategy used in the mechanistic understanding of life and cognition consists in simulating the dynamics of cell components and processes by the traditional tools of systems science, such as partial differential equations, cellular automata, or Petri nets. Because the power of computers progressively increases, it is believed that soon it will be possible to process much more complex models that handle all the details of living systems at multiple scales. This computational approach to life is often used as grounds for claiming that life and cognition are not qualitatively different from nonliving processes; the differences are only quantitative and can be measured by the complexity of algorithms. In particular, variational principles of optimization are successfully applied to both physical systems (the shape of a hanging chain or path of light beam) and to biological systems (natural selection or learning) (Menon and Ranganathan, 2022; Simidjievski et al., 2019). But despite the similarity with computational methods, there is an important difference: in physics, the least action principle is fixed and based on universal laws, whereas learning of a living system tends to minimize its internally developed cost function (Friston, 2010), which is not fixed but can include subjectively chosen sensorial inputs, new categories of objects, or tools, and new sources of energy and chemicals. In other words, organisms develop and preserve their own laws of action that do not contradict the laws of physics, but that are not prescribed by it.

Also, mechanistic modeling of living cells at the molecular level is a difficult task because of the large number of interacting components. Such models have been implemented for certain parts of bacterial cells, such as a gene regulatory network of membrane dynamics (Boags et al., 2017; Skalnik et al., 2023). But detailed models of whole bacterial cells are still incomplete (Karr et al., 2015; Macklin et al., 2014). There are plans to build models of whole eukaryotic cells (Netz and Eaton, 2021; Szigeti et al., 2018), but the number of molecular components of these cells is by several orders greater than in bacteria, which poses considerable challenges. In addition, our knowledge of protein functions is still incomplete, and thus, such models cannot be expected to make valid predictions in the long run. To avoid these problems, models are simplified by limiting analysis to input–output relations and presenting a cell or organism as just a black box. For example, models of evolution assume that genotypes are mapped into phenotypes. Each phenotype is then associated with its fitness value, which is a function of average survival rate combined with reproduction rate. Then the relation between genotypes and their fitness is visualized as a smooth "fitness landscape" function, and evolution by natural selection is presented as the shortest walk to the summit of this landscape, which corresponds to the most reproductively "successful" genotype.

This approach, however, is oversimplified far beyond available facts. There is no evidence that new random mutations can be beneficial in natural populations. Every mutation is a disturbance that is harmful when it occurs in a functional region of the genome (e.g., in a coding or regulatory region) or neutral if it occurs in a nonfunctional location. No facts indicate that new random mutations in nature can encode proteins with more efficient and robust functions than already existing proteins. Even if a better protein structure is possible theoretically, it would require multiple mutations in the coding region of a gene with no mutations elsewhere; the probability of such a precise and complex event is negligibly small. Thus, there is no direct path uphill in the fitness landscape, as claimed by the model; and the waiting time for a new more useful genotype via random mutations can be as long as millions of generations.

In those rare cases, when some low-frequency genotypes have a higher fitness, these genotypes are not new, which challenges the idea that natural selection is a mechanism of evolutionary innovation. This can be illustrated by two classical examples of natural selection: resistance of bacteria to antibiotics and the emergence of new coloration in the peppered moth. Many species of bacteria and fungi compete for utilization of the same natural organic resources, and thus, some of these species produce toxic chemicals to deter or kill their competitors. Naturally, bacteria developed various adaptations to survive in toxic surroundings, which include avoidance, inactivation on the surface, internal enzyme-dependent degradation, active export, and modification of a target site of affected protein. In other words, many bacteria are already "experts" in chemical warfare and have readily available tools for developing resistance to new toxic chemicals. What is needed is to adjust these tools to new compounds (which often resemble natural toxins encountered earlier). Thus, some bacteria appear resistant to certain antibiotics even before first exposure (Wright, 2010), and in the presence of antibiotics (especially, in low concentrations), these resistant bacteria grow in numbers and replace the original nonresistant individuals.

The change of wing color in peppered moths (*Biston betularia* L.) in England is the best documented case of natural selection in wild populations (True, 2003). The white form successfully hides on birch trunks and avoids bird predation, whereas the dark "carbonaria" form is normally rare, but it became abundant in industrial polluted areas where it could hide on the surface of tree trunks covered with soot. How did the peppered moth managed to produce new coloration so fast? The dark color of its wings is caused by the pigment melanin, but no mutation was found in melanin-related genes that could explain the difference between white and dark moths (Van't Hof and Saccheri, 2010). Thus, the authors of this study suggested that moth coloration is instead caused by mutations in some regulatory genes. After six years of additional research, they identified a 22-Kb genomic insert that was present in 95% of dark moths (N=110) and in 0% of white moths (N=283). Surprisingly it was a large tandemly repeated transposon (a virus-like self-reproducing unit) in the first intron of the gene *cortex* (Van't Hof et al., 2016). The function of this insert is still not clear; the authors showed slightly higher, but statistically significant, expression of *cortex* in dark moths with transposon in comparison to white moths, but it is not known if this difference can change the coloration of wings. Many species of moths can become darker in special conditions (e.g., after temperature drop in the fall) even without mutations. Thus, dark-colored peppered moths might have been present as a rare epigenetic variation in the population even before the industrial revolution in England. Correlation of the transposon with wing color could have emerged later, during genetic accommodation together with other minor genetic changes that increased the heritability of the dark color phenotype.

These two examples show that evolutionary events in nature associated with fast phenotype change employed the *already existing capacities of organisms*. In other words, these events have been well "rehearsed" in the long-term evolution of ancestral lineages, and thus, there is no surprise that these changes occurred so fast (within a few generations). Without these preexisting capacities, it would take millions of generations to develop these phenotypes from scratch via new random mutations. Thus, the analysis of the *agential* capacities of organisms[2] is of primary importance for predicting potential evolutionary changes in the future, whereas the analysis of mutation rates is much less informative.

Other theories and concepts in science also rely on input–output analysis, such as Shannon's theory of information, which considers transmission of abstract symbols stripped of their meanings. Although this simplified approach has benefits for designing computation algorithms, it brings confusion when the computational metaphor is applied to the human brain (e.g., by Dennett, 1991). Another example of the input–output approach is behaviorism, which considers a stimulus as a *cause* of behavior. As a result, many studies in physiology and psychology ignore such aspects of cognition as perception, representation, free will, and the role of social interactions.

Finally, all such attempts to define life through physics are short-coming because organisms have their own laws of change (i.e., goal-directedness) that are not prescribed by physics and not fully known. Features of agents that are so far not explained by physics include capacities to grow, differentiate, and reproduce, generate memory signs, interpret signs creatively, produce subagents at multiple levels of hierarchy, get integrated into superagents of higher levels, establish cooperative relations with other members of the superagency, and sustain essential functions and sign processes, despite unfavorable conditions or damages.

6.2 The Essential Feature of Life Is Meaning-Making

To overcome the limitations of physics and input–output analysis in the study of life and cognition, it is necessary to recognize that organisms are agents—autonomous goal-directed entities, whose activities are meaningful in the sense of being intentionally acquired, optimized, and utilized on demand. The purpose of each living activity is solving a particular problem such as finding food, avoiding predators, or surviving life-threatening conditions. Organisms learn how to solve problems either within their lifespan or in a long-term evolutionary way, and hence convert even accidental problem-solving into a reliable habit that can be used on demand.

By acquiring a habit, an organism changes its identity to some extent, a process which Gilbert Simondon called *individuation* (Scott, 2014). Simondon emphasized the similarity between biological and psychological individuation and physical processes (e.g., crystallization) because both emerge via phase transition (in a physical sense) that brings the system into a new metastable state. But he did not fully realize that the difference between living and nonliving systems is substantially more fundamental than the difference between self-organizing and not self-organizing physical systems. Hurricanes, tornados, and flames are self-organizing systems that temporarily persist far from the equilibrium, but they are not alive because they do not have heredity and cannot accumulate new habits (Sharov and Tønnessen, 2021). Self-organization is thus necessary, but not sufficient, for a system to be considered living. Instead, the main event in the origin of life was the emergence of *expandable heredity*, together with *self-reproducing catalytic agency* capable of duplicating hereditary units and delivering them to progeny systems. In contrast to tornados and flames, such systems can develop new kinds of recurring self-organizing processes that did not exist before, and this capacity is functionally equivalent to memorizing a new habit.[3] Thus, living systems are capable of meaning-making, whereas physical self-organizing systems do not carry or engage with meaning. Meaningful habits emerge for a purpose—to solve a certain problem of living, even if the organism does not realize it consciously; and thus, living systems are goal-directed (i.e., purposeful) agents.

The notion that life is a meaning-making process is the central idea of biosemiotics (Anderson et al., 1984; Hoffmeyer, 2008; Kull et al., 2009), which is focused on integrating biology with semiotics, to result in a theory of biological signification and meaning. It would be a mistake to consider biosemiotics narrowly as just application of semiotic terminology to living systems, however. The true novelty of biosemiotics is that it proposes a cardinal rethinking of the scope and methods of both disciplines: biology *and* semiotics. Dominating mechanistic research in biology should be complemented by the study of biological meanings, and semiotics needs to be expanded beyond the study of human-centered communication. According to biosemiotics, life is coextensive with semiosis (Anderson et al., 1984, p. 9; Sebeok, 1976, p. 69) in the sense that all living functions require the interpretation of signs and the production of signs—and that all sign processes belong to living organisms, although these processes can be extended from organisms to their nonliving components (functional proteins, ribosomes) or the external artifacts (computers, robots) that are products of organisms.

In the biosemiotics context, a sign is a semiotic tool that is perceived and interpreted by certain organisms, and the result of interpretation (interpretant or meaning) prompts the organism to start or change its activity in a habit-dependent way. In contrast to physical forces, the effect of signs on organisms is not deterministic. A sign provides an option that can be either taken or rejected by an organism, depending on its current goals. Sign processes, collectively known as *semiosis*, include the phenomena of *signification* (connecting a sign with its meaning), *representation* and *manipulation* of signs (producing a sign to keep it in memory or to inform others), and *communication* (exchange of meanings) with other organisms (Sharov and Tønnessen, 2021, p. 192). The actions of agents are thus always sign-mediated and draw on memory and/or perception. Most signs are passive due to their lack of autonomous goal-directedness, whereas organisms and cells are goal-directed, and therefore, their role in semiosis is mostly active. Although signs can change the activity of agents, it is organisms that are ultimately in charge of sign relations and semiosis, not the other way around (Sharov and Tønnessen, 2021, pp. 206–207).

6.3 Semiotic Agency Is a Generalization of the Notion of Organism

In the previous section, I argued that sign processes are essential for living organisms. However, sign processes also exist in other organism-like entities. For example, organs and cells within multicellular organisms are autonomous goal-directed systems with their own networks of sign processes. Similarly, super-organisms such as colonies of bees, ants, and termites are highly integrated and display colony-level sensing, foraging, digestion, and reproduction. Thus, it makes sense to find a term that is applicable to all these organism-like units. The most relevant term is "semiotic agency," which implies autonomy, goal-directedness, *and* capacity for semiosis (Sharov, 2010; Sharov, and Tønnessen, 2021, pp. 153–154; Tønnessen, 2015). This term has been already used in semiotics, philosophy, and biosemiotics (Hoffmeyer, 1998; Kockelman, 2017; Nöth, 2009), but more recently Sharov and Tønnessen (2021) have revitalized this term and slightly modified its meaning, arguing that, as shown in figure 6.1, organisms (core agents), as well as organisms' subagents and superagents, are all—semiotic agents.

Living organisms are hierarchically structured agents[4] containing subagents at multiple levels, and while it is not always clear which level is the most highly integrated and competent to be called the "organism," each higher level of agential hierarchy emerged gradually in evolution. For example, free-living eukaryotic cells in their early evolution began cooperating and formed colonies, which initially were not sufficiently integrated enough to be called organisms. But later they developed structural integration (e.g., tissues), division of labor, and the communal use of resources, and eventually became multicellular organisms. Other organisms emerged via integration of several unrelated ancestors, through symbiogenesis. Most common examples are lichens (the integration of fungi and algae) and eukaryotic cells (the integration of archean cells with bacteria that became mitochondria or plastids).

The reproduction of organisms always restructures the tree of agential hierarchy. Most multicellular organisms originate from a single specialized cell (egg or spore) within the

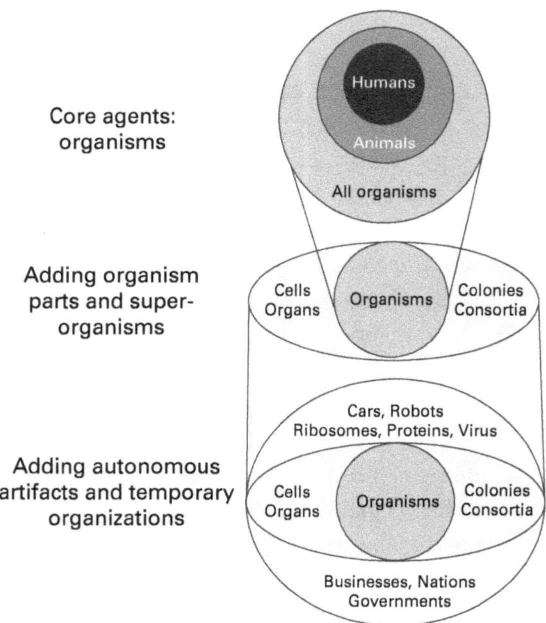

Figure 6.1
Semiotic agency is a generalization of the notion of organism. Organisms are core agents; organism's parts and super-organisms are added next; autonomous artifacts and temporary organizations are added last.

parental organism. Thus, a progeny organism always emerges first as a subagent within a parent, and this is an important reason for recognizing agency in parts of organisms, such as organs and cells. In viviparous animals, the embryo remains within the parent and grows like any other organ. Then at birth, the baby organism finally emerges from the parent's body, but it may still need some interaction with the parent until it is fully functional on its own.

The hierarchy of semiotic agents is extended in both directions: down to subagents and up to superagents that integrate multiple organisms. Examples of superagents are colonies, families, populations, evolutionary lineages, and ecological consortia that integrate evolutionary unrelated organisms. Some of these superagents emerge transiently, such as in slime mold (e.g., *Dictyostelium* spp.), whose amoebae-like cells live separately, but under some conditions aggregate, form a fruiting body, and produce spores. Humans also form various types of superagents, such as cultures, states, governments, and businesses, which are qualitatively different from superagents of other species because of humans' unique interpersonal relations, such as symbolic language, education, artificial living environment, monetary economy, laws, and ethics. Some of these superagents exist transiently during office hours and reemerge on the following workday.

Yet all semiotic agents are capable of some level of autonomous sign-driven activity. Autonomy here means a degree of independence from the environment and a capacity to interpret signs and select actions accordingly. All living cells include a functionally closed network of self-repair and self-production (Maturana and Varela, 1980). Functional capacities of organisms strongly depend on their local environment. Thus, additional layers of autonomy can be gained by niche construction—the process by which organisms improve their environments, making them more suitable for survival and reproduction (Laland and

O'Brien, 2010; Laubichler and Renn, 2015; Odling-Smee et al., 2003). Organisms of different taxonomic groups vary in the degree of their autonomy. For example, parasites are less autonomous than free living species because they cannot live without their host organism.

To emphasize the semiotic aspect of autonomy, Jesper Hoffmeyer and Norbert Wiley independently proposed the term *semiotic freedom*. Hoffmeyer (1992, p. 109) defined it as "the depth of meaning," and Wiley (1994, p. 15) wrote that it is "in the creative act of interpretation." The hallmark of semiotic freedom is the capacity to make choices at the level of *goals* or *representations* of objects rather than directly choosing objects or actions. Advanced levels of autonomy include *anticipation* of future events and *proactive* behavior, both of which rely on signs. Humans (and, possibly, some animals with advanced cognition) can facilitate their behavior by manipulating their cognitive environment, making it easier to navigate and rearrange when necessary, a phenomenon known as "cognitive niche construction" (Bertolotti and Magnani, 2017; Magnani, 2009).

Besides autonomy, semiotic agents also require goal-directedness (teleology), which is the capacity for utilizing historically accumulated experience to plan and regulate activities in the future (Barandiaran, 2017; Barandiaran et al., 2009; Sharov and Tønnessen, 2021, pp. 166–168). Goals are needed for setting priorities in both perception and action, and therefore, teleology is not limited to conscious agents but applies to any organism since all organisms pursue goals at least non-consciously, even prokaryotes. For example, the sensitivity of bacterial membrane receptors to glucose increases in environments where glucose is scarce (Porter et al., 2011). This regulation is a product of adaptive evolution, and it helps bacteria to find sources of food via chemotaxis.

Although the execution of goal-dependent activities is usually fast (e.g., by signal transduction in cells or conditioning in the nervous system), the acquiring of goal-directed habits by accumulation of prior experience may take a much longer time. It would be too restrictive to require that all semiotic agents acquire their goals within their lifetimes, especially in bacteria whose generation time is measured by minutes. Thus, it is reasonable to accept that goal-directed habits can be built into semiotic agents without direct experiential learning, but externally, by parental agents (Sharov and Tønnessen, 2021, p. 166). Innate habits acquired without learning exist in all organisms, including humans, but in bacteria, all habits are innate just because their lifespan is too short to learn. Of course, there are additional reasons why learning is not possible in bacteria, such as the lack of sufficient integration and flexibility in their signaling networks. The assumption that goal-directedness can be supplied by parental agents allows considering semiotic agents that span multiple generations. For example, extant organisms can utilize experiences and habits of organisms that lived million (or more) generations ago.

Moreover, even autonomous mechanisms produced by cells (e.g., proteins and ribosomes) and autonomous artifacts produced by humans (e.g., computers, robots, cars) can be considered semiotic subagents or dependent agents (figure 6.1) because they are programmed by their parental agents to act in a goal-directed way. Thus, the notion of semiotic agency transcends the boundary between life and nonlife. For example, a ribosome is a nonliving semiotic subagent in a living cell, which is programmed by the cell to produce polypeptides (precursors of proteins) with a specific sequence of amino acids encoded genetically. In fact, a ribosome within a living cell is functionally equivalent to a human-made robot because both of them produce components that are useful for parental agents.

Subagents, however, need to be distinguished from mere tools, which have no autonomy and cannot respond to signs. Although semiotic agency exists in both, living systems and nonliving autonomous artifacts, all agents are rooted in life because autonomous artifacts are always made by organisms.

6.4 Ontological Status and Types of Semiotic Agents

The notion of *semiotic agency* is an attempt to clarify the qualitative differences between life and nonliving objects such as atoms, rocks, planets, and stars. Usually, life is defined in biological terms as entities capable of "eating, metabolizing, excreting, breathing, moving, growing, reproducing, and responding to external stimuli."[5] But it is not clear why all these attributes need to be combined. In contrast, the notion of semiotic agency points to a more fundamental difference: semiotic agents exist in a self-designed form and therefore belong to a distinct category of being with an extended ontological status (Sharov and Tønnessen, 2021, p. 165; Walsh, 2018, p. 182). Agents exist to preserve and expand their identity, whereas nonliving objects are not self-defined and exist in the form determined externally by the laws of physics and the environment. All the other properties of life listed above are just the necessary *means* for preserving and propagating the organism's identity.

Because of their special ontological status, semiotic agents rely on their historically accumulated experience and knowledge. Thus, they cannot emerge spontaneously by pure chance; instead, they are produced by parental agents with at least a comparable level of semiotic competence (Sharov, 2010, p. 1054). The only exception would appear to be the origin of life, when noncellular agents, much simpler than viruses, are thought to have emerged spontaneously—an event with an extremely small probability, perhaps, once in a billion years in any given galaxy.

The gene-centric theory of evolution (neo-Darwinism) erroneously assumes that all evolutionary changes occur in the genome and that organisms are just "survival machines" carrying genomes (Dawkins, 1976). Biosemiotics overturns this idea of gene primacy and considers semiotic agents (including organisms) as the main players in evolution, whereas genomes have a supplementary role as hereditary sets of signs. In short, *the process of life has two channels: informational and agential.* Informational components provide a repository of passive hereditary signs (including genes), which are organized in such a way as to ensure integrity and global access by agents, whereas the role of agents is to interpret the hereditary and sensory information and use it for managing their activities, which include constructing and repairing the physical bodies of the agents and subagents, as well as maintaining and copying hereditary signs (Sharov, 2014). The informational channel is not self-explanatory, and thus, genomic information is meaningful only thanks to the interpreting capacity of proteins, ribosomes, and other subagents in living cells. Although the protein sequence of interpreting subagents is encoded in the genome, the genome alone would not help to preserve life. An organism requires *physically functioning agents* in the form of properly folded proteins or ribonucleoproteins in order to interpret genes.

Semiotic agents vary in their complexity and level of autonomy, ranging from simple functional mechanisms in living cells to complex organisms and super-organisms. There is a continuous scale of agential capacities instead of a binary choice between rational agents

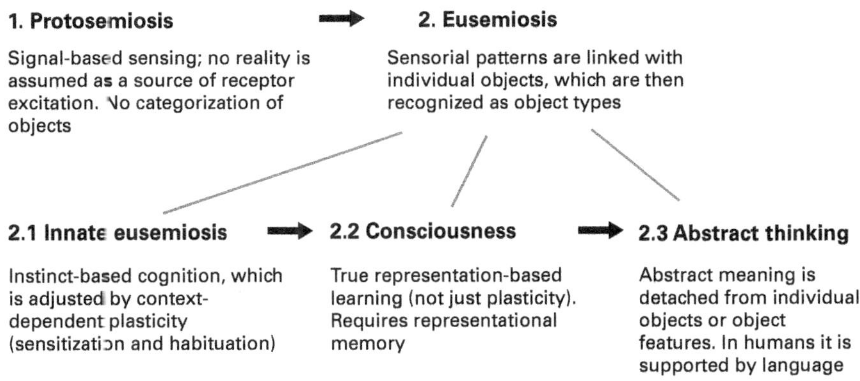

Figure 6.2
Major types of semiosis in agents.

(humans) and mechanistic non-agents as in the doctrine of Cartesian dualism. In terms of hierarchy, agents include organisms, organs, tissues, cells, cell components, families, colonies, populations, species, holobionts (multicellular organisms with microbiota), consortia (mutually supporting organisms from different species), societies, organizations, and autonomous artifacts. In terms of reproduction, semiotic agents can be *primary*—that is, self-reproducing (most organisms, cells, many super-organisms)—or *secondary* if they do not self-reproduce and are instead produced by other kinds of agents (e.g., cell components that are produced by the cell or worker bees that are produced by the queen bee) (Sharov and Tønnessen, 2021, pp. 166–167). Some agents are continuously active, whereas others may become inactive (dormant) for a certain period (e.g., sleep or diapause of organisms). An important feature of dormant organisms is that they can get reactivated on their own, either in response to some environmental factor or spontaneously. In comparison, a virus capsid can get activated only with the help of another agent—the host cell. Thus, it is a *potential agent* that does not have all components necessary for activation (Sharov and Tønnessen, 2021, pp. 180).

Semiotic agents likewise differ in their types of semiosis. There are two basic levels of semiosis: *protosemiosis* and *eusemiosis* (figure 6.2) (Sharov and Tønnessen, 2021). In protosemiosis, which is a rudimentary form of semiosis typical for bacteria, archea, and basic metabolism in eukaryotic cells, individual chemical signals are interpreted by cells as signals to change their activity *directly*—that is, without inferring cause–effect relations. As a result, bacteria have no "representations" of external objects, and they do not "know" that there are external objects that are the sources of the molecules detected by their chemoreceptors. The evolutionary transition from prokaryotes to eukaryotes, and especially to multicellular eukaryotes, is associated with the emergence of eusemiosis, which is the capacity to associate sensorial inputs with external objects and the capacity to recognize (i.e., categorize) these objects. Despite many publications on the origin of eukaryotes, biologists are largely unaware of this fundamental transition from protosemiosis to eusemiosis.

Eusemiosis emerged in its primitive innate form, where the capacity to detect and classify external objects is formed unconsciously based on hereditary information. Such *innate eusemiosis* exists even in single-cell eukaryotes which can classify particles into edible and nonedible before starting phagocytosis. Emergence of the nervous system in multicellular animals helped them to develop more efficient innate semiosis by chemically targeted neurite growth, synapse formation, and synapse plasticity.

Nervous systems were reused later in evolution for another major transition in the semiotic organization of animals—that is, the emergence of *consciousness*, which is eusemiosis by true learning and representational memory. Some biologists use the term "learning" in a broad sense as any effect of past events on future changes, a phenomenon which exists in all living systems and even in some self-organizing nonliving systems (Baluška and Reber, 2021; Shapiro, 2011). The term "learning" is very important for the behavioral studies of animals and humans, and thus, it would be a great loss for science to redefine it in such nonspecific way. Thus, it makes sense to preserve the traditional definition of "learning" as *the capacity to establish new mental associations of stimuli with representations of important objects or expected events*. Such terminology is consistent with Ginsburg and Jablonka (2019), who understood consciousness as unlimited associative learning.

The next major transition in semiotic competence occurred with the appearance of *abstract thinking* which includes extrapolations, contextuality, and the mental planning of future actions. This transition is characterized by Ginsburg and Jablonka (2019, p. 405) as imagination and dreaming. When organisms develop some processes of higher-level semiosis, they keep more primitive forms of semiosis at lower hierarchical levels. For example, such innate forms of behavior as the walking of a newborn calf are adjusted and improved by learning balance and coordination.

An essential component of semiosis is *semiogenesis*, which is the emergence of new kinds of sign relations in agents (Sharov and Tønnessen, 2021, pp. 192, 225–226). In living organisms, new sign relations are established mostly as a result of adaptive evolution, development, and learning. The degree of novelty can range from minor modification of a sign relation to major restructuring or repurposing. Semiogenesis in evolution can be triggered by environmental change or disturbance—for example, by mutations—but such disturbances are not the primary causes of adaptive evolution, as discussed above (subsection 6.1.1). The development of immunity toward infection, for example, is a vital component of somatic semiogenesis. Nervous systems provide mechanisms for rapid development of new connections between neurons via synapses, which is important for learning, both conscious and subconscious. Subconscious learning includes sensorimotor reflexes that support balance and precise action that is needed for walking, running, and other kinds of movement. In cognitively developed animals and humans, the major path for semiogenesis is the conscious learning of new behaviors, concepts, and models of reality.

6.5 Indeterminacy of Semiotic Agents

Determinacy is abundant in the nonliving physical world, where many objects remain self-identical for long times (e.g., most atoms, minerals, liquids, gases, tectonic plates, and planets). The movements of material bodies, for example, obey deterministic laws of physics, which remain true even for molecules. Thus, the discovery of fundamental indeterminacy in quantum mechanics was a surprise, and there is still no consensus among physicists on its interpretation.

Living organisms are mostly dynamic because they have to perform living functions to survive and reproduce. And yet, we believe that these changes do not affect certain deep levels of identity associated with heritable organization and individual traits. This belief is supported by the fact that each species makes progeny of the same kind: cats give birth to small kittens that grow into adult cats resembling their parents, and elephants give birth to baby elephants that grow into big adult elephants. The development of insects is a little more complicated because they keep transforming into distinct stages: egg, larva, pupa, and adult. But each stage (e.g., adults or larvae) is nearly identical across the sequence of generations, which allows us classify organisms into species and supports the idea of innate determinacy.

Moreover, determinacy is deeply embedded in the human language, where each kind of entity is named by a specific word. As a result, languages assume world determinacy by default and are not sufficiently flexible to describe indeterminacy and change. But the world includes indeterminacy both in the microscale (quantum mechanics) and macroscale (living organisms), and thus, indeterminacy needs to be conceptualized in science. Deacon (2011) assumes that nature itself is incomplete, and thus, absence is a part of the physical world. However, it seems better to characterize indeterminacy epistemologically—that is, as incomplete knowledge and/or incomplete control of the world by agents.

Let us consider several examples of indeterminacy in the living world. The first one is the indeterminacy of death. In this respect, a physicist would consider Schrödinger's cat, which is hypothetically isolated in a box together with a quantum-triggered killing mechanism (Schrödinger, 1935). Naturally, we do not know if such cat is alive or dead until the box is opened. But what will change if a technician forgot to put the killing mechanism into the box? Or what if the technician liked the cat and inactivated the killing mechanism? The result would be the same: we will still not know if the cat is alive or dead until the box is opened. Moreover, each of us is a Schrödinger's cat, and we never know if any other one of us is alive or dead until we open . . . the door and see that person. Obviously, the problem has little to do with quantum mechanics itself; indeterminacy is just a feature of life.

The second example is the indeterminacy of evolution. It is well known that some bacteria become resistant to antibiotics if they live in a medium with low concentration of antibiotics. However, we cannot know the mechanism of acquired resistance (avoidance, inactivation, degradation, export, modification of metabolism, or something else) until we identify it *post factum*. If we repeat the experiment, the mechanism of acquired resistance may appear different than in the first experiment, either grossly or in some details. And the third example is a hybrid agent, such as a bacteria cell with a latent viral infection. If the infected bacteria cell dies and produces a lot of virus particles, then we say that the infected cell has been already doomed, and its identity now is "virus." But if the bacteria cell manages to inactivate the virus, we say that its identity is "bacteria." This means that the identity of infected bacteria is fundamentally undetermined; it is a mixture of "virus" and "bacteria."

These examples indicate that indeterminacy results from the inability of agents to predict their own fate, as well as the fate of other things and agents—and this inability is fundamental: It cannot be fixed by collecting and analyzing more data, simply because the process cannot be described axiomatically. In this case, we have no other choice than to suspend our judgment and wait till "the box is opened." For each agent, there is nothing present in its umwelt (model of the world) beyond representation, and this is what reality is about. This does not imply solipsism, because representation is not just the result of imagination, but a summary of the agent's experienced interactions with the world, which is updated continuously.

Possibly, the agential approach can help to resolve paradoxes of quantum mechanics following the logic of Macháček (2019, p. 238), who suggested, "The unseen and unmeasured nature of things is not a sum of indices inductively symbolized by physics, but it is a continuum of possibilities of interpretation of indices." He also wrote, "The Peircean interpretation of quantum mechanics can show us a different path by simply discarding—on metaphysical grounds—the idea that there is anything but representation." I understand this statement as the notion that agents can interact with the world and gain experience (representation) even if the world is not axiomatically described.

These considerations imply that the indeterminacy of semiotic agency is unavoidable because the world is not a collection of true facts that simply wait for the time when they get identified by agents (e.g., humans), and it is not a formal system of axioms where truth is determined a priori. This topic is brilliantly discussed by Stuart Kauffman and Roli (2021) in the article "The World Is Not a Theorem."

The most amazing feature of living organisms, however, is their ability to persist and propagate *despite* such indeterminacy. Physics and mathematics provide recipes for finding exact solutions to various problems, which makes us think that organisms and humans should accept a deterministic way of life, where all factors and chances are taken into account. But this is a misguided approach because most of life's problems cannot be solved in a deterministic way. Moreover, striving toward determinism is a very expensive and inefficient strategy; it draws scientists and engineers to collect tons of data and build thousands of models that eventually appear useless. In contrast, organisms collect the minimum necessary information and take bets on various methods of survival and reproduction. Some of these methods have appeared successful and become accommodated genetically. Multicellular organisms do not try to exterminate all microbes in their body and instead develop immune systems to keep microbes under control. This strategy pays off because some of these microbes appear useful for digestion or other functions. Microbes contributed to various useful functions in eukaryotic organisms; for example, cell metabolism is supported by mitochondria and plastids originated from ancient bacteria, whereas viruses facilitated the evolution by horizontal gene transfer (Gontier, 2015) and by replicating binding motifs for transcription factors (Bourque et al., 2008). Thus, organisms use indeterminacy to their advantage instead of trying to get rid of it. Indeterminacy is also useful for cognition because it allows making choices and executing free will.

6.6 Discussion

In the past, the notion of agency has been criticized for being nonscientific because its terms (e.g., autonomy, goal-directedness, cognition, evolution, and function) are loosely defined. Daniel Dennett called it *folk psychology* (Dennett, 1991, p. 304) or *user illusion* (Dennett, 1991, pp. 311–312) and claimed it incompatible with science. An example of user illusion is the belief of a computer user that she moves the cursor around the screen and copies files from one location to another, whereas in fact, all these operations are done in quite a different manner by the computer's parts. Dennett argues that consciousness is likewise simply a "virtual machine, realized in the astronomically adjustable connections of a human brain" (Dennett, 1991, p. 431). Dennett's frustration with notions of agency and consciousness are understandable since human history is full of examples when people assumed the existence of conscious agents that were not there (e.g., angels and gods). But clear definitions exist only in axiomatic theories, which have their own limitations: (i) only a few phenomena are satisfactory described with axioms, (ii) the same phenomenon can be described by multiple sets of axioms, and (iii) an axiomatic system cannot be both complete and non-contradictive following Gödel's theorem. It is for these reasons that ordinary language remains the main communication tool in science, despite its ambiguity. Note, too, that Dennett uses terms such as experience, mind, reality, evolution without clear definitions, and his hypothesis that consciousness is a virtual machine is not elaborated. In biosemiotics, the notion of semiotic agency provides a clear separation between agents and borderline physical phenomena that resemble agency (e.g., tornado, river path, geothermal or hydrothermal flows, and tectonic plate movement). In the physical world, there is no reproduction with heredity, and thus, physical phenomena have no capacity to accumulate adaptive habits in evolution.

Semiotic agency is presented here as a hierarchy of subagents, which contradicts the common view that each human has a centralized control of its actions. The idea of a hierarchy is supported by the fact that a progeny organism is initially a subagent of the parent. Subagents can be criticized as a revival of the notion of homunculus, a medieval idea that agency is transferred from parents to offspring in the miniature form. Biology rejects the existence of homunculi as well as any other agents within an organism. But subagents of an organism are not homunculi for several reasons: (i) they are not a miniature form of the organism and have completely different structures, (ii) they do not take control of all organism functions, and (iii) subagents have their own autonomy despite following commands from other parts of the organism. The notion of multiplicity and autonomy of subagents is supported by notions of autopoiesis (Maturana and Varela, 1980), enactivism (Varela et al., 1991), and multi-agent cognition (Minsky, 1986).

The concept of semiotic agency is coherent with the Extended Evolutionary Synthesis, which assumes that living organisms and their parts are important players in the process of evolution (Pigliucci and Müller, 2010) rather than simple gene-copying machines, as claimed by neo-Darwinism. Organisms have sufficient plasticity to survive in nonfavorable conditions and mobilize their resources and variation to withstand harmful factors. These new and modified responses can be stabilized genetically via genetic accommodation. Neo-Darwinism explains evolutionary adaptations by new randomly occurring beneficial mutations, which have not been detected in nature. In contrast, the hypothesis of phenotypic plasticity combined with genetic accommodation is a more realistic pathway toward evolutionary adaptation because it does not require immediate specific beneficial mutations (Sharov and Tønnessen, 2021, pp. 230, 240).

Studying semiotic agency is aimed, first, at finding similarities of functions and semiotic processes in various kinds of agents, and second, at delineating qualitative differences between classes of agents with lower and higher semiotic competence. Considering the first approach, I discussed in this chapter that the hallmark of life is its capacity to develop meaningful goal-directed activities, use them repeatedly on demand in the original or slightly modified form, and combine them creatively. In other words, life is an evolving organic network of meaningful problem-solving habits, and the progress of life is seen in the extension of this network (e.g., by adding nodes and regulatory links) and in its reorganization toward increased efficiency, robustness, and creativity in changing environments. Because goal-directedness does not emerge from nothing, all agents are produced and programmed by parental agents with sufficient level of semiotic competence. All agents act in a goal-directed way, but not all of them can generate or modify their goals. Agents that are not capable of learning are fully programmed by their parental agents, whereas learning agents can combine their innate habits with newly learned action patterns and use them to thrive in changing environments.

Major qualitative differences within the domain of semiotic agency are demarked by evolutionary transitions (or transition zones), and especially by those transitions that restructure the organization of semiotic processes. Transition from protosemiosis (signal processing) to eusemiosis (perception of objects and categorization) coincides approximately with the emergence of eukaryotic cells. Eusemiosis is further partitioned into three subtypes: innate eusemiosis, consciousness, and abstract thinking.

The importance of studying organisms as autonomous goal-directed agents has been realized in biology and philosophy and has attracted the attention of scientists from several other transdisciplinary projects besides biosemiotics, such as second-order cybernetics, autopoiesis, enactivism, constructivism, and code biology. The theory of autopoiesis (Maturana and Varela, 1980) supports the notion of autonomy of living systems and their functional closure, and in this respect, it is similar to biosemiotics. However, the autopoiesis theory does not take into consideration semiotic processes and goal-directedness (other than the return to a steady state when perturbed) and describes living cells only in the terms of their physics. As a result, it ignores the essential property of life, which is meaning-making.

Enactivism and code biology are more closely related to biosemiotics than other theories. Enactivism explains the cognition of animals and humans as *meaning-making* and acknowledges the umwelt theory of Uexküll (Varela et al., 1991). It includes a theory of agency (Barandiaran, 2017; Barandiaran et al., 2009), and some researchers supported the integration of enactivism with biosemiotics (de Jesus, 2016, 2018; Tønnessen, 2016). The project of code biology assumes that functions of living cells are regulated by various kinds of codes, of which the genetic code is the main example (Barbieri, 2015). Yet some ideas about agents are better covered by enactivism and constructivism than by biosemiotics. One of them is the notion that mental models of the world in animals do not necessarily match to reality and instead are constructed subjectively as pragmatic tools. For example, Glasersfeld (1984) in his theory of radical constructivism assumes that human cognition is a biological adaptation that does not necessarily represent reality in iconic way.

However, alternative theories have certain drawbacks in comparison to biosemiotics. Most followers of autopoiesis and enactivism ignore the semiotic nature of life. Weak aspects of radical constructivism include a tendency toward excessive subjectivity (including solipsism), and lack of connection with the philosophy of pragmatism and with semiotics. Limitations of the code biology project include narrowing down the notion of biological meaning to the immediate products of molecular interactions in living cells. Further, coding is described as a mapping from a set of codes to the set of meanings (Barbieri, 2008). In such understanding, the functional role of proteins is not considered as the meaning of a gene. As a result, code biology appears relevant only for mechanistic biochemical relations between molecules in cells (i.e., protosemiosis). Considering these limitations, biosemiotics appears to be the most promising project in the study of semiotic agency.

Acknowledgments

I thank participants of Gatherings in Biosemiotics and participants of seminars, where I have presented, for helpful comments and discussion. Many thanks to editors of this volume, Kalevi Kull and Donald Favareau, and to anonymous reviewers.

Notes

1. Here, I do not consider aggregates of living cells, like in xenobots (Blackiston et al., 2021). In these cases, cells are living agents with extensive adaptive capacities, and their ability to self-organize is not surprising.
2. Agential capacities of organisms can be explored by experimental embryology, physiology in extreme environments, and studies of organism behavior in challenging conditions, and also by studying natural phenotypic variation within a population, species, or larger taxonomic unit—for example, homologous series (Vavilov, 1922).
3. Accidental problem-solving does not constitute a new self-organizing event if it is not memorized. The probability of such an event would be too low to be functional.
4. The term "hierarchy" can be replaced by "heterarchy" considering bidirectional dependencies between subagents (McCulloch, 1945). However, I prefer using the term "hierarchy," which is more familiar to readers.
5. Life, 2024, https://www.britannica.com/science/life2024.

References

Anderson, M., Deely, J., Krampen, M., Ransdell, J., Sebeok, T. A., and Uexküll, T. von. 1984. A semiotic perspective on the sciences: Steps toward a new paradigm. *Semiotica*, *52*(1–2), 7–47.

Baluška, F., and Reber, A. S. 2021. The biomolecular basis for plant and animal sentience senomic and ephaptic principles of cellular consciousness. *Journal of Consciousness Studies*, *28*(1–2), 34–59.

Barandiaran, X., Di Paolo, E., and Rohde, M. 2009. Defining agency: Individuality, normativity, asymmetry and spatiotemporality in action. *Adaptive Behavior*, *17*(5), 367–386.

Barandiaran, X. E. 2017. Autonomy and enactivism: Towards a theory of sensorimotor autonomous agency. *Topoi*, *36*, 409–430.

Barbieri, M. 2008. Is the cell a semiotic system? In M. Barbieri (Ed.), *Introduction to biosemiotics: The new biological synthesis* (pp. 170–207). Springer.

Barbieri, M. 2015. *Code biology*. Springer.

Bertolotti, T., and Magnani, L. 2017. Theoretical considerations on cognitive niche construction. *Synthese*, *194*, 4757–4779.

Blackiston, D., Lederer, E., Kriegman, S., Garnier, S., Bongard, J., and Levin, M. 2021. A cellular platform for the development of synthetic living machines. *Science Robotics*, *6*(52).

Boags, A., Hsu, P. C., Samsudin, F., Bond, P. J., and Khalid, S. 2017. Progress in molecular dynamics simulations of gram-negative bacterial cell envelopes. *Journal of Physical Chemistry Letters*, *8*(11), 2513–2518.

Bourque, G., Leong, B., Vega, V. B., Chen, X., Lee, Y. L., Srinivasan, K. G., et al. 2008. Evolution of the mammalian transcription factor binding repertoire via transposable elements. *Genome Research*, *18*(11), 1752–1762.

Dawkins, R. 1976. *The selfish gene*. Oxford University Press.

De Jesus, P. 2016. From enactive phenomenology to biosemiotic enactivism. *Adaptive Behavior*, *24*(2), 130–146.

De Jesus, P. 2018. Thinking through enactive agency: Sense-making, bio-semiosis and the ontologies of organismic worlds. *Phenomenology and the Cognitive Sciences*, *17*, 861–887.

Deacon, T. W. 2011. *Incomplete nature: How mind emerged from matter*. W. W. Norton and Company.

Dennett, D. C. 1991. *Consciousness explained*. Little, Brown and Co.

Friston, K. 2010. The free-energy principle: a unified brain theory? *Nature Reviews Neuroscience, 11*(2), 127–138.

Ginsburg, S., and Jablonka, E. 2019. *The evolution of the sensitive soul: Learning and the origin of consciousness*. MIT Press.

Gontier, N. (ed.). 2015. *Reticulate Evolution: Symbiogenesis, Lateral Gene Transfer, Hybridization and Infectious Heredity*. Springer.

Glasersfeld, E. von 1984. An introduction to radical constructivism. In P. Watzlawick (Ed.), *The invented reality* (pp. 17–40). Norton.

Hoffmeyer, J. 1992. Some semiotic aspects of the psycho-physical relation: The endo-exosemiotic boundary. In T. A. Sebeok and J. Umiker-Sebeok (eds.), *Biosemiotics. The Semiotic Web 1991*. Mouton de Gruyter, 101–123.

Hoffmeyer, J. 1998. Surfaces inside surfaces: On the origin of agency and life. *Cybernetics and Human Knowing, 5*(1), 33–42.

Hoffmeyer, J. 2008. *Biosemiotics: An Examination into the Signs of Life and the Life of Signs*. University of Scranton Press.

Karr, J. R., Takahashi, K., and Funahashi, A. 2015. The principles of whole-cell modeling. *Current Opinion in Microbiology, 27*, 18–24.

Kauffman, S. 1996. *At home in the Universe: The Search for the Laws of Complexity*. Penguin.

Kauffman, S., and Roli, A. 2021. The world is not a theorem. *Entropy, 23*(1467), 1–9.

Kockelman, P. 2017. Semiotic agency. In N. J. Enfield and P. Kockelman (eds.), *Distributed Agency*. Oxford University Press, 25–38.

Kull, K., Deacon, T., Emmeche, C., Hoffmeyer, J., and Stjernfelt, F. 2009. Theses on biosemiotics: Prolegomena to a theoretical biology. *Biological Theory, 4*(2), 167–173.

Laland, K. N., and O'Brien, M. J. 2010. Niche construction theory and archaeology. *Journal of Archaeological Method Theory, 17*, 303–322.

Laubichler, M. D., and Renn, J. 2015. Extended evolution: A conceptual framework for integrating regulatory networks and niche construction. *Journal of Experimental Zoology Part B: Molecular Development Evolution, 324*(7), 565–577.

Macháček, M. 2019. A Peircean interpretation of probability in quantum mechanics. In *Semiotics 2018: Resilience in an age of relation* (pp. 233–241). Semiotic Society of America.

Macklin, D. N., Ruggero, N. A., and Covert, M. W. 2014. The future of whole-cell modeling. *Current Opinion in Biotechnology, 28*, 111–115.

Magnani, L. 2009. *Adaptive cognition. The epistemological and eco-cognitive dimensions of hypothetical reasoning*. Springer.

Maturana, H., and Varela, F. 1980. *Autopoiesis and cognition: The realization of the living*. D. Reidel Publishing Co.

McCulloch, W. 1945. A heterarchy of values determined by the topology of nervous nets. *Bulletin Mathematical Biophysics, 7*, 89–93.

Menon, D., and Ranganathan, R. 2022. A generative approach to materials discovery, design, and optimization. *ACS Omega, 7*(30), 25958–25973.

Minsky, M. 1986. *The society of mind*. Simon and Schuster.

Netz, R. R., and Eaton, W. A. 2021. Estimating computational limits on theoretical descriptions of biological cells. *Proceedings of the National Academy of Sciences of the United States of America, 118*(6).

Nöth, W. 2009. On the instrumentality and semiotic agency of signs, tools, and intelligent machines. *Cybernetics & Human Knowing, 16*(3–4), 11–36.

Odling-Smee, F. J., Laland, K. N., and Feldman, M. W. 2003. *Niche construction: The neglected process in evolution*. Princeton University Press.

Pigliucci, M., and Müller, G. B. (Eds.). 2010. *Evolution: The extended synthesis*. MIT Press.

Porter, S. L., Wadhams, G. H., and Armitage, J. P. 2011. Signal processing in complex chemotaxis pathways. *Nature Reviews Microbiology, 9*(3), 153–165.

Schrödinger, E. 1935. Die gegenwärtige Situation in der Quantenmechanik. *Naturwissenschaften, 23*(48), 807–812.

Scott, D. 2014. *Gilbert Simondon's psychic and collective individuation: A critical introduction and guide*. Edinburgh University Press.

Sebeok, T. A. 1976. *Contributions to the doctrine of signs*. Indiana University Press.

Sebeok, T. A., and Umiker-Sebeok, J. E. (Eds.). 1992. *Biosemiotics*. Mouton de Gruyter.

Shapiro, J. A. 2011. *Evolution: A view from the 21st century*. FT Press Science.

Sharov, A. A. 2010. Functional information: Towards synthesis of biosemiotics and cybernetics. *Entropy, 12*(5), 1050–1070.

Sharov, A. A. 2014. Evolutionary constraints or opportunities? *Biosystems, 123*, 9–18.

Sharov, A. A., and Tønnessen, M. 2021. *Semiotic agency: Science beyond mechanism*. Springer.

Simidjievski, N., Bodnar, C., Tariq, I., Scherer, P., Andres Terre, H., Shams, Z., et al. 2019. Variational autoencoders for cancer data integration: Design principles and computational practice. *Frontiers in Genetics, 10*, 1205.

Skalnik, C. J., Cheah, S. Y., Yang, M. Y., Wolff, M. B., Spangler, R. K., Talman, L., et al. 2023. Whole-cell modeling of *E. coli* colonies enables quantification of single-cell heterogeneity in antibiotic responses. *PLoS Computational Biology, 19*(6), e1011232.

Szigeti, B., Roth, Y. D., Sekar, J. A. P., Goldberg, A. P., Pochiraju, S. C., and Karr, J. R. 2018. A blueprint for human whole-cell modeling. *Current Opinion in System Biology, 7*, 8–15.

Tønnessen, M. 2015. The Biosemiotic Glossary Project: Agent, agency. *Biosemiotics, 8*(1), 125–143.

Tønnessen, M. 2016. Agency in biosemiotics and enactivism. In J. Zlatev, G. Sonesson, and P. Konderak (Eds.), *Meaning, mind and communication. Explorations in cognitive semiotics* (pp. 69–82). Peter Lang.

True, J. R. 2003. Insect melanism: The molecules matter. *Trends in Ecology and Evolution, 18*(12), 640–647.

Van't Hof, A. E., Campagne, P., Rigden, D. J., Yung, C. J., Lingley, J., Quail, M. A., et al. 2016. The industrial melanism mutation in British peppered moths is a transposable element. *Nature, 534*(7605), 102–105.

Van't Hof, A. E., and Saccheri, I. J. 2010. Industrial melanism in the peppered moth is not associated with genetic variation in canonical melanisation gene candidates. *PLoS One, 5*(5), e10889.

Varela, F. J., Thompson, E., and Rosch, E. 1991. *The embodied mind: Cognitive science and human experience*. MIT Press.

Vavilov, N. I. 1922. The law of homologous series in variation. *Journal of Genetics, 12*, 47–89.

Walsh, D. M. 2018. Objectcy and agency: Towards a methodological vitalism. In D. J. Nicholson and J. Dupré (Eds.), *Everything flows: Towards a processual philosophy of biology* (pp. 167–185). Oxford University Press.

Wiley, N. 1994. *The semiotic self*. University of Chicago Press.

Wright, G. D. 2010. Antibiotic resistance in the environment: A link to the clinic? *Current Opinion in Microbiology, 13*(5), 589–594.

7 The Origins of Life, Agency, and Meaning

Stuart Kauffman

7.1 Overview

Organisms are open thermodynamic, self-reproducing, molecular reaction networks that are Kantian Wholes that achieve Catalytic and Constraint Closure. Good theory and data suggest that molecular reproduction arose in the universe as small molecule collectively autocatalytic sets. A plausible further pathway to prokaryotes involves the coevolution of such autocatalytic metabolisms with peptide and RNA autocatalytic sets. Ultimately, this pathway might lead to template replication and coding.

Kantian Wholes achieving catalytic and constraint closure constitute Molecular Autonomous Agents able to sense and act upon their worlds. This sensing and action are the onset of meaning. From these rudiments, coevolution of the biosemiotic web has flowered as coevolving organisms cocreate ever-novel niches for one another.

The emergence of life in the evolving universe remains a major mystery. With evolving life, agency and meaning arose. The universe evolved from matter to mattering. My aim in this brief article is to sketch a testable pathway from the origin of the universe to the emergence and early evolution of life.

7.2 A Brief Definition of Life, Agency, and Meaning

I. Living organisms are open thermodynamic self-reproducing chemical reaction networks that are Kantian Wholes that achieve Catalytic Closure, Constraint Closure and are spatially bounded, often by an enclosing membrane (Kauffman and Roli, 2023).

II. A Kantian Whole has the property that the Parts exist (in the universe) for and by means of the Whole. You are a Kantian Whole. You exist for and by means of your organs—heart and liver and genes. They exist for and by means of being Parts of you, the Whole.

III. The molecules in living cells form collectively autocatalytic sets. Each molecule has at least one last step in its formation catalyzed by some molecule in the set, or in the set of molecules that constitute its exogenous "food" (Hordijk et al., 2010).

IV. Living cells achieve a newly recognized and remarkable property: *Constraint Closure* (Kauffman and Roli, 2023; Montévil and Mossio, 2015):

  a. Thermodynamic work is the constrained release of energy into a few degrees of freedom. A cannon with powder and a cannon ball at its base is an example. The cannon is a boundary condition constraint on the release of energy when the power explodes and blasts the ball down the hollow bore of the cannon. Without a boundary condition constraint on the release of energy, no thermodynamic work can be done (Atkins, 1984).
  b. Thermodynamic work can construct boundary condition entities that can serve as constraints. For example, thermodynamic work was used to construct the cannon.
  c. In a Constraint Closed system, a set of nonequilibrium processes, [1, 2, 3], and a set of boundary condition constraints, [A, B, C], coordinate in such a way that the constraints, [A, B, C], constrain the release of energy in the processes, [1, 2, 3], such that the work done constructs the very same set of constraints, [A, B, C]. For example, A constrains the release of energy in process 1 to construct a B. B constrains the release of energy in process 2 to construct a C. C constrains the release of energy in process 3 to construct an A. The system literally constructs itself by doing thermodynamic work to construct the boundary conditions that constrain the release of energy to construct the same boundary conditions. This self-construction is entirely new. We construct our automobiles. These have organized constrained releases of energy that do work. Gas explodes; wheels turn. But automobiles do not construct their own constraints on the release of energy. Cells do. Living cells construct themselves (Kauffman and Roli, 2023; Montévil and Mossio, 2015).
  d. The union of Kantian Whole—catalytic and constraint closure—constitutes the mysterious *elan vital*, here rendered entirely nonmysterious.
  e. The evolving biosphere is a propagating diversifying *construction, not an entailed deduction.*

V. Living cells not only construct themselves, but as open thermodynamic systems must "eat" to survive. In general, cells have been able to evolve to "eat" because living cells are nonlinear dynamical systems with complex dynamical behavior that enables living cells, receiving inputs from their environment and acting on that environment, to sense and categorize their worlds, orient to relevant features of their worlds, evaluate these as "good or bad for me," and act based on those evaluations. This is the basis of *agency and meaning*. Agency and meaning are immanent in evolving life. The "biosemiotic meaning" is "I get to exist for a while." The capacity to "act" is immanent in the fact that living cells achieve constraint closure and do thermodynamic work to construct themselves. The same capacity enables cells to do thermodynamic work on their environment. A cell's action is embodied, enacted, embedded, extended, and emotive (Peil, 2014).

7.3 How Did Life Emerge?

A debate has endured for decades: Is life based on some form of template replication? Or is life based on some deeper and more general means of molecular reproduction seen in collectively autocatalytic sets? (Harrison et al., 2023; Kauffman and Roli, 2023). The debate is surely not settled. Progress in the first direction has been achieved by Philipp Holliger, who has in vitro evolved a long RNA polymer able to template replicate several hundred of its own nucleotides (Wachowius and Holliger, 2021). This is a triumph. In an RNA World, any such single replicating "gene" must somehow evolve some form of connected small-molecule metabolism catalyzed by RNA sequences that constructs the nucleotide building blocks needed by the single gene and that mobilizes energy resources to do so. This may well have been possible.

The alternative view, sometimes called "metabolism first," envisions the spontaneous emergence of collectively autocatalytic sets. Increasing theory and experimental grounds now provide favorable evidence for this latter pathway. Experimentally, DNA, RNA, and peptide collectively-autocatalytic sets have been constructed (Sievers and Kiedrowski, 1998; Vaidya et al., 2012; Wagner and Ashkenasy, 2015). Well-grounded theory has for years shown that in a system with an increasing diversity of molecular species of increasing molecular complexity, at some point, a *first-order phase transition* is reached at which collectively autocatalytic sets emerge (Farmer et al., 1986; Hordijk and Steel, 2018). Such a phase transition would be the emergence of molecular reproduction.

Parallel theory development supports the same theory. Years ago, Walter Fontana created "Algorithmic Chemistry" (Fontana, 1990). Here Fontana maintained 50,000 Lisp expressions on a model "vat" on his computer. Fontana fed novel Lisp expressions into his vat, and then randomly removed Lisp expressions from the vat. The system models a flow reactor. Lisp expressions can act on one another to produce new Lisp expressions. This is like a chemical reaction, but without considering thermodynamics and energy. By maintaining a constant total number of Lisp expressions in the vat, the system selects for Lisp expressions that increase in abundance. Fontana found, first, Lisp expressions that copied themselves and took over the vat. He disallowed these and reran his experiments. Fontana reliably found the spontaneous emergence of collectively autocatalytic sets of Lisp expressions.

Very recently, Aguera y Arcas et al. (2024) have extended Fontana's work with eight different base languages. They found the same spontaneous emergence of collectively autocatalytic sets in seven of the eight.

Perhaps the most powerful recent evidence that molecular reproduction is based on collective autocatalysis is in two papers with Joana Xavier as lead author (Xavier and Kauffman, 2022; Xavier et al., 2020). The first (Xavier et al., 2020) demonstrated small-molecule collectively autocatalytic sets with no DNA, RNA, protein, or lipid polymers. These sets, each with several hundred small molecules and metals, were found in archaea and bacteria from before oxygen was in the atmosphere. It is deeply interesting that the two small-molecule collectively autocatalytic sets have an intersection set of 175 small molecules and 172 reactions that is itself collectively autocatalytic. This suggests that the precursor to archaea and bacteria utilized this intersection small-molecule autocatalytic set. If so, in subsequent evolution, this precursor, small-molecule collectively autocatalytic set diverged into the set found in the archaea and in the bacterium. These sets became the basic metabolism of all subsequent life.

In a second paper, Xavier and Kauffman (2022) have shown that all 6,700 prokaryotes searched have small-molecule collectively autocatalytic sets. These sets range in size from a few dozen molecular species and reactions to a few hundred. As noted, presumably these evolved from the primordial intersection set between archaea and bacteria over two billion years ago. An important caveat is that these sets are identified computationally. It has not yet been demonstrated that they reproduce in vitro.

It now seems plausible, perhaps highly plausible, that such small-molecule collectively autocatalytic sets were the first form of "life" in the universe. Such sets are Kantian Wholes that achieve catalytic and constraint closure. Moreover, such sets typically form a number of amino acids and at least one nucleotide, ATP. In addition, via ATP and NAD, they couple the rudiments of the two major energy systems in living cells. Other sources of energy have been hypothesized as arising from wet dry cycles (Damer and Deamer, 2020) or from transmembrane proton and CO_2 gradients (Harrison et al., 2023).

The critical experiments now needed are based on the experimental creation of high-diversity small-molecule systems to test if collectively autocatalytic small molecular systems emerge. These experiments are feasible.

7.4 From the Big Bang to the Small-Molecule Collectively Autocatalytic Set

Briefly, the Big Bang was followed by the formation of hadrons, then the simple and complex atoms, then simple and ever more complex molecules. The Murchison meteorite, formed about five billion years ago, has tens of thousands of small and larger organic molecules (Schmitt-Kopplin et al., 2010). Space chemistry on solid surfaces, in general, may be very complex. Therefore, the basic emergence of high-diversity systems of small molecules occurred naturally in the evolution of the universe. If we can show that such systems do form small-molecule collectively autocatalytic sets readily, we will have a beginning experimental estimate of the expected frequency of molecular reproduction in a universe with 10^{22} stars, most of which have solar systems. At present, we have no experimentally based estimate.

7.5 Toward the Emergence of the Prokaryote

There is a plausible and testable pathway from the forgoing facts and ideas to prokaryotes, discussed in two papers (Kauffman and Lehman, 2023; Lehman and Kauffman, 2021). First, peptide collectively autocatalytic sets have been constructed, as have RNA collectively autocatalytic sets. Each is a Kantian Whole achieving catalytic and constraint closure. One can envision the small-molecule collectively autocatalytic set evolving to become the metabolism of a now coupled peptide-RNA collectively autocatalytic set. Such an evolutionary set merges these three distinct Kantian Wholes into a single *higher-order* Kantian Whole. Because the small-molecule autocatalytic set is already autocatalytic on its own, the evolution of RNA or peptide catalysts to take over catalysis of some, then all, of the metabolic steps can proceed piecemeal.

Further, the authors demonstrate how such a coevolution of Kantian Whole peptide RNA sets, together with a preference by each amino acid to bind to its current anticodon, can lead to the evolution of colinearity and coding. Similar processes might underlie the piecemeal evolution of template replication among transiently 3′–5′ stacked double-stranded sequences of RNA autocatalytic sets.

7.6 The Evolution of Agency, Function, Semantics, and Meaning

The very concept of a Kantian Whole affords a noncircular definition of "function." *The function of a Part in a Kantian Whole is that subset of its causal consequences that helps sustain the Whole.* The function of the heart is to pump blood, not jiggle water in the pericardial sac.

But the very concept of the function of a Part in a Kantian Whole is inherently semantic: *By virtue of my heart pumping blood, I get to exist for a while in the universe.*

Why *semantic*? The fundamental reason is that the universe is vastly nonergodic. The evolving universe will never make all possible sequences of 200 amino acids in vastly longer than the lifetime of the universe. Most complex things will never exist. The human heart weighs 300 grams. Why do human hearts exist in the universe? To answer this, we must move beyond physics alone. Physical laws must permit hearts to exist. But hearts exist in the biosphere because as organisms became larger, diffusion was not enough, and it was of enormous selective advantage to a lineage of organisms for hearts to evolve. Therefore, we cannot explain the existence of hearts in the nonergodic universe without making use of the function of the heart. Hearts exist in the universe by virtue of physics and also by virtue of the function of hearts in Kantian Wholes.

Getting to exist in the nonergodic universe by virtue of adaptive organs as Parts with functions sustaining the Kantian Whole is the heart of semiotics. Later, this becomes classification of the world and cognizing signs pointing to the good, and the foreboding of the meaning laden umwelt. Function yields semantics yields meaning. Matter has transformed to mattering.

And critically, this semantics and meaning inheres in physically embodied agents. As noted above, because cells are open thermodynamic systems, they must find means to eat to survive. To do so, cells have had to evolve means to *identify relevant features* of their world. Identification of relevant features involves some form of classification of the umwelt leading to action: Good for me, Bad for me, Irrelevant. A natural way for this to arise is in a nonlinear dynamical system with some diversity of dynamical attractors—say alternative steady states of its variables. The basins of attractions are "content addressable memories," and the attractors are the memories, or classes.

In order to act, cells must be able to do thermodynamic work of some sort. By virtue of Constraint Closure and further evolution, cells do work to construct themselves so they can evolve to do work on their world.

Molecular Autonomous Agents are reproducing embodied molecular systems able to do one or more thermodynamic work cycles and able to make choices between "Yuck" and "Yum" (Kauffman and Clayton, 2006; Peil, 2014).

7.7 The Collective Diversification of the Umwelt

Yuck and Yum, the basic "bad for me versus good for me" cognition and evaluation of an organism's umwelt (Uexküll, 2001), has diversified into the myriad biosemiotic umwelten of the cat world, the bat world, the worm world, the fungi world, and the bacteria world.

How is this so? The basic reason is that organisms coevolve, so they create ever-new niches that enable one another to exist. If you are bacteria, my flat stomach affords you a highway. You can crawl over my flat stomach to get to food. We are learning more what the myriad ways are by which soil organisms make livings with one another. Organisms form coevolving *Collective Affordance Sets*. As noted by Gibson, "a flat surface affords me a place to sit" (Gibson, 1979). Coevolving soil organisms have for billions of years created ever-novel ways of making their livings with one another. Organisms are the niches for one another, as well as is the abiotic umwelt. The explosion of the human global economy from a few tens of goods 100,000 years ago to billions today is merely another expression of the global economy among all species finding ways to make livings with one another. Novel species afford novel niches into which other species evolve. Once the swim bladder emerged, a worm or bacterium might evolve to live only in swim bladders. This is biosemiotics.

Yet more: The coevolution of biosemiotic networks cannot be deduced. The fundamental reason rests on the fact that all organisms really are Kantian Wholes, where *the function of a Part is that subset of causal consequences of the Part that sustains the Whole.* But this implies that the same Part, that has myriad subsets of causal consequences, may evolve novel uses that now function to sustain the Whole in novel ways. These are Darwinian preadaptations or Gould and Vrba's (1982) exaptations. In general, it is not possible to deduce one use, or function, of a Part from its other uses. From the use of an engine block to drill eight holes and make an engine, it is not possible to deduce that the sharp rigid corners of the engine block can be used to crack open coconuts. This implies that we cannot deduce the ever-new functions that arise in the evolving biosphere (Kauffman and Roli, 2023).

The evolving biosphere is a propagating non-deducible co-construction, not an entailed deduction (Kauffman and Roli, 2023). We cannot deduce the evolving biosemiotic web. Consider again an ecology of hundreds of soil bacteria and fungi. They are coevolving to "make a living" with one another. What goods, services, signs, signals, and behaviors are they exchanging? The umwelt of each organism is all the other organisms with whom it makes its living and the abiotic surrounding world.

A final major point: Selection acts on the Kantian Whole, not directly on its Parts. Whole organisms with Parts that function better in sustaining the Whole have more offspring. The improving Parts are selected indirectly. Here is the implication: While each Part may afford myriad novel causal uses each of which might become a novel function, because selection acts on the Whole, any novel functions among the Parts must always *integrate* to achieve the functional integration required to sustain the Whole. Emergence is not engineering (Kauffman and Roli, 2023).

The same emergent Holism arises when a set of species coevolves novel functions by which they manage to make a living together. As they do so, they become more functionally integrated. So too do the goods and services coevolve as complements and substitutes in our global economy. The biosemiotic web coevolves piecemeal but remains self-consistent.

The global umwelt expands, and with it what is next possible, the Adjacent Possible, expands. The expanding biosphere self-consistently and non-deducibly constructs itself.

We the Emergent.

Acknowledgments

I am truly grateful to Philip Clayton, David Deamer, Walter Fontana, Wim Hordijk, Katherine Peil Kauffman, Kalevi Kull, Andrea Roli, Roberto Serra, Mike Steel, and Eörs Szathmáry for wonderful discussions over the years.

References

Aguera y Arcas, B., Alakuijala, J., Evans, J., Laurie, B., Mordvintsev, A., Nicklasson, E., Randazzo, E., and Versari, L. 2024 *Computational life: How well-formed, self-replicating programs emerge from simple interactions*. arXiv: 2406.19108v2.

Atkins, P. W. 1984. *The second law*. Scientific American Library.

Damer, B., and Deamer, D. 2020. The hot spring hypothesis for an origin of life. *Astrobiology, 20*(4), 429–452.

Farmer, J. D., Kauffman, S. A., and Packard, N. H. 1986. Autocatalytic replication of polymers. *Physica D, 2*, 50–67.

Fontana, W. 1990. Algorithmic chemistry: A model for functional self-organization. SFI Working Paper 1990–011.

Gibson, J. J. 1979. *The ecological approach to visual perception*. Houghton Mifflin.

Gould, S. J., and Vrba, E. S. 1982. Exaptation—A missing term in the science of form. *Paleobiology, 8*(1), 4–15.

Harrison, S., Rammu, H., Liu, F., Halpern, A., Palmeira, R., and Lane, N. 2023. Life as a guide to its own origins. *Annual Review of Ecology Evolution, and Systematics, 54*, 327–350.

Hordijk, W., Hein, J., and Steel M. 2010. Autocatalytic sets and the origin of life. *Entropy, 12*(7), 1733–1742.

Hordijk, W., and Steel, M. 2018. Autocatalytic networks at the basis of life's origin and organization. *Life, 8*(4), 62.

Kauffman, S., and Clayton, P. 2006. On emergence, agency, and organization. *Biology and Philosophy, 21*(4), 501–521.

Kauffman, S., and Lehman, N. E. 2023. Mixed anhydrides at the intersection between peptide and RNA autocatalytic sets: Evolution of biological coding. *Interface Focus, 13*, 20230009.

Kauffman, S., and Roli, A. 2023. A third transition in science? *Interface Focus, 13*, 20220063.

Lehman, N. E., and Kauffman, S. A. 2021. Constraint closure drove major transitions in the origins of life. *Entropy, 23*(1), 105.

Montévil, M., and Mossio, M. 2015. Biological organisation as closure of constraints. *Journal of Theoretical Biology, 372*, 179–191.

Peil, K. T. 2014. Emotion: The self-regulatory sense. *Global Advances in Health and Medicine, 1*(2), 80–88.

Schmitt-Kopplin, P., Gabelica, Z., Gougeon, R., and Hertkorn, N. 2010. High molecular diversity of extraterrestrial organic matter in Murchison meteorite revealed 40 years after its fall. *Proceedings of the National Academy of Sciences, 107*(7), 2763–2767.

Sievers, D., and Kiedrowski, G. von. 1998. Self-replication of hexadesoxynucleotide analogues: Autocatalysis versus cross-catalysis. *Chemistry—A European Journal, 4*(4), 629–641.

Uexküll, J. 2001. An introduction to umwelt. *Semiotica, 134*(1–4), 107–110.

Vaidya, N., Manapat, M. L., Chen, I. A., Xulvi-Brunet, R., Hayden, E. J., and Lehman, N. E. 2012. Spontaneous network formation among cooperative RNA replicators. *Nature, 491*(7422), 72–77.

Wachowius, F., and Holliger, P. 2021. RNA replication and the RNA polymerase ribozyme. In S. Müller, B. Masquida, and W. Winkler (Eds.), *Ribozymes* (Vol. 1) (pp. 359–386). Wiley-VCH.

Wagner, N., and Ashkenasy, G. 2015. How catalytic order drives the complexification of molecular replication networks. *Israel Journal of Chemistry, 558*, 880–890.

Xavier, J., Hordijk, W., Kauffman, S. A., Steel, M., and Martin, W. 2020. Autocatalytic chemical networks at the origin of metabolism. *Proceedings of the Royal Society B, 287*, 20192377.

Xavier, J., and Kauffman, S. A. 2022. Small-molecule autocatalytic networks are universal metabolic fossils. *Philosophical Transactions of the Royal Society A*, 380, 20210244.

8 Inverse Darwinism: An Abductive Complement to Natural Selection?

Matt Kalkman and Terrence Deacon

8.1 Overview

Darwin's 1838 insight that gave birth to his concept of natural selection was stimulated by his realization that Malthus's analysis of competition for limited resources could be applied to the natural history of species. It explained how the subtractive logic of competitive elimination could hone the fit of organism traits to their local environments. But with the advance of molecular and cell biology at the dawn of the current millennium, we have become aware that a somewhat different logic characterizes evolution *within* organisms. In this context, the duplication of genes, molecules, and cells provides functional redundancy that allows degenerate duplicates to accumulate as though neutral. This vastly increases the probability of the evolution of otherwise improbable synergistic effects. We show that this Inverse Darwinian effect is formally analogous to the iconically based generative mode of inference that Peirce called abduction. This exemplifies the semiotic nature of the evolutionary process.

8.2 Introduction

At its core, biology is not merely a physical, chemical, or even an information science. It is fundamentally a science of semiosis, with evolution as its central unifying theme. Its most basic molecular processes are all organized to record, inform, and preserve a system of material relationships that is highly improbable thermodynamically. One implication of taking this perspective is that it treats evolution as the material instantiation of an inferential process. But what is the form of this inference?

Utilizing the terminology and semiotic analysis of logic developed by Charles Sanders Peirce, we attempt to reconstruct the inferential logic implicit in the evolutionary process. Peirce felt that the theory of logical inference of his time was fundamentally incomplete because it lacked an explicit logic of hypothesis generation. He argued that hypothesis generation is more than just unstructured guessing. He called this "abduction" and endeavored to articulate its formal structure.

Current neo-Darwinian theory treats the introduction of novel information as unstructured accident (e.g., genetic mutation). This accidental modification is retained or lost over succeeding generations with respect to its functional utility—that is, by natural selection. Critics of this neo-Darwinian paradigm tend to invoke some version of a "directed variation" perspective to counter the apparently vast improbability that purely accidental modification can provide functional improvement in a living organism. The source of "direction" that is invoked in these criticisms can range from the extremes of intelligent design to subtle variations on a Lamarckian theme, such as in theories of epigenetic inheritance.

We argue that strict neo-Darwinian theory leaves itself open to these criticisms because it focuses almost exclusively on one half of the process—the subtractive logic of "selection"—while ignoring the physical constraints and biases of the processes that generate "variety"—the materially embodied raw materials that are ultimately "tested" by natural selection.

In this essay, we depart from both perspectives. We will demonstrate that Peirce's abductive logic of hypothesis generation has its counterpart in a similarly structured evolutionary hypothesis generation process. Like abduction, this bio-logic is neither just a consequence of random fiat nor is it directed by some prescient anticipation of utility. It is a process that is ultimately analogous to what Peirce called abduction, because it too has its basis in iconism.

In the biological world, the counterpart to iconic interpretation is duplication of function or material form, which results in functional redundancy. In its simplest form, gene duplication has been well documented in molecular biology since the 1970s (Ohno 1970). A duplicate gene is iconic to the organism in which it arises to the extent that it produces a functionally indistinguishable physiological interpretation. In recent decades, a range of related processes has been described at the level of molecular biology and given the name Constructive Neutral Evolution (CNE) (Stoltzfus 1999). But this process and its evolutionary effects can be generalized to apply to evolutionary processes at all levels of biology, such as applying to the theme and variation structure of multicelled body morphology or codependence relations between organisms. We refer to this general evolutionary logic as *Inverse Darwinism* (Deacon, 2022, n.d.) because of its independence from the influence of selection while retaining the replicative and variational aspects of the Darwinian logic.

In this essay, we will argue that Inverse Darwinism is a necessary complement to the logic of neo-Darwinian theory, analogous to the way that abduction is a necessary complement to the logic of inductive and deductive forms of inference. In the next sections, we will first review the basic logic of abductive inference, then review the differences between Darwinian theory and its inverse, and then demonstrate the deep analogies between Inverse Darwinian processes and variations on the theme of abductive inference, before finally exploring some of the broader implications of this synthesis.

8.3 Abductive Inference

Charles Peirce revolutionized logic by recognizing that deduction and induction by themselves did not provide a complete system of logical inference. He realized that the generative aspect of inference was not well explained. By reflecting on the reasoning processes that contribute to creative and innovative thinking, he identified a mode of inference that he believed to be essential for hypothesis generation. Although he argued that it was a prior mode of inference that all other modes depended on, he also recognized that it was more fallible than its more recognized cousins. He called it abduction.

He noted that while deductive inference could provide predictions that necessarily follow from known general premises and inductive inference could provide predictions of probable general principles supported by examples, there was (at the time) no formal analysis explaining how hypotheses tested by these inferential tools could be initially generated. He introduced the concept of abductive inference to demonstrate that there is a definite structure to this process, not just random fiat. Taken together, he showed that the interdependence of these three modes of inference is responsible for building a well-grounded and well-integrated system of knowledge, beginning from merely possible to probable to necessary truths.

Another way to describe this relationship is that deductions predict necessary properties of individual cases; inductions predict probable relations between properties; and abductions use the similarity of properties to suggest other possible common properties. The centrality of similarity was highlighted by Peirce when he describes an abductive inference as an "inference through an Icon."[1]

Peirce more generally describes abduction as "the process of choosing a hypothesis"[2] and further argues that it is "the invention, selection, and entertainment of the hypothesis."[3] He also notes that previously, abduction was often called retroduction because it inverts the logical structure of deduction. He commented, "I consider Retroduction (a poor name) to be the most important kind of reasoning, notwithstanding its very unreliable nature, because it is the only kind of reasoning that opens up new ground."[4] One of his early examples involves beans in a bag. If it is known that all the beans in a particular bag are white, and one is provided with a white bean, then it is a reasonable hypothesis to guess (abduce) that this bean came from that bag, because of that shared property.

In the past, abductive reasoning has often been overlooked because it appears to be a logical fallacy in comparison to deductive inference.

In general, Peirce emphasizes the importance of abduction by suggesting that "[not] the smallest advance can be made in knowledge beyond the stage of vacant staring, without making an abduction at every step."[5] This introduction of novelty is therefore a key attribute of abductive inference. In this respect, Peirce claims that "it is the only logical operation which introduces any new idea" and that "all ideas of science come to it by the way of Abduction."[6] This is because abduction is "the inference which starts a scientific hypothesis."[7] Ultimately, where one assumes "something to be true that we do not know to be true, these facts would arrange themselves luminously. That is abduction."[8] In this way, Peirce argues that all inferential processes begin with an abductive inference. And despite its fallibility, all other inference rests upon this foundation of identifying categorical similarities.

To see the systematic interdependence of these three modes of inference, consider the classic syllogistic examples in figure 8.1.

Each is composed of the same three propositions arranged in one of three possible orders, where two are used to infer the third. In each case, it can be seen that the inference to the third proposition is due to being able to use a redundant element present in the two prior propositions to link their nonredundant elements. The redundant elements are canceled and can be ignored. But the antecedent-consequent order of propositions and which elements of the propositions are redundant and cancelable determines the difference in mode of inference. In an abduction, the same properties are attributed to different individuals; in induction, different

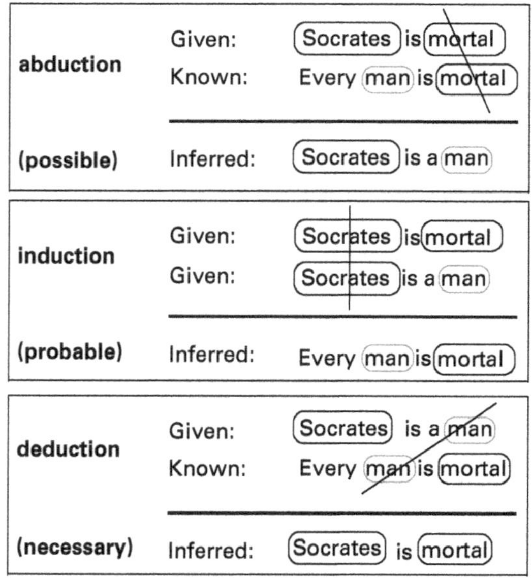

Figure 8.1
Peirce's three canonical modes of inference exemplified in the form of syllogisms. Redundant terms in the premises are marked as canceled, leaving the remaining nonredundant terms linked by virtue of this redundancy. The different inferential "strengths" of abduction, induction, and deduction (i.e., possible, probable, and necessary connections, respectively) are determined by the relative syntactic "positions" of these different redundancies. Notice that each of these forms of inference is specified only by the syntax linking the terms across sentences. This is, of course, what makes abstract symbolic logic possible (effectively an algebra of these relationships in which words are replaced by variables like a, b, c, etc.). It is also why logical inferential processes can be modeled by physical processes, such as in the switching dynamics of computing and the molecular interactions in cells (see below).

properties are attributed to the same individual; and in deduction, the subject property of one proposition is redundant with the object property of the other proposition. Because each is in the form of a syllogism, there is also a distinction between individual cases and general rules and between what information is known or assumed and what is "given" as new.

Peirce concludes that careful reasoning typically involves utilizing all three forms of inference in complex sequences, depending on the problem at hand and on one's purpose. Typically, one begins with a hypothesis (abduction), seeks other examples (induction), and tests its application (deduction) before rejecting or adopting it as a rule of thumb. Depending on the context, and what is and is not known in advance, the order of chaining inferences may be different, but only abduction makes a claim assigning a property to a subject and suggests a feature for characterizing an instance or defining a type.

This demonstrates a sort of inferential recursion that constitutes a form of meta-inference. It is an inferential process[9] that draws its premises not just from the conclusions of other inferences, but from prior inferential cycles. In this one-two-three repeat process, future abductions can be generated from prior inductions and deductions and can initiate yet further abductive-inductive-deductive cycles. This recursive dynamic enables higher-order inferences to build on prior inferential analyses in such a way that semiotic complexity systematically increases.

Over the course of his life, Peirce explored variations on this theme. In particular, he was fascinated by the recursive chain of reasoning that led Johannes Kepler to his discovery of the elliptical regularity of planetary orbits. According to Peirce, Kepler was "convinced that the hypothesis of Copernicus ought to be accepted."[10] This prior assumption established a general abductive frame within which Kepler would generate and test a series of subsequent hypotheses. From that starting point, Kepler "proceeds to modify this theory, after the most careful and judicious reflection, in such a way as to render it more rational or closer to the observed fact"[11] until he arrived at a final answer by "blundering along in the most inconceivable way (to us), from one irrational hypothesis to another, until, after trying twenty-two of these, he fell, by the mere exhaustion of his invention, upon the orbit."[12] And yet, as Peirce notes, Kepler generated each hypothesis while "never modifying his theory capriciously, but always with a sound and rational motive for just the modification he makes."[13]

Kepler was likely first drawn to the heliocentric model because of its platonic elegance. As an astrologer and mystic, it would have been natural to assume that the order of the celestial realm embodied geometric ideals. So, with this principle as a guide, he generated hypothesis after hypothesis, each modeled on platonic geometry—circles inscribed between regular polygons that did not quite fit the measurements, then circles inscribed between platonic solids that also did not fit—each consistent with his ideal geometric premise. Yet even when finally relaxing this criterion of mathematical simplicity to accept that the orbits were elliptical, this intuition led him to look for and discover an additional mathematical regularity: planetary orbital velocities precisely varied with respect to their distance from the sun. Because of this abductive bias, each succeeding hypothesis resembled those proposed previously with a slight modification motivated by deviations from expected predictions. These hypotheses were not generated randomly, and yet their adoption was not systematic either. There was a theme and variation pattern to their generation. Over time, the original constraint of platonic symmetry was progressively relaxed until it was reduced to abstract mathematical regularity.

This pattern of hypothesis exploration by testing increasingly divergent variations on an original theme convinced Peirce that abduction is a distinctively structured form of inference. Moreover, it suggests that different patterns of hypothesis generation may be more or less likely to lead to a useful result. Even the choice of an original theme and the pattern of generated variant forms may also be the products of prior abductive explorations. Nevertheless, as Peirce recognized, abduction alone provides no assurance of usefulness despite being the ground upon which all other forms of inference—and hence all knowledge—must rest.

8.4 Variations on the Theme

Charles Peirce's efforts to articulate an inferential form that characterizes the logic of hypothesis generation has given rise to many parallel approaches. In many respects, they can be understood as variations on his basic concept of abduction. Here, we will briefly describe three that can also be described as inference by iconism. They are Arthur Koestler's concept of *bi-sociation*, Gregory Bateson's concept of *double-description*, and Fauconnier's and Turner's concept of *conceptual blend.*

Arthur Koestler introduces his concept of bi-sociation in his 1964 book *The Act of Creation.* Bi-sociation involves combining two previously unrelated ideas or concepts to create a new, novel idea, perspective, or experience. Figure 8.2 redraws his way of visualizing the

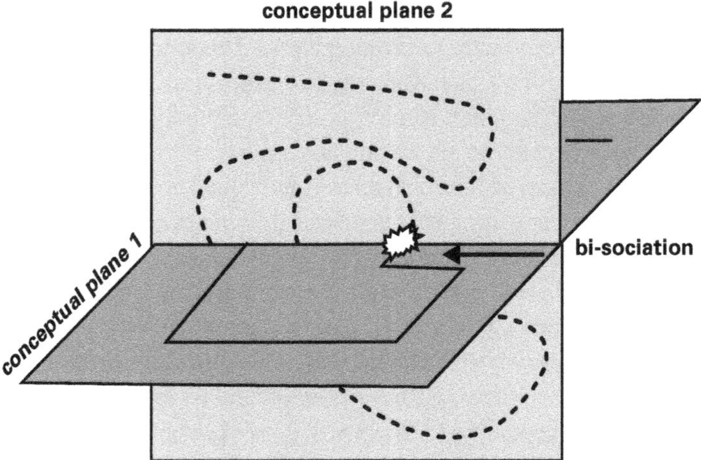

Figure 8.2
A redrawing of Arthur Koestler's (1964) depiction of the logic of bi-sociation in which two independent conceptual domains are depicted as intersecting planes/dimensions and within which separate strands of reasoning take place (solid and dashed lines). A bi-sociation that enables their fusion can occur if the two trajectories of thought intersect.

intersection of two "planes" of thought. Koestler argues that bi-sociation is the mental operation that is common to humor, art, and scientific discovery. It involves recognizing a novel synergistic relationship between previously unrelated processes. Examples from science and technology include Gutenberg's marrying the idea of a wine press with moveable type or Kepler's hypotheses about planetary motion by comparing the face of a clock with the orbits of the planets. An example from the arts is impressionist painting in which painters endeavored to capture a more direct sense of visual experience by focusing on the play of light and color and less on the objects being painted. Even a joke exemplifies bi-sociation. Consider this example: *What do you call a bear with no teeth?—A gummy bear*. This joke juxtaposes two incongruous domains linked by a pun in which the term "gum" refers to two radically different things–that is, the effect of being toothless (gums without teeth) and the popular candy "gummy bears."

Gregory Bateson introduces the concept of double-description in his 1979 book *Mind and Nature: A Necessary Unity*. Double description results when "two or more information sources come together to give information of a sort different from what was in either source separately" (Bateson, 1979, p. 21). Bateson exemplifies this with the example of binocular vision. Because both of our eyes face forward, they have overlapping fields of view. But each image is only two-dimensional. It is only when the images of the two eyes are combined that the brain creates the additional sensation of depth in the forward direction. The redundant information from each retina becomes fused to form a single image in our experience, but the systematic differences between the two images generate the experience of depth perception, thus creating the experience of an additional dimension. According to Bateson (1979, p. 87), "the two-eyed way of seeing is itself an act of comparison" (see figure 8.3).

Bateson expands the relevance of double description to apply to animal morphology when he asks, "What pattern connects the crab to the lobster and the orchid to the primrose

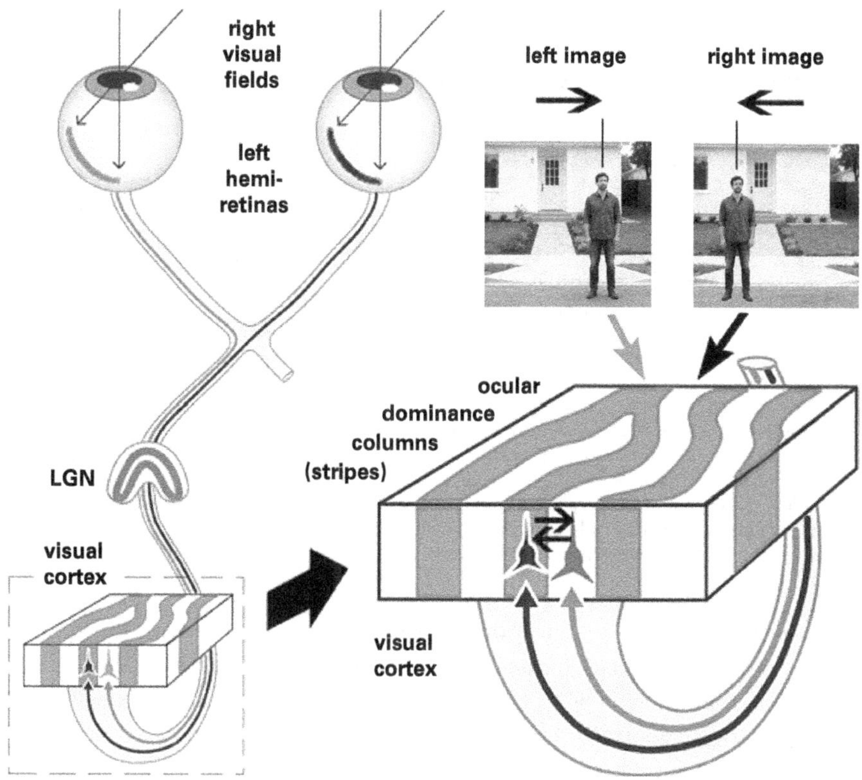

Figure 8.3
A graphic depiction of the neuroanatomical logic underlying binocular vision. Light arriving from the same spatial source in each half retina is relayed to neighboring positions in adjacent strips of visual cortex. The spatial geometry of the two-dimensional images from each eye are nearly identical (iconic) for distant objects and divergent for close objects (shown in the upper right). Because of the zebra stripe-like geometry of the two maps, neurons receiving signals from the same spatial and retinal locations are near each other and interact to resolve signal incompatibilities. Similar interactions occur for each point in the perceptual field and resolve the discrepancies via double description as virtual third dimension.

and all the four of them to me? And me to you?" (Bateson, 1979, p. 8). He is basically asking the reader to consider the biological significance of what has been traditionally called serial homology: the theme-and-variation pattern that characterizes animal and plant body plans. As is shown in figure 8.4, the individual legs of a crab or lobster are each variants of one another, distinguished both by their position along the body axis and by the ways that the different segments of each limb are relatively enlarged or reduced compared to the corresponding segment on each leg. This is particularly exemplified by the hypertrophy of the front leg and its greatly enlarged pincers. Not only does he point out the multiple-description logic within the body, but he then invokes a higher-order double-description comparison (at a phylogenetic level) by drawing attention to the same global morphological pattern in both crabs and lobsters, despite systematic differences of each limb between the two species. He goes on further to point out that this multilevel theme-and-variation morphological pattern characterizes plant bodies too.

Gilles Fauconnier and Mark Turner provide a detailed review of their concept of conceptual blends in their 2002 book *The Way We Think: Conceptual Blending and the Mind's Hidden*

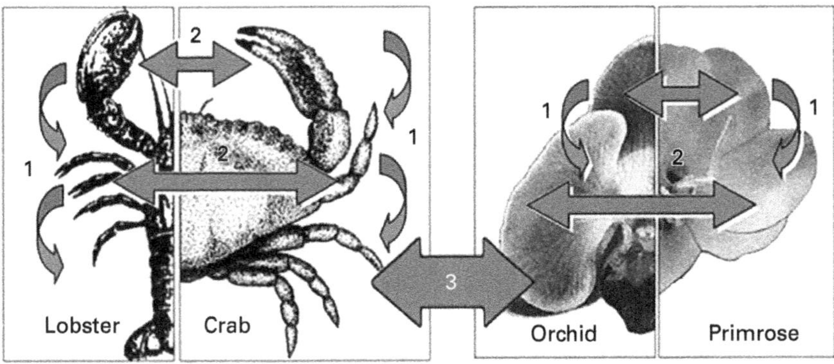

Figure 8.4
Three levels of theme-and-variation double description of morphological homologies/analogies as described by Gregory Bateson. 1. Redundant but systematically varying features of duplicate body structures that have differentiated so that they provide a complementary "division of labor." 2. homologies between global morphological organization of related species in which their global iconisms have been inherited from common ancestry. 3. analogies between the theme-and-variation organization of plant and animal body plans that exemplifies the shared abductive process that was independently evolved in plants and animals due to a common abductive evolutionary logic (described in the next section).

Complexities. Conceptual blending is a way of characterizing the cognitive process combining different mental "spaces" or concepts to create new meaning. It involves taking elements from separate mental spaces and integrating them into a new, blended space. Fauconnier and Turner argue that this is a critical source of creative thinking and the generation of new ideas. A classic example involves a riddle (depicted in figure 8.5): "One morning, a monk climbs a mountain to reach a monastery at the top that evening. The next day he descends back down to the bottom. Is there a location on the mountain where the monk will be at the same time both days?" One way to solve this riddle is to imagine both ascent and descent happening on the same day—one man walking up and one walking down, as though two men who must inevitably pass by one another at a specific location and time (indicated by the arrow).

Notice that the discovery, the joke, and the riddle all involve bi-sociation, double description, and conceptual blends. So, how are these forms of creative juxtapositions variations on the theme of abductive inference? Notice first that all are based on iconic comparisons. The syllogistic abduction is based on recognizing shared properties. In Peirce's example of beans in a bag, described in the previous section, the shared properties of being white supported the hypothesis that the bean in question was another member of a general type (beans from the bag). Each of these extensions of abductive inference also involve an initial isomorphic comparison–that is, iconic interpretation. The shared properties basically cancel, leaving the remaining differences unresolved and potentially highlighting any systematicity or regularities in their differences. In the case of binocular vision, it is the regularity of parallax by which progressively closer objects are more divergent in their left-right positioning than more distant objects. This regularity is interpreted as an additional dimension of depth. In the case of the riddle, the imaginary juxtaposition of the two separate processes highlights what they have in common.

Below, we will argue that this iconically based form of inference can also be embodied by molecular interactions interpreted by a living evolving system. In particular, we will demonstrate that there is an evolutionary analogue to abductive inference in what can be described as evolutionary hypothesis generation.

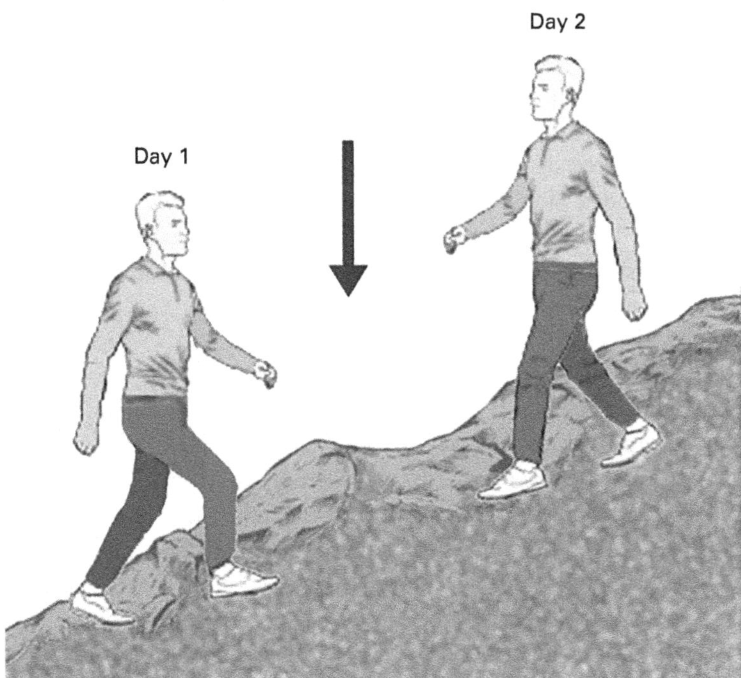

Figure 8.5
Depiction of the conceptual blend solution of the mountain climbing riddle, in which the two different trips, up one day and down the next, are imagined as happening on the same day as though two different individuals.

8.5 Inverse Darwinism

The field of molecular evolution has begun to challenge the view that evolution is primarily due to the incremental accumulation of useful chance variations. And yet it does not challenge the basic logic of Darwin's theory, nor has it provided evidence that any Lamarckian-like process provides directionality to mutations or anticipatory biases in the generation of useful variations.

It was initiated in the late 1960s and early 1970s by two controversial suggestions about evolution at the genetic level that appeared to challenge the neo-Darwinian status quo. One was the proposal by Motoo Kimura (1968) that a large fraction of genetic changes in evolution might be effectively neutral (i.e., with neither positive nor negative effects with respect to selection). The other was a proposal by Susumu Ohno (1970) that a significant source for the emergence of novel traits in evolution might be due to gene duplication.

These were controversial claims in the context of a neo-Darwinian paradigm that assumed that most traits had evolved as a consequence of natural selection. By definition, neither natural selection nor Lamarckian inheritance are relevant for explaining the production of neutral traits. So, in many respects, neutral effects tended to be ignored, except in the special case of genetic drift due to population bottlenecks and possibly correlated traits that result from the interdependencies of other selected effects. More troubling, however, was the possibility that, as Ohno suggested, the evolution of novel functions might be the result of degenerative effects following after gene duplication, and not the result of "positive" mutations.

So long as neutrality was a miniscule feature of biological systems and remained irrelevant to those features subject to the pruning shears of natural selection, it could be set aside. But by the mid-1990s, the widespread presence of selectively neutral traits was becoming more and more apparent. Evidence was also accumulating to show that they often influenced other non-neutral features of organism function. This caused many to suspect that incorporating these effects might require radical changes to evolutionary theory.

One of the most influential analyses of the evolutionary effects of neutrality at the intracellular level was introduced by Arlin Stoltzfus in 1999 known as Constructive Neutral Theory (CNE).[14] Muñoz-Gomez and colleagues[15] list five key concepts that characterize CNE with respect to molecular evolution: (1) excess capacities, (2) epistasis, (3) random genetic drift, (4) biased variation, and (5) purifying selection. Excess capacity refers to those situations where there happens to be more capacity in the system to complete a function than necessary. Epistasis refers to the notion that the context a mutation finds itself in, affects its selective effect. Certain contexts therefore allow certain mutations to remain neutral even when harmful in others. Random genetic drift allows mutations to spread by statistical chance throughout a population without any correlation with the environment, typically in small populations. Biased variation reflects the fact that variation is with respect to a previous function and because degeneration is more likely than improved function. Characterizing natural selection as "purifying" recognizes its role in stabilizing any new functional relationships that cannot be allowed to disappear and thereby preserves any useful epistatic interactions that might develop. Describing it as "purifying" also stresses the purely negative or eliminating role that natural selection is relegated to play, rather than any "creative" role. Finally, the word "constructive" in CNE refers to the increase in complexity that often results, irrespective of whether this is useful or just uncorrelated complexity.

A review by Gray and colleagues[16] describes CNE as "a neutral ratchet-like process that results in irremediable complexity." The complex features that CNE produces have also been described as the consequence of "runaway bureaucracy" or "biological Rube Goldberg machines which are unnecessarily complex and over-engineered to perform a single task."[17] It has also been argued that CNE allows for "diffusion of function" in the sense that the same function tends to evolve to depend on more part types or interactions. So, in effect, CNE allows for a structure to "degenerate its way into complexity."[18] These characterizations all emphasize increases in complexity that do not add novel capacity, but this is not necessarily always the case.

Despite its over twenty-year development, many evolutionary biologists still appear to be unaware of this explanatory framework for the origins of biological complexity. One possible reason is that the term "neutral" is somewhat misleading. A more accurate term might be *neutralized*. The words "neutral theory" and "constructive neutral theory" suggest to a casual reader that there are three forms of mutation: negative, neutral, and positive. However, it would be more accurate to describe most mutations as likely to be destructive or deleterious, with only a tiny minority producing positive effects. So, it would be more accurate to think of the effects of redundancy as rendering any potentially negative effects "tolerated." In this way, the error tolerance provided by redundancy ensures that potential negative effects are buffered or neutralized.

Proponents of CNE called this effect *pre-suppression*, as though corrective action is implemented prior to the appearance of a potentially deleterious mutation, as opposed to correcting a mutation after it has occurred. An example often cited is the evolution of chaperone proteins that prevent the deleterious deformation of other proteins in stressful conditions, such as high temperature, by binding to them and preventing deformation.

Their evolution not only maintains certain protein functions despite potentially harmful conditions, but it can also prevent miscoded proteins (due to mutation) from malfunctioning because they are also incidentally prevented from misfolding. In retrospect, it is as though the solution preceded the problem. But of course, chaperone proteins evolved in response to a different adaptive need and just happened to have a related affordance with respect to mutational damage. But once they enable the neutral fixation of otherwise malfunctioning gene products in the population, they become indispensable and will tend to be maintained by purifying selection.[19]

The term "presuppression" is in many respects problematic because of the connotations of evolutionary anticipation that it suggests. Yet the term "exaptation," coined in a paper by Gould and Vrba to avoid this same connotation for the older term "pre-adaptation" (once commonly used by neoDarwinians), does not convey the additional epistatic basis of this neutralizing effect. According to CNE, neutralization is provided by epistatic redundancy, but (as described above in the case of chaperone proteins) there are other possible ways that this can be accomplished besides simple redundancy. In a related analysis, also published in 1999, Allan Force and colleagues (1999) provided an explanation for the unexpected tendency for duplicated genes to persist. Prior theory had predicted that negative epistatic effects between duplicated genes should be common and often result in natural selection eliminating a duplicate gene. To explain the widespread presence of duplicate genes, they argued that minimizing direct epistatic interactions can also provide tolerance of negative epistatic effects. Through a process they described as duplication followed by degeneration followed by complementation (DDC) they reasoned that duplicate genes can, in effect, get out of each other's way. This might result from mutations affecting promoter or other regulatory sites, rather than protein coding regions. The result would be differences in either the timing or location of expression. In this way, independent variations in protein coding features can be independently selected for different effects. In this way, any potentially conflicting consequences of gene duplication are minimized, and complementary interactions are more likely to be exposed.

This suggests that there are likely to be multiple possible strategies for achieving genetic neutralization and error tolerance. Both involve the initial production of redundancy and excess function. Both also are preserved due to epistatic effects. And ultimately, both are the result of mutually supportive functions with respect to one another. It is just that CNE focuses on the stabilizing effects of codependence, and DDC focuses on the effects of a division of labor. It was quickly realized, however, that these are variants of a common evolutionary logic, in which redundancy provides error tolerance, which allows variations (often degenerate) to accumulate that are thereby more likely to fall into complementary interactions.

One of us (Deacon[20]) has argued that this general logic is an inversion of Darwin's original insight because of the way such processes are enabled by excess rather than constrained by limiting resources (as in Malthus's account). It also inverts the expectation that evolutionary complexification requires the accumulation of variants that each provide a slight improvement of some function. In this way, the processes described as CNE and DDC invert central tenets of neo-Darwinism (or as Deacon, 2022 and n.d. describes it "Inverse Darwinism").

More significantly, the causal logic of these evolutionary processes is general. It can apply to evolutionary changes at any level of the biological hierarchy. Consider, for example, the iterative theme and variation structure of insect legs, plant leaves and petals, or quadrupedal vertebrate limbs (as described in the previous section). These morphological partial redundancies are now understood to be correlated with the duplication and variation of genes for transcription factors expressed in separate body regions.[21] In the case of insect legs, their partially redundant structures likewise minimize any negative consequences for locomotion should one or more lose some feature optimized for that function. But this redundancy also makes it less costly for one or more of these partial duplicate structures to become exapted for some entirely different function, as are the enlarged forelimbs and pincers of lobsters. Although natural selection is invoked to explain the fine-tuning of these morphological variations for their distinctive functions, the evolution of complementary interdependence and division of labor is made more probable due to this redundancy and the tolerance of functional deviance it allows.

This effect of redundancy in evolution is paralleled by insights developed independently in another quite different field: information science. In a technical report for the Bell Telephone Company in 1948, Claude Shannon showed that redundancy in a communication channel is both the basis of useful information and a means for error correction and noise tolerance. This is because redundancy makes it possible to recognize and ignore noise. Accidental errors (like typos) corrupt a signal in effectively random (uncorrelated) ways, whereas intended signal elements will be systematically located in the same position in each copy. Shannon was therefore able to show that any level of noise can be compensated by a corresponding level of redundancy.

So, provided with a highly redundant signal (e.g., like written English), it tends to be relatively easy to correct spelling errors and to tolerate a considerable frequency of typos while still correctly interpreting the meaning that was intended. Thus, the error tolerance provided by redundancy also provides resistance to misinterpretation. But in biological systems, this flexibility due to redundancy provides a significant enhancement of what can be described as *evolvability*. In particular, it can greatly increase the probability that otherwise highly improbable complementary interactions will be uncovered.

Redundancy in a biological system has an effect that is analogous to the effect of redundancy in communication. It preserves the interpretability of the genetic signal despite mutational error. But instead of correcting errors in the genetic signal, it allows them to accumulate and persist as long as there are redundant sources of functional information still available. Though variant forms will share considerable family resemblances inherited from their common precursor, the probability that mutations will by themselves be improvements is miniscule because "mutations that reduce function are more common than those that confer new functions" (Force et al., 1999; Levasseur and Pontarotti, 2011; Proulx, 2012; Stoltzfus, 1999). These family resemblances are the major source of genetic bias that occurs when mutational changes are more common than others irrespective of the influence of selection. This bias is distinct from another long-recognized source of genetic bias—drift—though the effects are similar. Drift is most common when the population size on a sexually reproducing species fall to a very low level. When the number of potential mates is small, reproduction can only sample a small fraction of the possible genetic combinations. As a result, even slightly maladaptive traits can become fixed in the population. Although the effects of redundancy-based neutrality are similarly random with respect to selection and can also allow degraded duplicates to persist and become fixed in a population, it is not population size dependent and does not have maladaptive consequences.

The retention of biased variants that have been fixed by neutral drift can create or enhance the (inter-)dependencies between excess capacities and their substrates. In this way, effectively neutral biased variation provides directionality, not selection. Traditionally, natural selection has served as the ultimate driver of evolutionary change. It has been assumed that selection is the only factor that determines the directionality of evolution and the fitness of phenotype to environment. However, the generality of Inverse Darwinian effects demonstrates that purifying selection only purges deleterious mutations from populations and thus only serves to prevent dissociation of neutrally evolved (inter-)dependencies. Purifying selection "locks in complexity rather than producing it."[22]

The recognition of the critical role provided by redundancy (a form of iconic relationship) in both evolutionary biology and information science provides a useful context in which to reformulate evolutionary processes in biosemiotic terms. This can help make explicit the generative and emergent nature of semiosis that is implicit in semiotic theories, but seldom precisely articulated. We will explore the possibility of such a transdisciplinary synthesis in the remainder of this essay.

8.6 Evolutionary Hypothesis Generation

We often describe the appearance of a creative insight in science or art as though it popped into consciousness from nowhere, without precedence, as though just by random chance. Yet, as the French microbiologist and chemist Louis Pasteur once said, "Chance favors the prepared mind."[23] This suggests that there is a definite structure to insightful hypothesis generation that is not just random, even if it is not driven by deduction from known facts. This is the point of the analysis of abductive inferential processes; to understand what constitutes this relative "preparedness" to make informed guesses. The comparison with evolutionary theory highlights the way that neo-Darwinian theory often appeals to something like "lucky accident" as the source of favorable variations and serendipitously useful mutations. According to Inverse Darwinism it is neither the case that variations and mutations are generated by processes that anticipate their functional value (as neo-Lamarckians might claim), nor is it the case that they are generated in a purely random and unconstrained biological context (as strict neo-Darwinians might claim). In contrast, the examples of Inverse Darwinian processes described above demonstrate that, like creative thinking, the generation of serendipitously adaptive biological "hypotheses" is made more probable in a context of partial redundancy (a biological analogue to iconic interpretation).

To see this, consider a simple gene duplication effect, such as in the case of the evolution of trichromatic vision. Color reception by the cones of the retina is enabled by a class of light-sensitive proteins called opsins. Most mammals have two variants of this protein each produced by a distinct variant of an originally duplicated opsin gene. This makes most mammals dichromats that do not see the range of colors that we humans and many other monkeys and apes experience. Early in the lineage leading to most monkeys and apes, an additional duplication of an opsin gene led to the evolution of our trichromatic vision. Dupli-

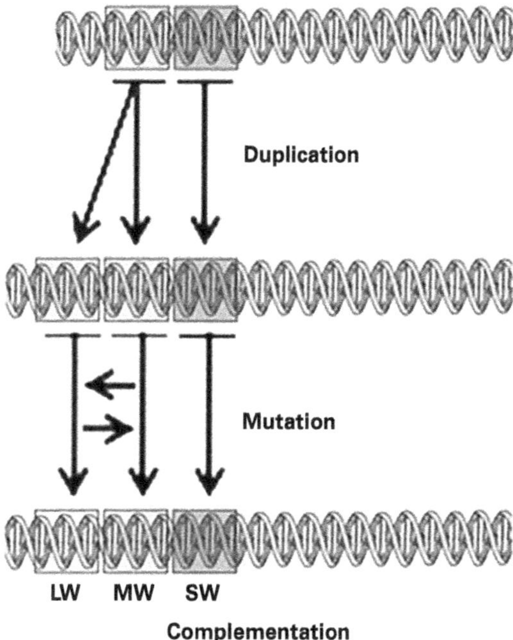

Figure 8.6
A graphic depiction of the opsin gene duplication and subsequent degeneration events that led to the evolution of trichromatic vision in monkeys and apes. LW = long wavelength (red) opsin gene; MW = mid wavelength (green) opsin gene; SW = short wavelength (violet) opsin gene.

cation and slight degradation of the middle-wavelength-sensitive opsin gene allowed the duplicate gene's protein to shift its wavelength sensitivity toward the red end of the spectrum, while the mid-wavelength opsin maintained its original light sensitivity (see figure 8.6). The systematic difference that resulted, enabled comparison of these signals within the visual cortex. But it did not simply result in an additional band of light reception. Rather, the fusion of this wavelength-shifted image with the other wavelength-specific images expanded the dimensionality of color complexity of visual experience. It is the color analogue to the double description that generates binocular vision—a "perceptual blend."

In a similar way, the duplication and divergence of the Hox family of regulatory genes, which are responsible for the segmental theme-and-variation morphology of insect and vertebrate body segments and limbs, (discussed above) can be seen to exemplify another variant of abduction. In this case, the DDC logic of the division of labor into nonoverlapping expression domains is involved. As noted in the previous section, this redistribution of expression has enabled each duplicate to also diverge in function without significant negative effects. This has effectively shifted the Inverse Darwinian effect up a level to involve the redundancy-facilitated sampling of synergistic interactions between serially organized body segments and limbs. Bateson's discussion of the serial homologies of crab and lobster limbs characterizes it as a version of double description. By additionally bi-sociating this abductive account with the DDC mechanism, we are provided with a deeper understanding of both.

What has been missing from these various abductive theories is an account of the processes by which these relations come into existence in the first place. Details of such semiotic processes are provided by analogy to these Inverse Darwinian mechanisms. So, rather than a retrospective description of the consequences of abductions, it may now be possible to outline some general mechanisms for how abductions emerge.

This highlights a common dynamic logic shared by abductive inference and Inverse Darwinian processes that undergirds the connection between biological and cognitive abductions. Ultimately, both biological and cognitive abduction rely on redundancy as a basis for structured exploration. Biological redundancy has traditionally been treated as mere excess, but as we have shown, it is the foundation for innovation. It is at once a buffer and affordance for exploration without immediate functional loss. This common logic suggests that hypothesis generation is not just a cognitive tool, but is a process embedded in the very fabric of evolutionary dynamics. It is a fundamental organizing principle that guides the emergence of complexity in both biological and symbolic domains.

8.7 Conclusion

In this essay, we have sought to show that just as abduction is a generative complement to induction and deduction, so too is Inverse Darwinism the generative complement to the subtractive logic of natural selection. This comparison has exposed a deeper underlying logic linking the structure of inference to the dynamics of evolutionary change. It demonstrates that at its very core, biology is a science of semiosis and evolution its unifying principle.

Historically, two opposed schools of thought were assumed to be the only contenders for evolutionary logic. One school argued that life is driven in some sort of progressive manner through a form of Lamarckian Inheritance, whereas neo-Darwinism treats the introduction of novel information as merely unstructured accident. But research in recent decades has highlighted a third alternative. Rather than either directed or accidental mutation, novel biological functions can arise due to a kind of functional iconism, such as is provided by gene duplication. The functional redundancy that results enables a kind of playful exploration of potential synergies that is both biased by past functionality and yet also an expression of blind chance. Like an unconsciously gestated spontaneous flash of creative insight, evolutionary innovations often appear as though emerging by pure fiat, but analogous to their mental counterparts, it turns out that a distinctly structured generative logic is involved.

In this essay, we have highlighted the parallel structure of this abductive logic as it is exhibited in both mental and evolutionary hypothesis generation. It is reflected in a mental process that occurs unconsciously but can be refined through conscious effort, and likewise, it is reflected in spontaneous neutral molecular processes that can become subsequently fine-tuned by purifying selection. It is a dynamic process that moves from possibilities (abductions), to probabilities (inductions), and ultimately to functional necessities (deductions), and can be recursively revisited in a meta-inferential process which we call (*anaduction*) to produce progressively more complex results.

In a very direct sense, the logic of this essay has itself involved a double description. We have taken advantage of the juxtaposition of semiotic and evolutionary analyses to provide a meta-inferential process that is more than guesswork "struck out at random."[24] It exemplifies commonalities with each of the variants on the theme of abduction that we have surveyed—bi-sociation, double description, and conceptual blending—and builds on the meta-inference that this enables.

Uncovering this underlying relationship between logical inference and biological evolution allows an inversion of what the field of biology has traditionally done in anthropomorphizing evolution as analogous to technological progress. This analysis demonstrates that mental processes grow out of a deeper generative logic found in biological processes preceding human reasoning. Mental inference is the product of a long recursive evolutionary logic that enables many cycles of higher-order abductions to be generated from prior lower-order abductions. It has been an incremental transition from "falling up" to "falling awake."

In the end, we hope we have demonstrated that it is more realistic to see both evolutionary biology and cognitive science as expressions of a deeper semiotic logic: A perspective that may help us to reimagine mental experience as "what it feels like to be evolution" (Deacon, 1997).

Notes

1. Charles S. Peirce, Pragmatism Papers, 1903, 276–77.
2. Charles S. Peirce, Collected Papers, 1901, para. 7.219.
3. Charles S. Peirce, Harvard Lectures on Pragmatism, 1901, 895.
4. Charles S. Peirce, Notebooks and Manuscripts, 1911, para. 3:206.
5. Charles S. Peirce, Harvard Papers, 1901, para. 2:899–900.
6. Charles S. Peirce, Collected Papers, 1903, 144–45.
7. Charles S. Peirce, Carnegie Application, 1902, letter L75; Charles S. Peirce, The New Elements of Mathematics, vol. 4, 1902, para. 62.
8. Charles S. Peirce, Pragmatism as a Principle and Method of Right Thinking, 1903, para. 282–283.
9. Following Peirce's terminological strategy, this upward directionality or meta-inference could be called *anaduction*, from ana-(upwards) and ducere (to lead). As opposed to ab- (away), in- (toward), de- (from).
10. Charles S. Peirce, Collected Papers, vol. 1, section 72.
11. Charles S. Peirce, Collected Papers, vol. 1, section 73, para. 3.
12. Charles S. Peirce, Collected Papers, vol. 5, section 362.
13. Charles S. Peirce, Collected Papers, vol. 1, section 74.
14. Stoltzfus, 1999.
15. Muñoz-Gomez et al., 2021.
16. Gray et al., 2010.
17. Lukeš et al., 2011.
18. Zimmer, 2013.
19. This effect has been compared to so-called toxin-antitoxin effects first described in the case of bacterial plasmids that become obligate in a bacterial lineage because of the way toxic effects only occur when the plasmid is not reproduced with a paired modified bacterial gene. See Ogura and Hiraga (1983).
20. Deacon, 2022, and n.d.
21. The transcription factors responsible for this mereological patterning differ for plants and animals, though in both plants and animals, duplication of transcription factor genes is involved.
22. See Wideman et al., 2019, and Zhang, 2018.
23. In a lecture given at the University of Lille on December 7, 1854, Louis Pasteur is quoted as saying "*Dans les champs de l'observation le hasard ne favorise que les esprits préparés*," translated as "In the fields of observation chance favors only the prepared mind."
24. Peirce, CP 1.74.

References

Bateson, G. 1979. *Mind and nature: A necessary unity*. Bantam Books.

Boole, G. 1854. *An investigation of the laws of thought on which are founded the mathematical theories of logic and probabilities*. Macmillan and Co.

Brunet, T. D. P., and Doolittle, W. F. 2018. The generality of constructive neutral evolution. *Biology & Philosophy*, *33*(1), 1–25.

Church, A. 1936. An unsolvable problem of elementary number theory. *American Journal of Mathematics*, *58*(2), 345–363.

Deacon, T. W. 2022. A degenerative process underlying hierarchic transitions in evolution. *Biosystems*, *222*, 104770.

Deacon, T. W. n.d. *Falling up: How inverse Darwinism catalyzes evolution*. (In preparation)

Fauconnier, G., and Turner, M. 2002. *The way we think: Conceptual blending and the mind's hidden complexities*. Basic Books.

Force, A., Lynch, M., Pickett, F. B., Amores, A., Yan, Y., and Postlethwait, J. 1999. Preservation of duplicate genes by complementary, degenerative mutations. *Genetics*, *151*(4), 1531–1545.

Gould, S. J., and Vrba, E. S. 1982. Exaptation—A missing term in the science of form. *Paleobiology*, *8*(1), 4–15.

Gray, M. W., Lukeš, J., Archibald, J. M., et al. 2010. Irremediable complexity? *Science*, *330*, 920–921.

Kimura, M. 1968. Evolutionary rate at the molecular level. *Nature*, *217*(5129), 624–626.

Kimura, M. 1979. Model of effectively neutral mutations in which selective constraint is incorporated. *Proceedings of the National Academy of Sciences*, *76*(7), 3440–3444.

Koestler, A. 1964. *The act of creation*. Penguin Books.

Levasseur, A., and Pontarotti P. 2011. The role of duplications in the evolution of genomes highlights the need for evolutionary-based approaches in comparative genomics. *Biology Direct*, *6*, 11.

Lukeš, J., Archibald, J. M., Keeling, P. J., Doolittle, W. F., and Gray, M. W. 2011. How a neutral evolutionary ratchet can build cellular complexity. *IUBMB Life*, *63*(7), 528–537.

Muñoz-Gómez, S. A., Bilolikar, G., Wideman, J. G., and Geiler-Samerotte, K. 2021. Constructive neutral evolution 20 years later. *Journal of Molecular Evolution*, *89*, 172–182.

Ogura, T., and Hiraga, S. 1983. Mini-F plasmid genes that couple host cell division to plasmid proliferation. *Proceedings of the National Academy of Sciences of the United States of America*, *80*(15), 4784–4788.

Ohno, S. 1970. *Evolution by gene duplication*. Springer-Verlag.

Ohta, T. 1973. Slightly deleterious mutant substitutions in evolution. *Nature*, *246*, 96–98.

Osawa, S., and Jukes, T. H. 1989. Codon reassignment (codon capture) in evolution. *Journal of Molecular Evolution*, *28*(4), 271–278.

Pasteur, L. 1939. Discours prononcé à Douai, le 7 décembre 1854, à l'occasion de l'installation solennelle de la Faculté des lettres de Douai et de la Faculté des sciences de Lille. In L. Pasteur, *Oeuvres de Pasteur, vol. 7* (L. P. Vallery-Radot, Ed.) (pp. 135–138). Masson and Co.

Proulx, S. R. 2012. Multiple routes to subfunctionalization and gene duplicate specialization. *Genetics*, *190*, 737–747.

Shannon, C. E. 1948. A mathematical theory of communication. *The Bell System Technical Journal*, *27*(3), 379–423; (4), 623–656.

Stoltzfus, A. 1999. On the possibility of constructive neutral evolution. *Journal of Molecular Evolution*, *49*(2), 169–181.

Turing, A. M. 1937. On computable numbers, with an application to the Entscheidungsproblem. *Proceedings of the London Mathematical Society*, series 2, *42*, 230–265.

Wideman J., Novick A., Muñoz-Gómez S., and Doolittle W. 2019. Neutral evolution of cellular phenotypes. *Current Opinion in Genetics & Development*, *58–59*, 87–94. https://doi.org/10.1016/j.gde.2019.09.004

Zhang J. 2018. Neutral theory and phenotypic evolution. *Molecular Biology and Evolution*, *35*, 1327–1331. https://doi.org/10.1093/molbev/msy065

Zimmer, Carl. 2013. The surprising origins of life's complexity. *Scientific American*, *309*(2), 84–89.

III DRAFTS

9 Evolution by Habits: Organismal Organization Beyond Natural Selection

Jana Švorcová and Karel Kleisner

9.1 Introduction

A *habit* can be understood as a characteristic of living systems that is often (but not necessarily) acquired, that is often (but not necessarily) behavioral, and that tends to be repeated, thus eventually leading to teleological behavior. In biological theories, habits are typically associated with learning, memory, and behavior. However, habit *per se* is a more universal concept: It is applicable at various levels of organismal organization and, crucially, it enables the evolution of complex traits. Evolutionary theories often emphasize processes such as selection, genetic drift, migration, or gene flow, while largely overlooking the role of habitual evolution—to their own detriment.

This perspective is not entirely novel. The concept of habit has deep historical roots in evolutionary thought, especially in Lamarck's ideas about how behavioral changes influence the development and evolution of organisms. Recent discussions have revived the notion of habitual evolution through concepts such as phenotypic plasticity, epigenetics, and niche construction. This notion has also attracted attention in biosemiotics, where its proponents draw upon the work of Charles Sanders Peirce (1839–1914) and emphasize the role of habit as a key semiotic process that governs the interaction between the organism and the environment (Bernardi della Rosa and Zengiaro, 2025). This renewed focus on habit underscores its significance not only in evolutionary processes but also in a wider philosophical and ecological context, shedding light on the dynamic interplay between behavior, environment, and (epi)genetic changes.

Traditionally, the study of organismal forms and functions has been dominated by reductionist approaches that explain biological phenomena within deterministic and mechanistic frameworks, especially within the theory of natural selection represented by the Modern Synthesis. Such models do not account for how organisms can actively communicate and shape their presence in ecosystems through their appearance and behavior. The agency of organisms, as expressed through their appearance and self-display, transcends traditional evolutionary theories, which focus narrowly on survival and reproduction (Kleisner, 2024). Nevertheless, these expressions are integral to how organisms navigate their relationships with the environment, communicate within their ecological niche, and engage in social interactions. Importantly, all these processes are inherently habitual.

In the following, we begin by exploring the role of *habit* in the evolution of scientific theories, tracing the development of the concept from Lamarck's foundational ideas to Peirce's refinements and contemporary reinterpretations in fields such as epigenetics and biosemiotics, where habit emerges as a crucial concept that links the process of interpretation to recurring patterns of behavior and adaptation. We examine how habits function at various levels of biological organization (cellular, supracellular, developmental, symbiotic) and investigate their role in facilitating evolutionary processes through ongoing reinterpretation. Using the example of organismal appearances, we highlight the interplay between choice, perception, and internalized experience, emphasizing how these factors are often overlooked due to evolutionary science biases such as selectionism, competition, and molecular-genetic centrality.

Finally, we address human mate choice and show that although adaptive theories dominate the mate choice discourse, arguments such as Prum's "Pleasure Happens" hypothesis propose that aesthetic preferences and pleasure function as powerful evolutionary drivers, reflecting a shift toward individual autonomy and challenging the historical role of parental influence in mate selection.

9.2 Habits in (Evolutionary) Natural Philosophy

Within natural philosophy,[1] which already includes some seeds of evolutionary thinking, the concept of *habit* features most prominently in the philosophy of Jean-Baptiste Lamarck (1744–1829), who was a proponent of both "transformism" (i.e., the view that species adapt to their environment and as it changes, so do they) and gradualism, like Darwin after him. He believed that habits provide a fundamental mechanism involved in the "use and disuse" of certain organs and behavioral patterns, and as such, they play a role in the inheritance of acquired traits.

According to Lamarck (2012 [1809]), an organism's behavior in response to its environment that gradually solidifies into a habit through repeated use (or disuse) is a vital component of adaptation. He also posited an organizing physical force that directs a gradual increase of complexity across all forms of life. This process, however, does not lead to a consistent gradualism because it is shaped by the environment in which organisms live and by their capacity to adapt to changing conditions (Corsi, 1989). Evolution is thus always shaped both by this organizing physical force and by the environmental conditions in conjunction with how organisms respond to them.

This view of evolution later strongly influenced Peirce, and our interpretation will proceed by comparing the views of these two thinkers (drawing mostly on Švorcová et al., 2023). Peirce explored evolution in essays such as *Man's Glassy Essence* (1892b) and *Evolutionary Love* (1893). Like many other thinkers of his time, he did not believe natural selection to be the only factor of evolution (Peirce, 1877); instead, he proposed three evolutionary principles—*tychasm*, *anancasm*, and *agapism*—which represented different evolutionary modes.

Peirce's concept of tychasm (from Greek *týchē* meaning "chance" or "fortune") involves chance and was inspired by Darwin's idea of random variation (i.e., "fortuitous" variation), although Peirce did not see it as particularly fortuitous (Peirce, 1893). Instead, he believed that the initial chaos is the primary source of spontaneity, which generates laws over time.

Chance, according to Peirce, operates within rules that can be articulated through statistical laws. It should not be viewed as chaos but rather as a form of spontaneity from which laws emerge (Peirce, 1892a).

Anancasm (*anánkē* meaning "necessity" or "inevitability"), or evolution by law, reflects mechanical necessity and thus also causality and physical, deterministic laws. Rigid cause-and-effect causality in anancastic evolution therefore amounts to development by "the mere force of circumstances or of logic" (Peirce, 1893) and logically leads to predictable outcomes, as in chemistry or the laws of gravity (Švorcová et al., 2023).

The concept of agapism (from *agápē* meaning "love"), strongly influenced by Lamarckian philosophy, focuses on the formation of habits. The laws of habits are fundamentally psychic. Unlike physical laws, they therefore concern final rather than efficient or material causation, especially when it comes to the relation between the psychic and the physical. This precedence of the psychic is related to Peirce's pan-semiotic vision, where the primordial mind precedes its material embodiment (Lacková, 2024) but mind and matter nevertheless form a continuum (Brier, 2019). Peirce sees the agapistic evolution as a progressive, purposeful process of gradual improvement driven by "creative love," the force of attraction between people, ideas, and even things. According to Peirce, habits can apply to both organic and inorganic nature: All things have a tendency to develop habits. We do not endorse this view in this paper, as we attribute habits solely to living beings.

For Peirce, habits accumulate and reinforce one another, resulting in a gradual growth and improvement within a process of "evolution by creative love" (Peirce, 1893). This teleological principle thus became the third complementary principle to chance and law. Chance and law are always fulfilled by habit. This is in line with Peirce's monistic perspective, where the mechanistic, material, and teleological aspect are ultimately all expressions of the same underlying processes. All these modes of evolution operate alongside each other and jointly account for the complexity of evolution.

According to Peirce, habit takes precedence over law. In the context of evolution, laws or automatic behaviors emerge only when habits solidify as temporarily final interpretants (i.e., the response that a sign elicits). Habit is inherently historical: its essence lies in repetition, ingraining, and reinforcement, whose action eventually links events into a seamless, continuous flow (Peirce, 1994). Over time, habits evolve into laws; the two are thus inherently linked. As Kull (2014) observed, another aspect of this process is semiosis, "meaning-making," which according to Peirce begins already in the lawless chaos of pure possibility, and which thus over time becomes organized. This is the process that drives nature's progression from chaos to an organized cosmos.

Peirce challenges the conception of universal and absolute laws we find in modern physics because he does not differentiate between laws and rules. An element of indeterminism, spontaneity, or absolute chance is always present (Hoffmeyer, 1996), and this indeterminism plays a significant role also in interpretation (see further).

Peirce's interpretation of habit differs from Lamarck's mechanistic view. According to Lamarck, organisms adapt in both form and function to environmental changes. New conditions lead to new experiences, which in turn drive the development of new organs and habits, while obsolete ones fade. Physiological changes associated with this process, such as muscle growth or behavioral shifts, can be inherited—and this drives the evolution of species (Lamarck, 2012 [1809]). Lamarck thus thought of habits in terms of repetition of beneficial traits associated with response to environmental changes (akin to what we would call behavioral plasticity), but this habitual causality was purely mechanistic.

Both Lamarck and Peirce viewed habit as central to evolution, but Peirce emphasized its role in evolution from chaos to structured laws, which contrasts with Lamarck's deterministic causality. Unlike Lamarck, Peirce moreover includes teleology and psychic attributes among habitual processes. Lamarck was a thorough materialist, and his scientific explanations relied heavily on concepts from physics and chemistry. In his view, evolution was thus ultimately driven only by environmental influences reflected in the behavior of organisms and by the organizing force, which has often been interpreted by neo-Lamarckians as some sort of vital or teleological force (Butler, 2014 [1910]; Pauly, 1905). However, Lamarck actually believed this organizing force to be a purely physical phenomenon (Lamarck, 1815; see Corsi, 2011),[2] and his evolutionary causality was a direct mechanistic cause-and-effect process. For instance, in the case of wading birds, the cause is the experience of the parents (who fish in shallow waters), and the effect is the elongation of legs in the offspring. In short, while Lamarck views habit as a mechanistic repetition of beneficial traits (and can dispense with teleology), Peirce's concept of habit involves an end-directed formation of law-like regularities, whereby mechanism and teleology belong to the same category as these law-like regularities.

After Lamarck, the concept of evolution by habit was further developed and refined by his followers through an analogy between organic memory and heredity, which drew parallels between memory and inheritance on the one hand and trait acquisition and learning on the other. According to this view, all traits (hereditary, acquired, and even initially conscious) are transmitted to subsequent generations through learning and repetition, ultimately leading to instinctive and automated behaviors. This idea was adopted by many thinkers,[3] including Peirce, in the late nineteenth and even early twentieth century, and the theory of "organic memory" became almost a paradigm of the era in about 1870 to 1918 (Otis, 1994). It claimed that, if sufficiently repeated, any (even initially conscious) activity can become a habit or instinct, whereby heredity and memory were thought to be governed by essentially the same processes. The downfall of this theory stemmed from the proponents of organic memory failing to understand the material basis of heredity (Bowler, 1983; Otis, 1994).

9.3 Further Development of the Concept

Similar ideas were developed in Germany by Theodor Eimer (1843–1898) and other adherents of the orthogenetic theory (see, for example, Bowler, 2013; Popov, 2018). While Eimer, unlike Lamarck or Peirce, did not explicitly focus on the concept of habit, his work on orthogenesis brings together Lamarckian inheritance, the direct influence of environment, and the iterative nature of evolutionary variation, which somewhat resembles a broader conception of habit as the stabilizing and creative principle in the evolutionary process. Eimer, the most prominent proponent of the theory of orthogenetic evolution, struggled with two issues that hampered a wider acceptance of his ideas. First, he was rather rude to his scientific opponents (Eimer, 1898) and second, Ernst Mayr (1904–2005), a prominent neo-Darwinian evolutionist, somewhat prematurely and without further ado dismissed orthogenesis altogether and used his position of editor of the *Evolution* journal to prevent any discussion or criticism of Darwinism from the orthogenetic perspective (Popov, 2008).

Somewhat curiously, Mayr later explained in his famous book *The Growth of Biological Thought* (1982) that the evolutionary trends so favored by orthogeneticists have two possible causes—namely, "consistent changes in environment" and "the internal cohesion of the genotype which places severe constraints on the morphological changes." He thus in effect repeated the explanation proposed by orthogeneticists, who claimed that evolution is driven by a combination of environmental influences and internal developmental causes. The sole difference was that Mayr avoided expressions such as "natural law" or "regularity." He concluded his argumentation with the following reassurance: "Hence, evolutionary trends are readily explained within the explanatory framework of the Darwinian theory and do not require any separate 'laws' or 'principles'" (Mayr, 1982, p. 530f).

In any case, the ideas and empirical evidence brought together under the umbrella of orthogeny were dismissed before they could be fully evaluated. The vision of evolution proposed by Eimer and partly by other adherents of the orthogenetic theory does not invoke any mysterious forces acting upon organisms. Orthogenetic explanations stressed the importance of developmental and other law-like processes immanent to living organisms, the variational constraints, and the internalization of environmental conditions (Bowler, 2013; Kleisner and Saribay, 2019; Popov, 2018). One of Eimer's insights, valuable also for later biology, is that many traits are passed onto the subsequent generations without being beneficial for the bearer and thus adaptive. That is also why he criticized August Weismann for excessive adaptationism (Eimer, 1890).

Eimer's arguments in support of Lamarckian inheritance are unusual but interesting. According to Eimer, however, noninheritance of injuries is not an argument against the inheritance of acquired characters because in many species, we commonly observe regeneration. If an injury is not visible in the next generation, it is thus due to a delayed effect of regeneration. In a similar vein, fledglings cannot fly immediately after hatching, and the art of playing a musical instrument is not directly inherited as a such because babies cannot play musical instruments. Nevertheless, birds eventually learn to fly, and it is generally acknowledged that some musical skills are passed down in certain families (Eimer, 1890, p. 10). Eimer's inheritance of acquired characters thus does not have the nature of some absolute and immediately manifested act.

After Lamarck, the neo-Lamarckians, Eimer, and Peirce, the concept of habit did not play a notable role in subsequent theories of evolution. Although Darwin, when commenting upon Lamarck's *Philosophie zoologique*, remarked, "Habits becoming hereditary form the instincts of animals—almost identical with my theory" (Burkhardt, 2011, p. 41), he did not develop this idea any further. He readily admitted that instinct and certain behaviors play a role in survival (Darwin, 1872), but habits as such did not have a specific role in his theory of evolution. He was firmly convinced that evolution is primarily driven by natural selection and that habits, with their inherited results, were thus rather treated only as material for natural selection to work upon (Burkhardt, 2007).

Still, in Darwin's *On the Origin of Species* (1859), we do find a theory of inheritance based on Lamarck's theory of inheritance of acquired characteristics. In fact, Darwinism and Lamarckism ideologically diverged only with the advent of the Modern Synthesis. After the discovery of Mendel's laws, neo-Lamarckism could no longer explain how habitual traits could be passed to the subsequent generations via genetic inheritance, and Mendelian genetics quickly prevailed in the 1930s and 1940s and became the dominant mode of explaining evolution (Gilbert, 2011). Phenotypic plasticity, the effects of environment on the evolution of organisms, and soft inheritance all fell into obscurity (Gissis and Jablonka, 2011). Save for a few exceptions, these subjects were ignored and received no institutional funding, while the genetic program was strongly supported and institutionalized (Gissis and Jablonka, 2011).

Yet even within the framework of Modern Synthesis, one can imagine that behavioral habits change the course of evolution—for example, a shift in food preferences (or some other behavioral change) of some individuals can eventually lead to isolation and a speciation event. Along similar lines, James Baldwin (1861–1934), in his article "A New Factor in Evolution" (1896), highlighted the role of behavior in evolution, proposing that behavior can facilitate selection.[4] For instance, when certain individuals can adapt their behavior to an environmental challenge, this can lead to their selective advantage in that environment. Such adaptations can take the form of building a more solid nest structure or finding a new food source. Fast-learning individuals—that is, those whose behavior quickly adapts to new conditions—are more likely to survive and reproduce.

From a contemporary point of view, random mutations, epigenetic changes, or other shifts in developmental regulation can lead to genetic changes that encode or enhance certain beneficial traits or behaviors initially learned or developed through plasticity. These mutations are then favored by natural selection, gradually making the behavior more stable and widespread in the population. Phenotypic plasticity thus can facilitate natural selection depending on the reaction norms of genotypes and selection pressures. In other words, if certain reaction norms result in phenotypes that survive in the new conditions, the corresponding genotypes have a chance to persist (West-Eberhard, 2003).

Thus, the concept of habit has reemerged with phenotypic plasticity, epigenetics, and genotypic accommodation—key ideas in the Extended Evolutionary Synthesis (Laland et al., 2015; Pigliucci and Müller, 2010). Niche construction theory (Matthews et al., 2014; Odling-Smee et al., 2003) also highlights how organisms shape their environments through habitual behaviors, influencing their own evolution and that of others. While this relates to ecology and behavior, biology shows that repeated interactions can alter epigenetic patterns, affecting gene expression transgenerationally (or even leading to biased mutations, as in the case of DNA methylation). Such habits can ensure developmental continuity but may also be lost (Švorcová et al., 2023). Before exploring habits at the cellular level, let us first examine their role in the explanatory framework of biosemiotics.

9.4 Habit in Biosemiotics

Peirce introduced the concept of a sign relation consisting of an irreducible triad of a representamen, an object, and an interpretant. This has become a widely used model of semiosis in biosemiotic studies (Favareau, 2015). *The representamen* (sign vehicle) is a perceptible form of the sign referring to an object (for instance, a deer skull in the woods, the word "deer," an image of a deer on the wall, or deer tracks in the snow). *The object* is that which is represented by the sign (a deer or deers), and the *interpretant* is the understanding or effect which the sign generates in the perceiver (the idea of a deer, or the habituated reaction to attack for food, or to pull the trigger).

Thus, the world can only be known through signs. Furthermore, Peircean sign triads continuously concatenate because the interpretant of one sign relation can serve as the sign vehicle for the next (Favareau, 2015). Across the biosemiotics literature, however, the application of this model is often inconsistent: The interpretant (the effect or change in the organism) is often replaced by the interpreter (a cell, an animal, a person), and the sign relation or its object are often substituted by the concept of its "meaning."

When we form a habit, we tend to regard our interpretation of an object as perfectly defined and final—until the habit changes and the meaning of the object collapses and is reinterpreted. It is the existence of a habit as *an anchored interpretation of an otherwise unknowable object* that enables intersubjective understanding (and with it the operation of objective science). Inspired strongly by Peirce, one of the pioneers of biosemiotics, Jesper Hoffmeyer, notes that life itself exemplifies nature's tendency to form habits: "The tendency of nature to generate regularities—or, as it is more commonly described today, to self-organize—may be understood as the very first exposition of the principle that will develop through cosmic evolution to become semiosis, the ability of living systems on planet Earth (and possibly many other places in the universe) to form interpretants—or self-maintaining and self-perpetuating habits, if you like" (Hoffmeyer, 1996, p. 612).

The habitual stage of the interpretant (an organism's reaction to a sign) is never fixed, never given once and for all. In principle, the organism can change it, but such change always depends on the history of the organism and its previous experience (Markoš, 2002). Sharov and Tønnessen (2021) similarly view habit as a vital evolutionary process that underpins the goal-directed nature of living beings formed as a self-sustaining, self-developing circuit between agents and their purposes.

For the purpose of this chapter, we do not need to endorse a specific model of semiosis. What we must do is acknowledge that organisms have the ability to engage in meaning-attribution/making, which implies that semiosis is an intrinsic property of living systems starting at the level of a single cell as a whole (Švorcová, 2024). This can be illustrated by the example of a single molecule interpreted differently by a cell depending on the cell's history and environmental conditions. Life evolves through a constant reinterpretation and co-option of the same informational toolkit (Caroll, 2005; Davidson and Erwin, 2006; Gilbert, 2016; see Švorcová, 2012, 2024; chapter by Markoš). Meaning as such can be defined as the existence of a mediated relation that is established by certain processes of life that constitute reality for the organism (Kull, 2023b). Meaning inherently presupposes interpretation (and misinterpretation), and that is possible only when some degree of indeterminacy is present—that is, the possibility of a choice. We can then say that even a cell can choose how to respond to a given molecule (Kull, 2018).

9.5 Cellular Habits

Our assertion that habits can be identified even in entities such as cells is based on the premise that, while the fundamental cellular genetic and molecular components remain the same across organisms, they can be, and are, utilized in many different ways—that is, there is a variety of habitual interpretations (Švorcová, 2012; Gilbert, 2016).

For example, a molecule of cyclic adenosine monophosphate (cAMP), which functions as a second messenger (a messenger of signals within cells that another signal messenger from outside the cell initiates), can be involved in various cellular processes. In the cells of skeletal muscles, cAMP activates enzymes such as protein kinase A (PKA), which enhances glycogen breakdown to release glucose for energy (Nielsen et al., 2002). In the cells of cardiac muscles, cAMP modulates calcium ion channels, thus increasing the strength and frequency of heart contractions (Hove-Madsen et al., 1996). In fat cells, cAMP triggers the enzymes responsible for lipolysis—that is, for the breakdown of triglycerides into free fatty acids and glycerol (Lass et al., 2011). In neurons, cAMP serves as a secondary messenger in signal transduction, contributing to synaptic plasticity, learning, and memory by influencing the proliferation and removal of neurotransmitter receptors (Kennedy, 2013).

Thus, a molecule with particular physical and chemical properties can be used differently by different cells depending on their history, environment, and interpretative capacity. This allows us to say that the elements of life are not molecules but signs (Kull, 2023b).

Bacteria, as well as single-celled and multicellular eukaryotic organisms, too can be understood as fundamentally habitual beings. They are capable of forming habits that persist across multiple generations through epigenetic processes such as DNA methylation, various histone modifications, and a range of RNA molecules (e.g., Crews et al., 2007; Gangisetty et al., 2022; Morgan et al., 1999; Pembrey et al., 2006; Rassoulzadegan et al., 2006; Rechavi et al., 2014; Vassoler et al., 2013; Yu et al., 2018). Science tends to view such epigenetic processes as largely stochastic, as they often occur as a result of environmental changes in tandem with the organismal experience of these changes (longer day/more light, social interactions, stress, change of food habits).

From this perspective, particular experiences and consequent epigenetic modification may either be reinforced and embedded or forgotten—and that depends on various environmental factors (Ginsburg and Jablonka, 2009; Robinson and Barron, 2017; Švorcová, 2024). This plasticity (in the Peircean sense of developmental teleology) allows for a degree of spontaneity and indeterminism (Švorcová et al., 2023). Teleology in this context integrates chance (variability generated by mutation), necessity (normative code), habit and spontaneity (epigenetic interpretive interplay)—that is, all three types of Peircean evolution (Lacková, 2024).

9.6 Supracellular Developmental Habits

As we repeatedly emphasized in our work (Švorcová, 2012, 2024), one can observe in development and evolution certain memory patterns that could be interpreted as semiotic habits. They take the form of developmental networks continuously reinterpreted throughout phylogeny. Homological elements are retained, but they are in the course of evolution repurposed and give rise to new phenotypic structures. A notable example is the *Hox* gene set, which—alongside other conserved elements, such as genes and their products (transcription factors and RNAs)—contributes to the formation of flexible gene regulatory networks.

Changes, or reinterpretations, of these networks during development tend to take the form not of mutations in conserved gene regions but rather their duplication and subsequent divergence (Davidson, 2006). Further factors, which contribute to the formation of flexible gene regulatory networks, are alterations in their *cis* regulatory elements and their overall configuration—in particular, the combination of transcription factors in a given tissue at particular times and locations (Davidson and Erwin, 2006). Further changes are induced by environmental influences; they, too, can be epigenetically regulated and stabilized (Švorcová, 2024). Such interactive networks operate as dynamic systems shaped by environmental experiences as well as maternal input, structural inheritance, genetic information, biophysical forces, and other factors. These networks, and the transformations they undergo, are fundamentally historical, and alterations in their architecture ultimately play a decisive role in the evolution of phenotypes (this dynamics is akin to what Hoffmeyer calls *semiotic scaffolding*; Hoffmeyer, 2007; see also the theory of facilitated variation; Gerhart and Kirschner, 2007). Habits such as those described above eventually become highly automatic. It might seem that their operation is highly law-like. Nevertheless, while it is beneficial for living organisms to entrust certain functions to automated mechanisms, we do not endorse the ontological claim made by contemporary biology according to which such automatisms are the fundamental nature of organismal reality.[5]

9.7 Symbiotic Habits

Symbiotic relationships can likewise be viewed as habitual. In many cases, the role that the symbiont plays for the host could be fulfilled by a symbiont of another species (as with bacteria in the human gut; Eisenstein, 2018) or achieved by other means (different species of termites digest cellulose either with the help of their gut symbionts or with the help of a fungus they grow in their nests; Martin and Martin, 1978). In other instances, symbiotic interactions occur sporadically or only under particular conditions (facultative symbiosis). But in many of these cases, such relationships are maintained across generations for millions of years, safeguarded by specific forms of symbiotic transmission, and the symbionts can be housed in specialized organs (e.g., the fungi and bacteria in insects are located in mycetomes or bacteriomes, respectively; Michalik et al., 2023).

If the symbiosis is to persist, the symbionts must be transmitted along with the egg's cytoplasm or positioned near it, allowing the emerging offspring to acquire them. Such symbiotic relationships can evolve into a deeper genetic and epigenetic dependence up to a point where we see the emergence of obligate symbiosis—that is, a state where the organisms cannot survive independently, without their symbionts. Such symbiotic bonds are then impossible to break. They often entail the provision of essential substances that neither the host nor the symbiont can produce independently, but they also affect functions such as digestion, detoxification, colonization of new niches (herbivory, survival in extreme environments, bioluminescence), defense against pathogens, mate selection, recognition of kinship, maintenance of social cohesion, and numerous other processes (for many examples, see Gilbert and Epel, 2015). Importantly, such relations are safeguarded by reciprocal sign-mediated communication, through which, for instance, our immune system recognizes commensal bacteria in our guts as familiar, which further influences homeostasis in our intestines and affects our immune system, brain development, digestion, and emotions (Carabotti et al., 2015).

The reader can imagine a plethora of such relationships throughout nature. Our message here is quite simple: When these relationships started to take form, there was no physical (anankastic) or chemical necessity, no inevitability arising from their configuration that bound the symbiont and its host. Instead, by establishing a new habit, a regularity repeated across generations, the host and symbiont created a new niche in which they could coexist, interact, and explore new possibilities (a food source, defense against predators, or some other evolutionary advantage).

This is exemplified, for instance, in the partnership between legume roots and the *Rhizobium* bacteria. While thousands of flavonoid types serve as a defense against pathogens, a particular group known as isoflavonoids, which are found only in legumes (Hirsch et al., 2001), play a different role. They signal to *Rhizobium* bacteria the presence of a compatible host. This shows how an initially stress-related signal evolved to be interpreted as an invitation to mutualistic interaction, effectively acquiring a new meaning within such habitual relationship.

Such relationships are ultimately enabled by the existence of an informational norm shared by all inhabitants of the biosphere (e.g., genetic coding, molecules of second messengers, genetic toolkit). Communication can lead to the establishment of habits: Adaptive advantages come later. In other words, habitual evolution together with such shared norms, as we shall see further on, typically leads to the formation of higher organic units, gestalts, or closures that enable the existence of organismal agents of a higher order (see also this volume's chapter by Markoš).

9.8 Biosemiotic Echoes of Organismal Forms and Functions

Choices influence future experiences and repeated choices–that is, certain kinds of habit create an internalized milieu that helps shape perception and cognition. Perception is selective and variable in terms of meaning attribution to objects in the organism's surroundings. In particular, perception based on certain experiences influences what objects enter the organism's subjective perceptual world, or *Umwelt*. In terms of the attractor field model (Brejcha et al., 2021; Kleisner and Saribay, 2019; Tanaka et al., 1998), perception literary warps the reality in which the perceiver lives. Objective reality can be only measured, but not fully perceived, experienced, or lived. That does not mean that we should resign on investigating general truths about the world around us because one can translate from the perceptual space and its objectified representation and then mathematically model differences between the two.

Perhaps not surprisingly, the accuracy and overall aptness of such modeling and any comparisons derived from it would critically depend on our prior knowledge of the system. The only difference from the approach of traditional post-Cartesian "objective science" is that the "perception" and "occurrence" of things and behaviors are treated as being of equal importance. There is no solid ground of "objective reality" (if life is included) that will one day be fully discovered, described, and explained (Deely, 2009)—that should be obvious to anyone who had even just a fleeting encounter with semiotics, phenomenology, structuralism, or any related strands of philosophy (and to anyone who ever had an argument with a spouse over a perfectly clear and self-evident fact).

The biosemiotic approach to morphology investigates how living organisms express their existence through their multimodal appearances. It links physical structures with agency—that is, morphology's ability to influence perception and interaction, and vice versa. That, in turn, is intimately linked to the philosophical and theoretical dimensions of biological organization because it highlights the unique ways in which living entities manage their appearances to establish their presence in ecosystems. In this sense, appearances are not superficial. They are essential to an organism's agency, serving as interfaces for communication, interaction, and survival.

One of the key arguments of the biosemiotic approach to morphology is that the agency and appearance of living organisms cannot be reduced to simple evolutionary adaptations (Kleisner, 2024). Living beings are unique in their ability to create appearances that reveal, conceal, or modify their attributes as part of repeated interaction with their environment. Unlike inanimate objects, which reflect light only passively, living organisms actively built their display during morphogenesis. This capacity for "self-display" distinguishes life from nonlife and is a foundational aspect of biological organization. From this perspective, an organism's surface is not merely a physical boundary: It is a medium of expression, an interface where life actively engages with the world.

In contrast, the molecular-genetic perspective reduced the study of evolutionary adaptations to a purely genetic perspective. While genetic research has discovered many essential details of evolutionary processes, it often fails to consider the wider context of morphology, behavior, and ecological interactions. The sum of genetic factors alone cannot account for the diversity of forms and behaviors observed in nature. Moreover, genetic selectionism that is based on the conviction of an exclusive status of the gene as the most important (if not the sole) unit of selection, one that is subjected to transgenerational transmission, underestimates the role of the female ovum and its cytoplasmatic factors. Genetic selectionism is thus actually a rather chauvinistic theory, and one can but wonder why no feminist movement has used this in a political argument as of yet.

Therefore, we advocate a renewed focus on morphology and a multidimensional view of biological appearance. Organic forms cannot be fully appreciated without considering their inherent agency and the communicative and aesthetic role they have in their ecosystems. Rather than viewing appearances solely through the lens of survival and reproduction, we suggest they should be examined also as expressions of organismal identity, autonomy, and interaction.

In short, we argue that life's expressive qualities are as central to evolution as the mechanisms of natural selection are. By focusing on organismal surfaces and their functioning within ecosystems, we advocate a rethinking of biology that transcends traditional Darwinian views. Instead of viewing appearances as merely adaptive, we claim they are intrinsic to the organism's agency—and this has far-reaching implications to our understanding of the complexity of organic life, including human affairs.

9.9 The Habitual Bloom and Adaptive Gloom of Evolutionary Psychology

Richard Prum aptly criticized evolutionary psychology and its approach to human mate choice, noting that "contemporary evolutionary psychology has a profound, constitutive, often fanatical commitment to the universal efficacy of adaptation by natural selection." Especially fitting is then his statement that "there is never any doubt what the conclusion of any evolutionary psychology study will be" (Prum, 2017, p. 227). In fact, this applies not only to evolutionary psychology but to mainstream evolutionary biology in general.

We agree with Prum that there are many odd biases in both the theory and empirical practice of evolutionary psychology. He is right to point out that women's mating preferences are less frequently studied than male preferences, and this is then reflected in the general theoretical framework of the subject. For instance, if we were to apply to humans the general animal analogy according to which females are the carriers of costlier gametes than males, then females should be the choosier sex, and that should lead to evolution of male ornamental traits.

In this scenario, females' choice is the power that drives the evolution of ornaments in the evolutionarily less important and more easily replaceable male; the process is reminiscent of little girls putting colorful clothing on their beloved but completely inert and mindless dolls. But to conclude on a more serious note, in humans, both sexes participate in mate choice, but this is apparent more on the habitual rather than genetic or gamete level.

Prum's alternative to the adaptive agenda around mate choice is the Beauty Happens hypothesis, which claims that human sexual ornaments, such as the female waist-to-hip ratio and permanently present breasts, originated due to coevolution of preferences. In other words, their development was driven by repeated choices of the opposite sex—that is, by specific habitual actions. But such traits do not necessarily advertise a better biological quality. It seems that the Beauty Happens hypothesis is but a special case of Fisherian runaway selection. One way or another, it is a nice slogan for research that studies how variation in habits can affect variation in phenotypes and give rise to adaptive elements such as honest signaling.

The most important quality of Prum's model is the fully acknowledged directness of the effects that influence the dynamics of evolutionary change. Neo-Darwinian explanations always assume some forces (or hidden reasons) acting behind the apparent effects, so that all manifestations of living organisms are in the service of some hidden reasons that are ultimately linked to reproductive success and the optimization of survival strategies. We should appreciate the role of pure beauty and pleasure as organizing forces that can change the behavior and, consequently, also the morphology (and its genetic underpinnings) of organisms. The simple but important innovation we find in Prum's proposal and also, analogically, in biosemiotics in general (compared to the standard neo-Darwinian model) is that these approaches allow the aesthetic dimension (beauty and pleasure) to function as a direct force of evolutionary change. In other words, aesthetic preferences and the tendency to fulfil them are granted the status of an immediate creative cause *per se*, one that is not viewed as just a proxy to some fitness-related biological quality such as good genes, fertility, viability, and the like.

Although Prum's argument that human sexual behavior is driven by female pleasure is convincing and can be supported by concrete evidence, we should remain slightly reserved. The history of science gives us plenty of reasons to be reticent. Just as, for example, Darwin, Wallace, Spencer, and others were in their day influenced by contemporary economic theories and the general social climate of early capitalism, so Prum's hypothesis fits strikingly well into the social climate of our times, characterized by efforts to achieve gender balance or even by active promotion of the feminine in all sorts of areas of the social establishment of Western societies. It should, however, be noted that Prum's model does not neglect male pleasure as a coevolved feature, so the overall image is balanced. This is no criticism of the aesthetic hypothesis, and Prum is fully aware of this dimension of his explanations when he writes that "in a delightful and unexpectedly feminist fashion, the Pleasure Happens hypothesis identifies women as the active agents in the evolution of their own capacity for orgasmic pleasure" (Prum, 2017, p. 379). The admission that women, and not only females of our species, but all living beings, whether they engage in sex or not, are active agents of their evolutionary capacities, is quite fundamental, and one cannot but agree with Prum on that point. Nevertheless, we should not forget that our theories and assumptions do not just hover in the air independently of the surrounding social events, as the history of science and technology convincingly demonstrates. It is thus little wonder that the open formulation of the Pleasure Happens hypothesis had to wait for our times.

9.10 Choices That Decide the Future

A necessary condition for semiosis is at least a minimal indeterminacy, which is provided by the simultaneous existence of possibilities and due to which exists a little freedom of choice.
—Kull, 2023c, p. 78

Janzen's (1985) idea that the coexistence of species in ecosystems is not due to some slow, long-term coevolution, but the result of a fast, real-time process that he calls "ecological fitting" may be in principle extended beyond the level of species. One such extension is the intraspecific individual level. If we take the human mate choice as the closest example, we see that it often manifests features that seem more in line with some kind of sudden fitting than with the assumption of choice made based on some ancient Pleistocene adaptations. Mate choice seems to be more about expectations, about the anticipation of the near future based on past experience, whereby the "past experience" is most likely a combination of individual experience and lineage (species) experience. Nowadays, many human intersex interactions are caused by sudden outbursts. After all, the so-called love at first sight (and also not at first sight) is actually a relationship that partners establish as a result of search and choice, and it corresponds to ecological fitting—or what Kalevi Kull, inspired by Janzen, would call semiotic fitting (Kull, 2023a).

Historically, however, it seems that in most human cultures, mate choice was not determined solely or mainly by individual choice. It has been argued that parental control, especially over daughters, was a major factor in human evolution. Parents tend to exert more control in traditional foraging societies, where they influence sexual selection through decisions about marriage and reproduction (Apostolou, 2007). This could be one of the greatest challenges to Prum's Pleasure Happens hypothesis. One could object that there is no pleasure (or at least no direct pleasure) to be derived from having to mate with someone who was selected not for physical attraction but, for instance, for social status and wealth. One could object that parental choice is an evolutionarily relatively new phenomenon that played a significant role for several thousand years exclusively in the human species, only to be recently replaced by free individual choice, which is, after all, the original evolutionary strategy for our species and one that evolved in the environment of evolutionary adaptedness (sometime in the Pleistocene, again). But do we really know this?

Parental choice can take on different forms in different cultures and times. The choice can be made only by parents, by the immediate family, or even by broader kin. In other cases, parents do not make the final decision but co-select the prospective spouse in collaboration with their children. In all cases, the mate choice is fully or partly delegated to the hierarchically superior unit above the level of individuals. Situations when parents only participate in the choice and do not take full control are more likely to occur in societies that are transitioning from a traditional to a modern style of life.

The recent appearance of individual mate choice and the associated weakening of parental choice seems to be associated with (or even is the consequence of) a decline of the traditional role of the family. This weakening of the role of the family is probably the result of social, cultural, and environmental changes that started with the Industrial Revolution and led to the development of living conditions that enabled individuals to break free from economic dependence on family clans and become an independent and self-reliant population unit. It is remarkable that in the evolution of complex organic units, we often observe a similar process in the opposite direction, one where lower-level units (individuals) are integrated into a higher-level unit and then restricted in freedom and forced to work for the good of this more complex, strongly integrated unit to which they now belong. We find this process exemplified in colonial polyps (e.g., the Portuguese man-of-war, *Physalia physalis*) or in the colonies of social insects. From the perspective of organizational systems, the disintegration of human family and the integration of insect colony are two sides of the same coin. Both are the products of the general principles of biological organization. In one case, we see a relegation of control to a higher level and the emergence of superorganism (colony), while in the case of the recent evolution of human sexual behavior, one can observe a weakening of the organizational powers of the higher level in favor of increased operational independence at the individual level. In our species, evidence shows that parental control over mate choice is more prominent in collectivist cultures. For example, a study found that Iraqi Kurds reported higher levels of parental influence on their mate choice than Dutch individuals from The Netherlands (Buunk et al., 2010). Still, although human individual mate choice tends to dominate in post-industrial societies, some parental influence persists. Especially mothers exert indirect control over their children's mate choices through various manipulation tactics, even in societies where individual autonomy in mate selection is high (Apostolou, 2007).

The adaptationist scenario suggests that humans have evolved a range of mating strategies, men and women have faced different adaptive challenges in mating, and individual mating choice should take on board the markers of biological quality (Buss, 2019; Buss and Schmitt, 1993). According to this neo-Darwinian theory, this ought to lead to a better quality of offspring and a general increase of fitness, but recent empirical evidence does not unequivocally support these theoretical expectations (Eastwick et al., 2014; Kleisner et al., 2024). Members of different species and of populations within species (demes, cultures, or subgroups) display different degrees of freedom of choice depending on particular environmental and historical societal situations (Henrich et al., 2010).

We view habit as a repetition of established sign relations, as something that reflects the repetitive aspect of sequential semiosis (Kull, 2023a). In the theory of biological signaling, a habit would correspond to a stereotyped response to a signal. In mate choice, a habit would correspond to a choice that constraints individual freedom and thus minimizes the risk of unexperienced decision. While freedom of choice increases potential risks, it also enables innovations and thus opens the way to other potential benefits.

9.11 Conclusion

In this chapter, we analyzed a number of examples that show that evolution by habit matters at all levels of biological hierarchy, from agency at the cellular level all the way to animal and human behavior. We have intentionally selected scenarios that contrast with the standard adaptationist description of the living world. We do not claim that all our examples are unbiased, neutral, or indisputably supported by repeatedly corroborated empirical results. Rather, we used the available empirical evidence and explored various theoretical perspectives of both contemporary and historical authorities to outline an alternative explanatory framework for a habit-centric view of the evolution of living entities.

We are convinced that life, at all levels, evolves through the formation, stabilization, and modification of habitual patterns of meaning. Habitual reinterpretation is one of the main principles of the living. Once we accept habit as a constitutive factor of evolution, organisms are no longer seen as merely the passive products of evolutionary pressures but as active agents capable of influencing their evolution. This argument has been made before (Bowler, 1983) and, at its core, highlights the fact that adoption of habits allows individuals to choose new habits when the environment changes, thereby shaping, at least to some degree, the course of their evolution.

Acknowledgments

We would like to thank Anna Pilátová, and the editors, for their thorough revisions. The publication was supported (JŠ) by the project of the Czech Science Foundation (GA23–05374S, Reframing Philosophical Anthropology: Searching for an Anthropological Difference Beyond the Nature/Culture Dichotomy) at the University of Hradec Kralove, Philosophical Faculty.

Notes

1. For the sake of brevity, we shall not go into some of the purely philosophical traditions, although the concept of habit has been investigated by thinkers across millennia, from Aristotle to Deleuze.

2. Lamarck defines life through organization of constitutive parts shaped by organic fluids like heat, electricity, or vapor—a view rooted in Cartesian philosophy and shared by eighteenth-century mechanists. Habit, in this context, is the repeated flow of subtle fluids through solid structures, reinforcing the same pathways over time (Corsi, 1989).

3. Such as Ewald Hering, Ernst Haeckel, Samuel Butler, Auguste Forel, Richard Semon, Patrick Matthew, Peter Willers Jessen, Adam Sedgwick, Thomas Laycock, Henry P. Orr, Edward Drinker Cope, Alpheus Hyatt, Fairfield Osborn, Francis Darwin, James Ward, Eugenio Rignano, and Ernst William MacBride (see Bowler, 1983; Otis, 1994; Schacter, 2001).

4. Baldwin's idea did not involve the inheritance of acquired traits; he did not consider himself a Lamarckian.

5. One might ask whether signal transduction or gene regulation dynamics are better understood as coded processes. Barbieri (2024) discusses such codes as conventional sets of rules linking two independent worlds, an arbitrary correspondence, and distinguishes them from semiosis, arguing that cells engage only in coding and decoding. We see code as an established interpretation, akin to a habit in Peirce's sense (Švorcová, 2012; Lacková, 2018). Rather than being frozen, codes arose through negotiation during life's emergence. Since semiosis requires interpretation, coding–decoding (as in Searls, 2002) is merely a by-product of semiotic negotiation, much like natural language. Meaning, therefore, emerges through interpretation rather than being inherent to a code.

References

Apostolou, M. 2007. Sexual selection under parental choice: The role of parents in the evolution of human mating. *Evolution and Human Behavior*, *28*(6), 403–409.

Baldwin, J. M. 1896. A new factor in evolution. *The American Naturalist, 30*(354), 441–451.

Barbieri, M. 2024. *Codes and evolution. The origin of absolute novelties*. Springer.

Bernardi della Rosa, S., and Zengiaro, N. 2025. The biosemiotic glossary project: Habit. *Biosemiotics, 18*.

Bowler, P. J. 1983. *The eclipse of Darwinism: Anti-Darwinian evolution theories in the decades around 1900*. Johns Hopkins University Press.

Bowler, P. J. 2013. *Darwin deleted: Imagining a world without Darwin*. University of Chicago Press.

Brejcha, J., Tureček, P., and Kleisner, K. 2021. Perception-driven dynamics of mimicry based on attractor field model. *Interface Focus, 11*(3), 20200052.

Brier, S. 2019. Peircean logic as semiotic expanded into biosemiotics. In E. Bisanz (Ed.), *Applied interdisciplinary Peirce studies* (pp. 59–84). Peter Lang.

Burkhardt, R. W. 2011. Lamarck, Cuvier, and Darwin on animal behaviour and acquired characters. In S. B. Gissis and E. Jablonka (Eds.), *Transformations of Lamarckism. From subtle fluids to molecular biology* (pp. 33–44). MIT Press.

Buss, D. 2019. *Evolutionary psychology: The new science of the mind*. Routledge.

Buss, D. M., and Schmitt, D. P. 1993. Sexual strategies theory: An evolutionary perspective on human mating. *Psychological Review, 100*(2), 204–232.

Butler, Samuel. 2014 [1910]. *Unconscious memory*. A. C. Fifield Press, Project Guttenberg re-edition of London.

Buunk, A. P., Park, J. H., and Duncan, L. A. 2010. Cultural variation in parental influence on mate choice. *Cross-Cultural Research, 44*(1), 23–40.

Carabotti, M., Scirocco, A., Maselli, M. A., and Severi, C. 2015. The gut-brain axis: Interactions between enteric microbiota, central and enteric nervous systems. *Annals of Gastroenterology, 28*(2), 203–209.

Caroll, S. 2005. *Endless forms most beautiful*. W. W. Norton.

Corsi, P. 2011. Jean-Baptiste Lamarck: From myth to history. In S. B. Gissis and E. Jablonka (Eds.), *Transformations of Lamarckism. From subtle fluids to molecular biology* (pp. 12–28). MIT Press.

Corsi, P. 1989. *The age of Lamarck evolutionary theories in France 1790–1830*. Oxford University Press.

Crews, D., Gore, A. C., Hsu, T. S., Dangleben, N. L., Spinetta, M., Schallert, T., Anway, M. K., and M. D., Skinner, M. K. 2007. Transgenerational epigenetic imprints on mate preference. *Proceedings of the National Academy of Sciences of the USA, 104*(14), 5942–5946.

Darwin, C. 1859. *On the origin of species by means of natural selection, or the preservation of favoured races in the struggle for life*. John Murray.

Darwin, C. 1872. *The expression of the emotions in man and animals*. John Murray.

Davidson, E. H. 2006. *The regulatory genome: Gene regulatory networks in development and evolution*. Academic Press.

Davidson, E. H., and Erwin, D. H. 2006. Gene regulatory networks and the evolution of animal body plans. *Science, 311*(5762), 796–800.

Deely, J. 2009. *Purely objective reality*. University of Scranton Press.

Eastwick, P. W., Luchies, L. B., Finkel, E. J., and Hunt, L. L. 2014. The predictive validity of ideal partner preferences: A review and meta-analysis. *Psychological Bulletin, 140*(3), 623–665.

Eimer, T. 1890. *Organic evolution as the result of the inheritance of acquired characters according to the laws of organic growth*. Macmillan.

Eimer, T. 1898. *On orthogenesis and the impotence of natural selection in species formation*. Open Court Publishing Company.

Eisenstein, M. 2018. Gut reaction. *Nature, 563*(7730), S34–S35.

Favareau, D. F. 2015. Creation of the relevant next: How living systems capture the power of the adjacent possible through sign use. *Progress in Biophysics and Molecular Biology, 119*(3), 588–601.

Gangisetty, O., Chaudhary, S., Palagani, A., and Sarkar, D. K. 2022. Transgenerational inheritance of fetal alcohol effects on proopiomelanocortin gene expression and methylation, cortisol response to stress, and anxiety-like behaviors in offspring for three generations in rats: Evidence for male germline transmission. *PLoS one, 17*(2), e0263340.

Gerhart, J., and Kirschner, M. 2007. The theory of facilitated variation. *Proceedings of the National Academy of Sciences of the USA, 104*(suppl. 1), 8582–8589.

Gilbert, S. 2011. The decline of soft inheritance. In S. B. Gissis and E. Jablonka (Eds.), *Transformations of Lamarckism. From subtle fluids to molecular biology*. MIT Press, 121–126.

Gilbert, Scott F. 2016. Ecological developmental biology: Interpreting developmental signs. *Biosemiotics, 9*(1), 51–60.

Gilbert, S. F., and Epel, D. 2015. *Ecological developmental biology: The environmental regulation of development, health, and evolution.* Sinauer Associates.

Ginsburg S., and Jablonka, E. 2009. Epigenetic learning in non-neural organisms. *Journal of Biosciences, 34*(4), 633–646.

Gissis, S. B., and Jablonka, E. (Eds.). 2011. *Transformations of Lamarckism. From subtle fluids to molecular biology.* MIT Press.

Henrich, J., Heine, S. J., and Norenzayan, A. 2010. Most people are not WEIRD. *Nature, 466*(7302), 29–29.

Hirsch, A. M., Lum, M. R., and Downie, J. A. 2001. What makes the rhizobia-legume symbiosis so special? *Plant Physiology, 127*(4), 1484–1492.

Hoffmeyer, J. 1996. *Signs of meaning in the universe.* Indiana University Press.

Hoffmeyer, J. 2007. Semiotic scaffolding of living system. In M. Barbieri (Ed.), *Introduction to biosemiotics. The new biological synthesis* (pp. 149–166). Springer.

Hove-Madsen, L., Méry, P. F., Jurevicius, J., Skeberdis, A. V., and Fischmeister, R. 1996. Regulation of myocardial calcium channels by cyclic AMP metabolism. *Basic Research in Cardiology, 91*(Suppl 2), 1–8.

Janzen, D. H. 1985. On ecological fitting. *Oikos, 45*(3), 308–310.

Kennedy, M. B. 2013. Synaptic signaling in learning and memory. *Cold Spring Harbor Perspectives in Biology, 8*(2), a016824.

Kleisner, K. 2024. Agency and appearance: Reading the face of life. In J. Švorcová (Ed.), *Organismal agency: Biological concepts and their philosophical foundations* (pp. 263–285). Springer.

Kleisner, K., and Saribay, S. A. 2019. The dual nature of mimicry: Organismal form and beholder's eye. *Biosemiotics, 12,* 79–98.

Kleisner, K., Tureček, P., Saribay, S. A., Pavlovič, O., Leongómez, J. D., Roberts, S. C., . . . and Varella, M. A. 2024. Distinctiveness and femininity, rather than symmetry and masculinity, affect facial attractiveness across the world. *Evolution and Human Behavior, 45*(1), 82–90.

Kull, K. 2014. Physical laws are not habits, while rules of life are. In T. Thellefsen and B. Sørensen (Eds.), *Charles Sanders Peirce in his own words: 100 years of semiotics, communication and cognition* (pp. 87–94). De Gruyter.

Kull, K. 2018. Choosing and learning: Semiosis means choice. *Sign Systems Studies, 46*(4), 452–466.

Kull, K. 2020. Semiotic fitting and the nativeness of community. *Biosemiotics, 13*(1), 9–19.

Kull, K. 2023a. Further considerations on semiosis in evolution: Arbitrarity plus semiotic fitting, and/or mutability plus natural selection. *Sign Systems Studies, 51*(1), 171–194.

Kull, K. 2023b. On the concept of meaning in biology. In P. A. Corning, S. A. Kauffman, D. Noble, J. A. Shapiro, R. I. Vane-Wright, and A. Pross (Eds.), *Evolution 'on purpose': Teleonomy in living systems* (pp. 161–174). MIT Press.

Kull, K. 2023c. Biosemiotic approaches. In R. Meylaerts and K. Marais (Eds.), *The Routledge handbook of translation theory and concepts* (pp. 77–93). Routledge.

Lacková, Ľ. 2018. A biosemiotic encyclopedia: An encyclopedic model for evolution. *Biosemiotics, 11,* 307–322.

Lacková, Ľ. 2024. *Language of life: A Peircean approach to living organisms.* Peter Lang.

Laland, K. N., Uller, T., Feldman, M. W., Sterelny, K., Müller, G. B., Moczek, A., Jablonka, E., and Odling-Smee, J. 2015. The extended evolutionary synthesis: Its structure, assumptions and predictions. *Proceedings of the Royal Society B: Biological Sciences, 282*(1813), 20151019.

Lamarck, J. B. 1815. *Histoire naturelle des animaux sans vertebres.* Déterville.

Lamarck, J. B. 2012 [1809]. *Zoological philosophy* (H. S. R. Elliott, Trans.). Forgotten Books.

Lass, A., Zimmermann, R., Oberer, M., and Zechner, R. 2011. Lipolysis—a highly regulated multi-enzyme complex mediates the catabolism of cellular fat stores. *Progress in Lipid Research, 50*(1), 14–27.

Markoš, A. 2002. *Readers of the book of life: Contextualizing developmental evolutionary biology.* Oxford University Press.

Martin, M. M., and Martin, J. S. 1978. Cellulose digestion in the midgut of the fungus-growing termite *Macrotermes natalensis*: The role ofacquired digestive enzymes. *Science, 199*(4336), 1453–1455.

Matthews, B., De Meester, L., Jones, C. G., Ibelings, B. W., Bouma, T. J., Nuutinen, V., et al. 2014. Under niche construction: An operational bridge between ecology, evolution, and ecosystem science. *Ecological Monographs, 84*(2), 245–263.

Mayr, E. 1982. *The growth of biological thought: Diversity, evolution, and inheritance.* Harvard University Press.

Michalik, A., Franco, D. C., Deng, J., Szklarzewicz, T., Stroiński, A., Kobiałka, M., and Łukasik, P. 2023. Variable organization of symbiont-containing tissue across planthoppers hosting different heritable endosymbionts. *Frontiers in Physiology, 14,* 1135346.

Morgan, H. D., Sutherland, H., Martin, D. I., and Whitelaw, E. 1999. Epigenetic inheritance at the agouti locus in the mouse. *Nature Genetics, 23*(3), 314–318.

Nielsen, J. N., Wojtaszewski, J. F., Haller, R. G, Hardie, D. G., Kemp, B. E., Richter, E. A., and Vissing, J. 2002. Role of 5'AMP-activated protein kinase in glycogen synthase activity and glucose utilisation: Insights from patients with McArdle's disease. *The Journal of Physiology, 541*(3), 979–989.

Odling-Smee, F. J., Laland, K. N., and Feldman, M. W. 2003. *Niche construction: The neglected process in evolution.* Princeton University Press.

Otis, L. 1994. *Organic memory: History and the body in the late nineteenth and early twentieth centuries.* University of Nebraska Press.

Pauly, A. 1905. *Darwinismus und Lamarckismus: Entwurf einer psychophysischen teleologie.* Ernst Reinhardt Verlag.

Peirce, C. S. 1877. The fixation of belief. *Popular Science Monthly, 12*, 1–15.

Peirce, C. S. 1892a. The doctrine of necessity examined. *The Monist, 2*(3), 321–37.

Peirce, C. S. 1892b. Man's glassy essence. *The Monist, 3*(1), 1–22.

Peirce, C. S. 1893. Evolutionary love. *The Monist 3*, 176–200.

Peirce, C. S. 1994. A guess at the riddle. In Peirce, C. S. *The collected papers of Charles Sanders Peirce* (electronic ed.). Vols. I–VI ed. by Ch. Hartshorne and P. Weiss, Vols. VII–VIII ed. by A. W. Burks. Harvard University Press.

Pembrey, M. E., Bygren, L. O., Kaati, G., Edvinsson, S., Northstone, K., Sjöström, M., Golding, J., and the ALSPAC Study Team. 2006. Sex-specific, male-line transgenerational responses in humans. *European Journal of Human Genetics, 14*(2), 159–166.

Pigliucci, M., and Müller, G. B. (Eds.). 2010. *Evolution: The extended synthesis.* MIT Press.

Popov, I. 2008. Orthogenesis versus Darwinism: The Russian case. *Revue d'histoire des sciences, 61*(2), 367–397.

Popov, I. 2018. *Orthogenesis versus Darwinism.* Springer.

Prum, R. O. 2017. *The evolution of beauty: How Darwin's forgotten theory of mate choice shapes the animal world—and us.* Doubleday.

Rassoulzadegan, M., Grandjean, V., Gounon, P., Vincent, S., Gillot, I., and Cuzin, F. 2006. RNA-mediated non-mendelian inheritance of an epigenetic change in the mouse. *Nature, 441*(7092), 469–474.

Rechavi, O., Houri-Ze'evi, L., Anava, S., Goh, W. S. S., Kerk, S. Y., Hannon, G. J., and Hobert, O. 2014. Starvation-induced transgenerational inheritance of small RNAs in *C. elegans. Cell, 158*(2), 277–287.

Robinson, G. E., and Barron, A. B. 2017. Epigenetics and the evolution of instincts. *Science, 356*(6333), 26–27.

Schacter, D. L. 2001. *Forgotten ideas, neglected pioneers: Richard Semon and the story of memory.* Psychology Press.

Searls, D. 2002. The language of genes. *Nature, 420*(6912), 211–217.

Sharov, A. A., and Tønnessen, M. 2021. *Semiotic agency: Science beyond mechanism.* Springer.

Švorcová, J. 2012. The phylotypic stage as a boundary of modular memory: Non-mechanistic perspective. *Theory in Biosciences, 131*(1), 31–42.

Švorcová, J. 2024. Plastic ontogenesis: Memory, closure, and habitual teleology in development. In J. Švorcová (Ed.), *Organismal agency: Biological concepts and their philosophical foundations* (pp. 143–164). Springer.

Švorcová, J., Lacková, Ľ., and Fulínová, E. 2023. Evolution by habit: Peirce, Lamarck, and teleology in biology. *Theory in Biosciences, 142*(4), 411–422.

Tanaka, J., Giles, M., Kremen, S., and Simon, V. 1998. Mapping attractor fields in face space: The atypicality bias in face recognition. *Cognition, 68*(3), 199–220.

Vassoler, F. M., Johnson, N. L., and Byrnes, E. M. 2013. Female adolescent exposure to cannabinoids causes transgenerational effects on morphine sensitization in female offspring in the absence of in utero exposure. *Journal of Psychopharmacology, 27*(11), 1015–1022.

West-Eberhard, M. J. 2003. *Developmental plasticity and evolution.* Oxford University Press.

Yu, R., Wang, X., and Moazed, D. 2018. Epigenetic inheritance mediated by coupling of RNAi and histone H3K9 methylation. *Nature, 558*(7711), 615–619.

10 Protective Mimicry as Potential: Does Its Realization Involve Active or Passive Perception, or Both?

Richard I. Vane-Wright

10.1 Introduction: Innate Potentialities

No characteristic of a living thing arises either wholly from nature or wholly from nurture. We are always dealing with systems which have certain *innate potentialities*.
—Conrad H. Waddington, 1960, p. 28, emphasis added

Mimicry is a logical necessity arising from the signal-based interactions that . . . exist in nature, *regardless of whether or not there are any real examples*.
—Richard I. Vane-Wright, 2019, p. 68, emphasis added

Waddington (1961) devoted a whole chapter of *The Nature of Life* to "innate potentialities"—contrasting the "atomistic approach" with "continuum theories" that "attempt to account for phenomena in terms of sets of relations" (Waddington, 1961, p. 22). Recently, the same distinction has been drawn between substance and process ontology (Dupré and Nicholson, 2018), and "object theories" and "agent theories" (Walsh, 2018). Such process or agency theories reflect Whitehead's rejection of dualism (the "bifurcation of nature": Whitehead, 1920), and his conclusion that all of reality is subjective (Whitehead, 1978 [1929]). "For Whitehead, the world conceived as stable is not reality; the only real things are the interlocking and diverging processual subjects, which in their realization presuppose structures, but always transcend them as well" (Schlehaider, 2023, p. 185).

Neo-Darwinians usually present mimicry in a genetic (i.e., substance) ontological framework (e.g. Fisher, 1930; but compare Turner, 1983). Inspired by Haukioja's (1982) distinction between process of life and process of change theories, I have tried to place mimicry in a process framework (Vane-Wright, 2019). In the context of strikingly altered forms, functional mimicry and "sematics,"[1] the need for separation was well understood by Portmann (1952, pp. 122–123): "It has not always been taken into account that, at best, . . . the [theory of evolution] might afford a reason for their preservation [e.g. functional mimics], but never provide the cause for their origin." A process-based, potentialities approach was inferred by Maran (2017, p. 1): "There are many resemblances and relations in nature that have not yet formed distinct mimicry systems, but have potential to do so in the future."

Mimicry is an epiphenomenon, in a causal sense, of meaningful, preestablished semiotic relations between two organisms: model and operator. Typically, the prior establishment of these relations is based on experiential learning by the operator (Poulton, 1890; Boppré et al.,

2017—although this is not simple: e.g., Skelhorn and Rowe, 2006). Mimicry is a "by-product" of an *embodied* biosemiotic system; it is not brought about by some disembodied "invisible hand" or diffuse "natural selection." Instead, functional mimicry is sustained and can be improved by *teleonomic* selection (Corning, 1983). Moreover, mimics are not "deceivers"—mimicry is brought into being by the actions of an operator, the agential selector, realized by means of a Jakob von Uexküll "functional circle":

> In this, the subject (as "self") embraces the object [in this context, model or mimic] like a forceps with its two claws: on the one hand as receiver (receptor), which receives stimuli and transposes them into signs, thus structuring the "nonself" with perceptual cues (properties of an object); on the other hand as operator (effector), which changes or erases the perceptual cues or properties of the "nonself" with operational cues. We can reduce this process to its simplest form by describing it as "self-stimulation" that is modified by the alternation of the "nonself" between stimulating operator (effector) and stimulated receiver (receptor). (Uexküll 1982, p. 18)

If any "deception" is involved, it is self-deception of the operator. Even though we can describe the sematic (exophenotypic) form of a mimic as "deceptive," to both human observer and operator (Boppré et al., 2017, table 1), classic Batesian mimics do not take conscious actions to deceive, or seek to do so. Of course, this has long been understood, but value-laden, emotive terms such as "deception" (including Georges Pasteur's "dupe" for operator) easily distort our understanding of causation. Once this is appreciated, more interesting questions arise regarding the circumstances in which "operator-self-deception" (the mimicry driver) can occur, and the nature of the underlying perceptive/cognitive processes involved.

10.2 Systems

To understand defensive mimicry, it is necessary to appreciate that, in reality, it involves more organisms than those simply implied by the paradigmatic tripartite schema of model, mimic, and operator (or signal-receiver: Wickler, 1965, 1968) that has tended to dominate discussion. Alternative prey are essential (Holling, 1965)—*reductio ad absurdum*, if the model were totally unacceptable and protective mimicry perfect, the operator would starve if there were no alternative prey. In addition, all organisms within the dynamic will have their own antagonists and synergists. Mimicry can only occur within a multispecies community, with all component populations subject to many regulating factors and resource needs.

Most examples of mimicry are found in speciose habitats, where the ecologies of many species overlap, and there is much ecological redundancy. Vane-Wright (1981, figure 12.8) proposed a *minimum* tri-trophic schema of eight species. Mimicry is an innate biosemiotic potentiality of (eco)systems, not of genes, and ultimately needs to be understood in terms of systems theory.

10.3 Batesian, Protective Mimicry: A Tri-partite System That Depends on Bipartite "Honest" Signaling

Mimicry systems depend on an honest, signal-based connection between two organisms that confers mutual benefit.[2] If an organism is well protected against specific predators, so that if attacked, its offensive, noxious, or nauseous qualities ensure the predator rapidly desists and the potential prey organism is not always fatally wounded, this is generally expected to lead to the evolution of aposematism ("away-sign"—Poulton, 1890; Loeffler-Henry et al., 2023). The well-defended organism develops a conspicuous warning signal (often in a multimodal manner) that the predator can perceive at a distance, understand its meaning, and thereby choose not to attack—or discontinue an attack commenced but not consummated.

Several factors must be satisfied for aposematism to be a successful primary defense:[3] mutual benefit; a "tuned" conspicuous signal; the key ability of the predator (agent) to detect, perceive, and associate the meaning of that signal (e.g., this is an unprofitable organism to eat); and then act on that knowledge (be an "operator" in the sense of Vane-Wright 1976; "operation" in the original sense of Jakob von Uexküll—see below). The signal-generator benefits by reduced risk of injury or death, and the operator by a reduction of wasted effort pursuing unprofitable prey. This can only come about if the operator is able to learn, "memorize" (pace James Gibson), the significance of the signal and then *generally* act appropriately—in most encounters, desist or withdraw from attack. This, in turn, implies some form of cognition by the predator—operating within its umwelt, its specific, subjective "world" that includes all meaningful aspects relevant to it (Kull, 1998).

Once signaling has been established, the potential for it to be "parasitized" (dependent, Batesian mimicry), or collectivized (cooperative, Müllerian mimicry) by other organisms (mimics), is established. Thus, the supposed dependent similarity of drone flies to stinging bees, and the collective similarity of local species of yellow jacket wasps, is understandable. Where all three components—model, operator, and mimic—belong to different species, the emergence of the "imitation" signal will typically entail evolution (Ruxton et al., 2018). Once some initial resemblance affecting the proto-mimic has occurred within an ecosystem,[4] this will typically undergo subsequent improvement by selection (automimicry[5] is an exception). Such a two-step model was explicitly proposed by Nicholson (1927) and discussed by Sheppard (e.g., 1975) and Turner (e.g., 1984).

However, processes such as mutation, recombination, chromosomal rearrangements, and accumulation of "modifiers" are not the drivers of this process—they are facilitators, "part of a regulatory system and not its directors" (Noble and Noble, 2023). As noted above, the predator (operator) is the prime-mover, its cognitive processes resulting in *teleonomic* (goal-orientated) selection. As Peter Corning has redefined it, "Teleonomic Selection is a purposeful (cybernetic) process (i.e. an act of choosing) that always occurs in the 'minds' of living organisms; it is living beings that do the selecting, and it is a process that is intimately related to meeting the basic survival and reproductive needs of a given organism in a given context" (Corning, 2014, p. 252). But with protective mimicry, there is a "catch." While the ongoing teleonomic selection process is entirely appropriate with respect to the original bipartite signaling system, faced with a functional Batesian mimic, it becomes disadvantageous—recalling, in a very different context, a statement made by moral philosopher Philippa Foot:

"A certain kind of behaviour can be judged, not of course as things work out in the particular case, where proper behaviour may by chance be disastrous, but as belonging to the 'how things should be' of an individual of this kind" (Foot, 2004, p. 10).

But before looking at cognitive "mind" processes, particularly with reference to perception (where James Gibson offered a very different interpretation), I need to address another issue, concerning the "survival needs" served by mimicry.

10.4 Mimicry and Population Regulation: A Historical Deviation

Here, it is necessary to address an issue that affected the theory of mimicry for many years—from the period when Charles Elton, Alfred Lotka, Vito Volterra, and Alexander Nicholson and others laid the foundations of population dynamics (Egerton, 2014). It has often been suggested that the evolution of defensive mimicry within a population enables it to exist, or exist at greater density or in higher numbers than if it were not mimetic. Nicholson (1927) challenged this idea, as he wanted to explain why Batesian mimicry is a comparatively rare phenomenon—if mimicry is good for some, why not all? "To explain why true mimicry, i.e., deceptive resemblance, is a comparatively rare phenomenon, the theory is put forward that natural selection must have operated more vigorously on the few forms which exhibit it than on other insects, and that the former owe their preservation solely to their great powers of variation. There is absolutely no evidence to support this theory and facts of common observation tell strongly against it. Why, for example, are mimetic insects no more successful than related non-mimetic forms?" (Nicholson, 1927, p. 81).

Nicholson's criterion of "success" equated to population numbers (Vane-Wright, 1981). Ultimately, he argued, numbers must be limited by the quantity of food available—but animals rarely exhaust their food resources to the point of starvation. So the quantity of food must somehow determine numbers *indirectly*. Maintaining stability would require a density-sensitive factor. For insects, he regarded larval parasitoids and other early-stage predators as the only suitable agents (he dismissed food, weather, and diseases). Nicholson's feedback model was mediated by early-stage-predator foraging becoming more and more efficient as prey density increased and less efficient as it decreased. Nicholson (1927, p. 85) next asked, "It is evident from what has been said that if the numbers of an insect are caused to vary from the normal by any cause there is a definite mechanism which will tend to bring the numbers back to normal. Suppose, then, that the major enemies, on which the numbers of a particular insect depend, attack the larval stage, and that the minor enemy attacks the adult. Does it follow that this increased attack will cause a diminution in the numbers of the insect?"

Nicholson concluded that it will not, because increased attack on the adults would reduce the total number of eggs laid, but this would then reduce the search efficiency of the early-stage predators (of eggs, larvae, and pupae), so that more insects reached the adult stage, such that "the numbers would remain practically unaltered." He next applied this logic to mimicry:

> From what has been said it follows that protection from attack in the adult stage can be of little, if any, importance to a species of butterfly: for reduction in the normal slight attack on the adults would only result in a slight decrease in the numbers of the earlier stages and the numbers of adults which emerged, the actual numbers of the species remaining practically unaltered. Therefore if a perfect mimetic pattern appeared suddenly in a non-mimetic species, giving complete immunity from attack, it would not increase the success of the species, which would be just as successful without the mimetic pattern. (Nicholson, 1927, p. 88)

Although the adults would suffer less attack, "less adults would be produced, on account of the increased severity of attack on earlier stages" (Nicholson, 1927, p. 89). Despite this, mimicry would still evolve, with the conclusion that "mimetic resemblance, therefore, simply serves to fit the [individual] possessors more perfectly to their natural environment, without conferring upon them any material advantage" (Nicholson, 1927, p. 90).

Being better fitted to "their natural environment" surely fulfills any reasonable definition of an adaptation. But, according to Nicholson, Batesian mimicry confers no "material advantage" on populations or species that adopt it—which seems counterintuitive, even paradoxical. The clue resides in Nicholson's use of the term *material*. Having dismissed food, weather, and diseases as factors capable of stabilizing insect numbers directly, he proposed early-stage predators as the density-dependent regulators necessary for his model. However, this was an indirect effect because Nicholson insisted that food availability was the ultimate regulator. Thus, resources are *material*, not relational, locating his account in a substance-based ontology. Mimicry would only be an advantage to the population if it somehow increased the amount of food consumed—which, he was convinced, it did not (see also Haldane, 1953, p. 19; Nicholson and Bailey, 1935).

Nicholson's view, endorsed by Haldane, became influential among those studying mimicry in adult insects from a neo-Darwinian standpoint. Thus, Philip Sheppard, an expert on the genetics of polymorphic butterflies and industrial melanism, stated,

> Mimicry is certainly advantageous to the individual; in addition, it is often claimed that mimicry is advantageous to a species, allowing it to exist or to exist at higher densities than it would if it were non-mimetic. However, it is unlikely that this is often the case, especially in butterflies, since it would be necessary for the predators maintaining the mimicry also to be controlling population size (Haldane 1953). It is probable that the population is controlled by parasites, diseases and little, if at all, by vertebrate predation on the adults. (Sheppard, 1961, p. 21)

Thus, it was argued, even if protective mimicry was granted the status of an adaptation, it was unimportant with respect to the use of fundamental (material) resources. But about this time, ideas on population numbers and their regulation began to broaden in the direction of process ontology and systems biology. Richard Levins (1975, p. 49) applied "loop analysis" to community dynamics, concluding that "in a predator-prey system . . . selection for more effective predator avoidance, even if accompanied by lower fecundity or reduced resistance to physical conditions, will increase its population size." Protective mimicry came to be seen, once again, as an adaptation concerned with that most fundamental of all biological imperatives—"simply staying alive" (Eldredge, 2004, p. 34).

10.5 Apparent Competition and Enemy-Free Space

By the 1970s, different views about the ecological dynamics of adaptations such as mimicry were emerging (antecedents included Lotka, 1925, and Williamson, 1957), most notably "apparent competition" (Holt, 1977) and "enemy-free space" (Lawton, 1978; Price et al., 1980). Jeffries and Lawton (1984) gave a contemporary summary: "Lotka, Williamson and Holt show that the abundance of species may be reduced, or species eliminated entirely from a community by populations of polyphagous enemies sustained by alternative prey species. Hence there may be competition between victim species for 'enemy free space.' By enemy free space we mean *ways of living* that reduce or eliminate a species' vulnerability to one or more species of natural enemies . . . predators may have profound effects in determining which, and how many species co-exist in ecological communities" (Jeffries and Lawton, 1984, pp. 270, 278, emphasis added).

Fryer (1986) offered a wry commentary on these developments, noting that the impact of predators on prey was well understood since ancient times. However, Fryer failed to appreciate how new theoretical analyses were necessary to counter the dominant, material-resource-competition models of population regulation that had held sway for decades (e.g., MacArthur, 1958; Nicholson, 1927). Moreover, when coupled with Holling's (1965) insistence on the need to include alternative prey (a paper perhaps alluded to by Jeffries and Lawton, but not cited by them), the ideas of apparent competition and enemy-free space offered new and complex insights into the significance of aposematism and protective mimicry in relation to community ecology (e.g., Kikuchi et al., 2022).

"Enemy-free space" has come to be seen by some ecologists as a resource (e.g., Resetarits et al., 2019; Valentine et al., 2020). The proposal of Dennis et al. (2003) that, for insects, "resources" be subdivided into "consumables" and "utilities," is noteworthy: "The resources required by each stage may be divided further into consumables (e.g. hostplant parts, adult resources) and utilities. The latter describe the conditions for existence and persistence, such as physical sites for various activities (e.g. mate location and pupation sites), and suitable conditions for development and activities (i.e. suitable local climates and microclimates) and enemy-free space" (Dennis et al., 2003, p. 417).

Although this division is not exclusive, "consumables" corresponds quite well to materials, and "utilities" to processes (conditions—cf. the concept of "ecosystem services"). However, just as the significance of process needs was (re)emerging, another rebranded and extended concept entered the field—the notion of affordances (Gibson, 1977, 1979). Broadly defined as the reciprocal relationships between agents and their environments, the idea of affordances was, in part, an independent formulation of Jakob von Uexküll's umwelt—a subjective model of the world (Kull, 2001) involving perception ("*Merken*") and action (operation, "*Wirken*") combined in the schema of the "functional circle" (Baedke et al., 2021, figure 1; Fultot and Turvey, 2019; Uexküll, 1982; see also Ginn, 2014; Pickering, 2007; cf. the "intentional arc" of Merleau-Ponty, 1962; Thompson, 2007, pp. 59, 77, 249).

> Most of Uexküll's work was devoted to the problem of how living beings *subjectively* perceive their environment and how this perception determines their behavior . . . he introduced the term 'umwelt' to denote the subjective (subjectivized, meaningful) world of an organism (Kull, 2001, p. 7; emphasis added)
>
> . . . each organism inhabits its own distinct perceptual universe, or umwelt, and acts not according to a blind stimulus-response but actively interprets signs appropriate to it. In the acting, organisms bring their world into being. Umwelt, therefore, breaks down distinctions between the creature and its environment. (Ginn, 2014, p. 130)

Like Jakob von Uexküll (and Maurice Merleau-Ponty), James Gibson was insistent that organism and environment were an inseparable coupling: An organism "is a perceiver *of* the environment and a behaver *in* the environment" (Gibson, 1979, p. 4). Although integration of the ideas of Uexküll, Merleau-Ponty, Gibson, Piaget, and others would be desirable (Dotov et al., 2012; McGann et al., 2020; Sonesson, 2006), here I "simply" consider resources versus affordances in the context of mimicry, prior to evaluating "mentalist perception theories" (specifically, Richard Gregory's) versus Gibson's "information pickup theory," in trying to understand how functional mimicry actually works. (In this context, and anticipating my conclusions, comparisons between the worldviews of Uexküll and Gibson are, at best, superficial.)

10.6 Mimicry as a Resource *or* an Affordance?

10.6.1 Resources

Aposematism creates an innate potential for enemy-free space that can benefit a poorly defended organism with respect to a particular predator if they become functional mimics of a defended, aposematic organism that the predator avoids. This involves inappropriate responses to the mimic by the operator (potential predator). Based on Gibson (1979, p. 133), we could say that the mimic presents *mis*information, resulting in *mis*perception by the operator (predator) (but see §6.2). However this comes about, there is a quantitative limit to the benefit of protective mimicry—an ever-increasing number of mimics within a system devalues the protection available. At equilibrium, any further increase in the number of mimics, or their density, will expose the mimics to increased rates of predation, rather than diminished rates. In effect, misperception by the predator will switch from the mimics to the models—actuarially, encounters will become rewarding rather than repelling.

Thus, as long understood (Poulton, 1890; Sheppard, 1975, p. 189), the enemy-free space offered by protective mimicry is strongly frequency- or density-attenuated. If the protection offered by a model is fully utilized by the combined populations of one or more mimetic species, then it is very unlikely that another species will be recruited as a mimic. There is little or no advantage for new individuals or species "late to the party." Moreover, if the enemy-free space (actually a probability space rather than a physical refuge) is fully "occupied," the devalued bipartite warning signal will become a source of "danger," not only to the models (Poulton, 1890, p. 243) but also to the predators. This reflexive property is surely one of the principal reasons why Batesian mimicry is relatively rare.

Protective mimicry can be seen as a resource of the enemy-free space "utilities" sort. However, in comparison to an organism that hides under fallen leaves to escape detection (a physical space), mimicry "space" is virtual. Unlike perceiving a hiding place and then making use of it, functional mimicry does not result from the actions of the mimetic organism itself. Instead, it arises indirectly, from the "misperceptions" of its potential antagonists—mimics "hide in plain sight." This is a special advantage of most mimicry systems: Mimics are largely free to move, without loss of mimetic protection, and thus continue with many vital activities that depend on extensive movement (e.g., foraging). In contrast, effective crypsis often requires long periods of inactivity. For Batesian mimics, the reflexive dynamic means that protection is progressively reduced by population growth, even to the point of becoming a disadvantage. Given all these peculiarities, is mimicry perhaps better seen as an affordance rather than a virtual process-dependent resource?

10.6.2 Affordances

Gibson's "*central* concept" of affordances (Dotov et al., 2012) has received enormous attention since first proposed (Gibson, 1977). Much of this interest has been within psychology and sociology. As with all fertile ideas, its meanings and applications have been subject to revisions and nuances (McGann et al., 2020). Originally "*the theory of affordances*" related to Gibson's "ecological" (as opposed to "mentalist") ideas about animals, their visual perception of the environment, and what the environment offers ("affords") different organisms for their needs—and how it does so. With respect to affordances and mimicry, here I largely confine myself to Gibson's original formulations (Gibson, 1977, 1979).

Gibson (1977, 67, footnote) stated that this paper was a "preliminary version" of what became chapter 8 in Gibson (1979). Although both have much in common, there are differences. Gibson (1977) sets out with a manifesto: "A description of what the environment *affords* the animal can be given in terms of a list . . . [that] includes features of the terrain, shelters, water, fire, objects, tools, other animals, and human displays. . . . What is being attempted is an explanation of how the 'values' or 'meanings' of things in the environment could be directly perceived" (Gibson, 1977, p. 67; emphasis original).

This is followed by a definition: "*The affordances of anything is a specific combination of the properties of its substance and its surfaces taken with reference to an animal.*" At this point in the later work, there is, instead, a statement: "The *affordances* of the environment are what it *offers* the animal, what it *provides* or *furnishes*, either for good or ill. . . . It implies the complementarity of the animal and the environment" (Gibson, 1979, p. 119). Returning to the 1977 paper, Gibson then went on to explore what he meant by "a specific combination": "The combination of properties is uniquely related to the animal or species being considered. *It is assumed* that if the properties of substance and surface are given in light the combination is given, and hence that if the properties are perceivable the special set of properties will be perceivable" (Gibson, 1977, p. 67; emphasis added).

Gibson seems to me in difficulty here—he sets out to explore how values and meanings could be directly perceived but then "assumes" the properties of a thing can be revealed directly by light reflected from its surface. This circularity is confirmed in the last section of chapter 8, "Misinformation for Affordances": "If there is information in the ambient light for the affordances of things, can there also be misinformation? According to the theory being developed, if information is picked up perception results; if misinformation is picked up misperception results" (Gibson, 1979, p. 133).

So "information" is that which, when received ("picked up"), provides reliable evidence about the properties and thus the affordances that something offers, but if "ambient light" does not provide reliable indications about what a thing affords, it is said to be misinformation, which results in "misperception"—a "heads I win, tails you lose" sort of argument. Is the information "picked up" from an object or place, in ambient light at least, directly perceived with respect to what it affords—or not? Gibson's answer seems to be sometimes it is, sometimes it is not.

If we set these difficulties aside, at first reading, there does seem to be much in Gibson's rich texts that could be applied to aposematism and Batesian mimicry. Chemically well-defended butterflies have "warning patterns" that are very conspicuous, and observations and experiments have shown that potential (experienced) vertebrate predators perceive these at a distance and regularly avoid trying to eat them. Even though a "negative affordance," one could say that ambient light reflected from the wing surfaces is "information" that the potential predator perceives directly to appreciate that "this object affords poisoning" ("First, consider substances that afford ingestion. Some afford nutrition . . . some afford poisoning, and some are neutral"—Gibson, 1979, p. 129). Then, the wing pattern of a chemically undefended Batesian mimic presents misinformation that results in misperception by the predator. As I put it shortly before Gibson's treatise was published, "In any given situation, confronted with a particular model or mimic, or both, an operator has a biological role with respect to each. Thus the role of the operator when confronted with a distasteful model, is normally to avoid it (negative, protective role). . . . The Batesian mimic, however, represents a potentially acceptable item, the true biological role of the operator being to eat it (positive, aggressive role)" (Vane-Wright, 1976, p. 30).

So what is the role of the mimic? Is it to "misinform" the potential predator—to cheat, dupe, or deceive? Does the mimic produce a different type of information on its illuminated wing surfaces, "misinformation," unlike the "information" produced by the (honest) model organism? Gibson would appear to be open to such an argument: "When Koffka asserted that 'each thing says what it is,' he failed to mention that it may lie. More exactly, a thing may not look like what it is" (Gibson, 1979, p. 134).

This is perplexing—Gibson was dedicated to a "radical hypothesis" proposing that "the 'values' and 'meanings' of things in the environment can be directly perceived . . . values and meanings are external to the perceiver" (Gibson, 1979, p. 119). Unlike Uexküll, values and meanings are objective, *not* subjective. But the (food) value of a model or a mimic is only realized by an operator—which chooses according to its perceptions of model and mimic as "the same" or "not the same." To allow for the fact that "things are not always what they seem" (a wonderful descriptor for mimicry—Byatt, 1995, p. 130), Gibson smuggled the contextual value of untruth into his system. He almost acknowledges this: "If the affordances of a thing are perceived correctly, we say it looks like what it *is*. But we must, of course, *learn* to see what things really are. . . . And this can be very difficult" (Gibson, 1979, p. 134). Another difficulty is that "misinformation" presented by a mimic at low density switches to "information" when the numerical limit of the protective enemy-free space is exceeded (§6.1)—emphasizing the relational rather than intrinsic nature of protective mimicry and aposematic and pseud-aposematic signs. However, according to Gibson (1979, p. 130), affordances are "invariant."

Gibson's theory locates the values and meanings of things externally to the perceiver. His analysis is based on an object ontology, emphasizing materials and structures, including substances, detached objects, and attached objects—which offer affordances such as resting places or hiding places (Gibson, 1979). Moreover, perceptions of "a real world" are supposedly not virtual (Gibson, 1966, p. 317). However, according to my interpretation, the value of mimicry for mimics is relational but unintentional, a virtual resource that is created from subjective "misperceptions" that arise from the actions of the perceiver (operator) itself, in trying to interpret the meanings and instrumental values (for it) of diverse objects within its umwelt—specifically and subjectively, what is "good" to eat, and what is not.

My conclusion is that protective mimicry is a phenomenon that cannot be regarded as an affordance—however seductive such a view might seem at first. With respect to mimicry (at least), the theory is incoherent. Instead, mimicry appears to be a virtual, relational resource brought into being by processes that can be described in terms of Uexküll's functional circle, driven by the embodied perceptions, interpretations, and actions of the operator. But what form of perception is it—passive (pick up) or active (mental)?

10.7 What Is the Nature of Perception in the Context of Mimicry?

Mimetic analogies, it is scarcely necessary to observe, are resemblances in external appearance, shape, and colours between members of widely distinct families: an idea of what is meant may be formed by supposing a Pigeon to exist with the general figure and plumage of a Hawk
—Bates, 1862, p. 502

Given that Gibson's "theory of affordances" has been so widely approved and used,[6] how can I have the temerity to suggest that, with respect to mimicry at least, it is "incoherent"? At the start of his most widely cited book, Gibson states, "The fact is worth remembering because it is often neglected that the words *animal* and *environment* make an inseparable pair. Each term implies the other. No animal could exist without an environment surrounding it. Equally, although not so obvious, an environment implies an animal (or at least an organism) to be surrounded" (Gibson, 1979, p. 4).

Do I disagree with that? Certainly not—not least because it is consistent with umwelt—as a reading of the delightful *Foray into the Worlds of Animals and Humans* (Uexküll, 2010 [1934]) will rapidly attest. Such ideas of a reflexive, self-creating, self-maintaining holistic relationship go back to Kant (Mensch, 2013) and Whitehead (e.g., 1920, 1978 [1929]) and appear foundational to such concepts as autopoiesis (Maturana and Varela, 1980) and enactivism (Varela et al., 1991). Thus, "like two partners in a dance who bring forth each other's movements, organism and environment enact each other through their structural coupling" (Thompson, 2007, p. 204).

But a fundamental, ontological separation in Gibson's thinking becomes apparent only a few sentences later (loc. cit.): "The animal is thought of as a highly organized *part* [emphasis original] of the physical world but still a part and still *an object* [emphasis added]." Animals are "animate objects" (op. cit.: 38). So again, we see here Gibson's commitment to an object-based ontology. This is the basis of his rejection of subjectivism and "mentalism" in his understanding of process-based relationships between organisms and their environments, and of perception. My critique here, regardless of other uses of the theory of affordances, relates to the impossibility of its application to mimicry systems. If one insists on Gibson's ideas about the direct nature of perception then, faced with mimicry (an enactive "dance" of three rather than two when all things are not always as they seem), one is forced to accept absurdities, such as the notion that some objects "lie."

10.8 Mind and Life: Active, Agential Perception in the Context of Mimicry

It is clear that the mind and life are intimately interconnected and essential to our understanding of the living world.
—Sanjuán and Viejo, 2023, p. 1

The organisms that bring functional protective mimicry into existence are the operators (typically predators) that have to negotiate their living world actively (as in Uexküll's umwelt). One of the great proponents of the "Active" nature of perception was Richard Gregory (1923–2010). Gregory criticized Gibson's ideas of direct "pick up" as "Passive" perception (e.g., Gregory, 1974, pp. xix, 351; 1997, p. 1121)—representing a debate that continues to this day (e.g., Ben-Zeev 1981; de Wit et al., 2015; Costall, 2017; Hatfield, 2021). Perception as an active, generative, and inferential process, at least in part, is a very old idea (Silva and Yrjönsuuri, 2014). A totally polarized view is untenable (Hatfield, 2021)—and Gibson's own views did embrace the significance of activity (Costall, 2017; "The perceptual systems, as it turns out, correspond to the organs of active attention with which the organism is equipped"—Gibson, 1966, p. 58). Even so, he remained very critical of "mentalist" theories (Gibson, 1979, p. 310, index "mentalism"), which would surely have included the ideas of Gregory at issue here.

In drawing this discussion to a close, I regret I cannot enter further into these complex matters—which include fundamental questions about holism vs. dualism, subjectivism vs. objectivism, agency vs. objectcy. Instead, I present one of Gregory's ideas that seems particularly relevant to mimicry—that perceptions can be seen as "predictive hypotheses" (Gregory, 1980) and thus be understood as an activity that operators undertake when faced with aposematic *potential* prey.

Gregory wrote several times about his idea that perceptions can be compared to scientific hypotheses. In his most well-known account, he explicitly claimed "that perceptions are essentially like predictive hypotheses in science," stating that the "approach is based on regarding perception and science as *constructing hypotheses* by 'fiction-generators' which may hit upon truth by producing symbolic structures matching physical reality" (Gregory, 1980, p. 182, emphasis original). A good part of his paper involves introspections regarding whether or not his arguments are valid. But he went on to propose a three-step process for perception that proceeds from *signals* (input stimulus patterns), to *data* (read from the signals by following a conventional code), and thence to the formation of *perceptual hypothesis* (Gregory, 1980, pp. 183–184).

At this point in his argument, he had to digress concerning the nature of hypotheses, reflecting that there was no general agreement. In order to proceed, he felt obliged to propose his own definition: "I suggest that *hypotheses are selections of signalled and postulated data organized to be effective in typical (and some novel) situations.* Hypotheses are effective in having powers to predict future events, unsensed characteristics, and further hypotheses. They may also predict what is *not* true. I shall assume that we accept that these are important characteristics of scientific hypotheses and perception" (Gregory, 1980, p. 184).

Gregory then spends most of the rest of his paper exploring similarities and differences between "hypotheses of science" and "perceptions," coming to a rather muted conclusion: "There are marked similarities and important identities between hypotheses of science and perceptions. It is these that justify calling perceptions 'hypotheses.' The differences are, however, extremely interesting, and I fear that I have not done them justice" (Gregory, 1980, p. 197).

There has been recent support for Gregory's "mentalist" position.[7] Cicchini et al. (2024) review work on "serial dependence" (Fischer and Whitney, 2014) to suggest that perceptual processes can be predictive, and the idea that perception can be generative: "The brain forms models of the world and tests these models against sensory input, rather than starting from the input itself" (Cicchini et al., 2024, p. 130). Here, I will have to be content with matching my understanding of mimicry to Gregory's (1980) definition:

i. *hypotheses are selections of signaled and postulated data*—for aposematism, the signal is decoded according to a rule, based on past (learned) experience, that an organism that has such an appearance (e.g., color, sound, odor) is well defended, leading to the postulate that it is best avoided;

ii. the *data* are *organized to be effective in typical (and some novel) situations*—typically, the data derived from the aposematic signal will offer reliable information for controlling action ("control information" in the sense of Corning, 2001)—and this behavior can sometimes be relied upon in a novel situation (e.g., a bird migrating from one ecosystem to another);

iii. *hypotheses are effective in having powers to predict future events, unsensed characteristics, and further hypotheses*—the "avoid potential prey exhibiting this signal" hypothesis offers valuable predictions about future encounters with potential prey, including their otherwise unsensed repellency or toxicity were they to override the avoidance rule (as is likely the case when, for example, the predator is sufficiently hungry and alternative prey are not readily available);

iv. *they may also predict what is not true*—encountering a protective mimicry system is where "untruth" enters—the "avoid potential prey exhibiting this signal" hypothesis will apply equally to a Batesian mimic as it will to the well-defended "model" that the perceptual hypothesis is founded upon.[8]

Timo Maran gave an excellent, pithy summary when he wrote, "In mimicry it is the receiver who learns through correct and mistaken interpretations about the actual properties of mimics and models. Thus mimicry becomes a learning device and the receiver becomes someone who aims to acquire better knowledge about the surrounding environment" (Maran, 2017, p. 67). I also agree with Maran when he states that "the recognition of mimics and models by the signal receiver is based on probability and [is] not an absolute" (Maran, 2017, p. 34). However, this is perhaps something for the future.

10.9 Conclusion

1. Protective mimicry cannot be seen as an affordance in Gibson's sense—it is seen here as a (virtual) biotic relationship that can better be included in the "enemy-free space" category of resources.
2. The selective agents (operators) that bring protective mimicry into being are the predators themselves; mimics do not intentionally become mimics. Batesian mimicry is the realization of the potential enemy-free space that arises as a byproduct of teleonomic selection (*sensu* Corning, 2014) affected by the operators with respect to the aposematic organisms they have actively learned to avoid.
3. On Gibson's theory of direct perception, the "misperceptions" (or better, "mistaken interpretations") that occur when operators fail to distinguish mimics from their models require either active intent on the part of mimics (deception) or that aposematic signals can switch from presenting "information" to "misinformation" due to shifting population dynamics. While the appearances (signal properties) of mimics are indeed "deceptive" (often to the human observer as well as the operator), neither of these interpretations makes biological sense.
4. Some element of active perception, for the study of which Gregory's notion of perceptual hypotheses is a useful starting point, is seen as essential for functional protective mimicry to be maintained, and to evolve. Passive perception could also be involved, perhaps in the form of instinctive behaviors, but that is not established here.

Thus, I suggest mimicry is a potential (biosemiotic) resource that, to be realized, necessarily involves some element of interpretation or active perception by the operator ("perceptual hypothesis" formation, or some nuanced equivalent); it could also involve other, direct or passive, perceptual processes. However, from my perspective, the existence of functional mimicry underscores the inescapably active role of all organisms in negotiating their umwelt—a world where choice and meaning-making (always with the attendant risk of some mismaking) is truly fundamental to life.

Acknowledgments

I am most grateful to the editors, Kalevi Kull and Don Favareau, for the opportunity to contribute to this work, and the reviewers for valuable and interesting suggestions. I dedicate this paper to the late Wolfgang Wickler and to my close friend and colleague Michael Boppré. All sparked the innate potentialities.

Notes

1. Sematics is the occurrence of phenotypic features intended to be detected and interpreted—cf. "Sematic Colours" of Poulton (1890, p. 338).

2. Although "honest signalling" raises many issues, "at a fundamental level, the concept of aposematism relies on the idea that conspicuous signals accurately reflect an underlying defence on the part of the signaller" (Summers et al., 2015, p. 1585).

3. The account as presented is very simplified—in reality, there are many complications regarding the origin and function of aposematism as a primary defence—including the vexed issue of the evolutionary process(es) involved. Ruxton et al. (2018), Rojas et al. (2017), and Kikuchi et al. (2023) offer discussion and further references.

4. It is often suggested this requires a mutation, but it could equally or even more plausibly entail an exaptation—for example, an independent change in behavior or in a species' distribution.

5. "Automimicry" (Brower et al., 1967) occurs when all members of a population share the same warning signal, but individuals vary widely in their defensive qualities—from well-protected to unprotected—with the latter benefiting from the established effectiveness of the former. The classic example is the monarch butterfly, in which the chemical defense of each individual can vary greatly, dependent on the heart poisons present or absent in its particular larval host plant.

6. Taking a cue from Dennis et al. (2003), I do wonder in what way the notion of affordances really differs from a sufficiently elaborated taxonomy of resources. Consider dividing resources into *energy* (e.g., sunlight, heat), *substances* (e.g., food, nesting materials), *situations* (e.g., vantage points, hiding places), *biotic relations* (e.g., endosymbiosis, mimicry), and *applications* (e.g., tools—including novel uses that a "quantum mind" can seize upon by "identifying, preferring, choosing and acting to do so": Kauffman and Roli, 2023, pp. 536–537). All these are "afforded" by the extraordinarily heterogeneous, unfolding, evolving "environment" within which each organism actively negotiates its own "life of meaning" (Goodwin, 2007)—invariably involving biosemiosis. As problem-solvers, organisms play a key role in creating and changing the resources available to them. Many resources are not "invariant"—they come and go, shift, morph, evolve—some due to the collective agencies of organisms themselves (e.g., the formation of an oxygen rich atmosphere by Cyanobacteria about 2.1 to 2.4 billion years ago—the "Great Oxygenation Event"). However, if resources were to be regarded, at least in part, the same as affordances, the meaning of the former would have to be nuanced so as not to lose the key idea that currently separates "passive" ideas of an environmental resource as just "stuff," opposed to the reflexive understanding of an affordance as "a joint property of a purposive system and the conditions with which it interacts" (Walsh, 2015, p. 163).

7. A reviewer has pointed out a parallel approach, stemming from Peirce's notion of "abduction" (the process of forming hypotheses), which should also be explored.

8. This is possibly what Henry Bates originally had in mind—he not only proposed the theory of protective mimicry that bears his name but also introduced the term "model" for the protected organism. Cf. "The brain forms models of the world and tests these models against sensory input" (Cicchini et al., 2024, p. 130).

References

Baedke, J., Fábregas-Tejeda, A., and Prieto, G. I. 2021. Unknotting reciprocal causation between organism and environment. *Biology & Philosophy*, *36*(5), 48 [29 pp.].

Bates, H. W. 1862. Contributions to an insect fauna of the Amazon Valley. Lepidoptera: Heliconidae. *Transactions of the Linnean Society of London*, *23*(3), 495–566, 2 pls.

Ben-Zeev, A. 1981. J. J. Gibson and the ecological approach to perception. *Studies in History and Philosophy of Science Part A*, *12*(2), 107–139.

Boppré, M., Vane-Wright, R. I., and Wickler, W. 2017. A hypothesis to explain accuracy of wasp resemblances. *Ecology and Evolution*, *7*(1), 73–81.

Brower, L. P., Brower. J. V. Z., and Corvino, J. M. 1967. Plant poisons in a terrestrial food chain. *Proceedings of the National Academy of Sciences*, *57*(4), 893–898.

Byatt, A. S. 1995. *Angels and insects* (vintage ed.). Random House.

Cicchini, G. M., Mikellidou, K., and Burr, D. C. 2024. Serial dependence in perception. *Annual Review of Psychology*, *75*(1), 129–154.

Corning, P. A. 1983. *The synergism hypothesis.* McGraw-Hill.

Corning, P. A. 2001. "Control information": The missing element in Norbert Wiener's cybernetic paradigm? *Kybernetes*, *30*(9–10), 1272–1288.

Corning, P. A. 2014. Evolution 'on purpose': How behaviour has shaped the evolutionary process. *Biological Journal of the Linnean Society*, *112*(2), 242–260.

Costall, A. 2017. 1966 and all that: James Gibson and bottom-down theory. *Ecological Psychology*, *29*(3), 221–230.

Dennis, R. L., Shreeve, T. G., and Van Dyck, H. 2003. Towards a functional resource-based concept for habitat: A butterfly biology viewpoint. *Oikos*, *102*, 417–426.

de Wit, M.M., van der Kamp, J., and Withagen, R. 2015. Visual illusions and direct perception: Elaborating on Gibson's insights. *New Ideas in Psychology*, *36*, 1–9.

Dotov, D. G., Nie, L., and de Wit, M. M. 2012. Understanding affordances: History and contemporary development of Gibson's central concept. *Avant*, *3*(2), 28–39.

Dupré, J., and Nicholson, D. J. 2018. A manifesto for a processual philosophy of biology. In D. J. Nicholson and J. Dupré (Eds.), *Everything flows: Towards a processual philosophy of biology* (pp. 3–45). Oxford University Press.

Egerton, F. N. 2014. A 1961 letter from A. J. Nicholson. *Bulletin of the Ecological Society of America*, *95*(2), 154–161.

Eldredge, N. 2004. *Why we do it: Rethinking sex and the selfish gene.* W. W. Norton.

Fischer, J., and Whitney, D. 2014. Serial dependence in visual perception. *Nature Neuroscience*, *17*(5), 738–743.

Fisher, R. A. 1930. *The genetical theory of natural selection.* Clarendon Press.

Foot, P. 2004. Rationality and goodness. *Royal Institute of Philosophy Supplements*, *54*, 1–13.

Fryer, G. 1986. Enemy-free space: A new name for an ancient ecological concept. *Biological Journal of the Linnean Society*, *27*(3), 287–292.

Fultot, M., and Turvey, M. T. 2019. Von Uexküll's theory of meaning and Gibson's organism–environment reciprocity. *Ecological Psychology*, *31*(4), 289–315.

Gibson, J. J. 1966. *The senses considered as perceptual systems.* George Allen & Unwin.

Gibson, J. J. 1977. The theory of affordances. In R. Shaw and J. Bransford (Eds.), *Perceiving, acting and knowing: Toward an ecological psychology* (pp. 67–82). Lawrence Erlbaum.

Gibson, J. J. 1979. *The ecological approach to visual perception.* Houghton Mifflin (consulted as Classic Edition 2015, Psychology Press).

Ginn, F. 2014. Jakob von Uexküll beyond bubbles: On umwelt and biophilosophy. *Science as Culture*, *23*, 129–134.

Goodwin, B. 2007. *Nature's due: Healing our fragmented culture.* Floris.

Gregory, R. L. 1974. *Concepts and mechanisms of perception.* Charles Scribner's Sons.

Gregory, R. L. 1980. Perceptions as hypotheses. *Philosophical Transactions of the Royal Society of London B*, *290*(1038), 181–197.

Gregory, R. L. 1997. Knowledge in perception and illusion. *Philosophical Transactions of the Royal Society of London B*, *352*(1358), 1121–1127.

Haldane, J. B. S. 1953. Animal populations and their regulation. *New Biology*, *15*, 9–24.

Hatfield, G. 2021. Gibson and Gestalt: (Re)presentation, processing, and construction. *Synthese*, 198(Suppl 9), 2213–2241.

Haukioja, E. 1982. Are individuals really subordinated to genes? A theory of living entities. *Journal of Theoretical Biology*, *99*, 357–375.

Holling, C. S. 1965. The functional response of predators to prey density, and its role in mimicry and population regulation. *Memoirs of the Entomological Society of Canada*, *97*(S45), 5–60.

Holt, R. D. 1977. Predation, apparent competition, and the structure of prey communities. *Theoretical Population Biology*, *12*, 197–229.

Jeffries, M. J., and Lawton, J. H. 1984. Enemy free space and the structure of ecological communities. *Biological Journal of the Linnean Society*, *23*(4), 269–286.

Kauffman, S. A., and Roli, A. 2023. What is consciousness? Artificial intelligence, real intelligence, quantum mind and qualia. *Biological Journal of the Linnean Society*, *139*(4), 530–538.

Kikuchi, D. W., Allen, W. L., Arbuckle, K., Aubier, T. G., Briolat, E. S., Burdfield-Steel, E. R., Cheney, K. L., and 39 others. 2023. The evolution and ecology of multiple antipredator defences. *Journal of Evolutionary Biology*, *36*, 975–991.

Kikuchi, D. W., Barfield, M., Herberstein, M. E., Mappes, J., and Holt, R. D. 2022. The effect of predator population dynamics on Batesian mimicry complexes. *The American Naturalist*, *199*(3), 406–419.

Kull, K. 1998. On semiosis, umwelt, and semiophere. *Semiotica*, *120*(3–4), 299–310.

Kull, K. 2001. Umwelt. In P. Cobley (Ed.), *The Routledge companion to semiotics and linguistics* (p. 281). Routledge.

Lawton, J. 1978. Host-plant influences on insect diversity: The effects of space and time. *Symposia of the Royal Entomological Society of London*, *9*, 105–125.

Levins, R. 1975. Evolution in communities near equilibrium. In M. L. Cody and J. M. Diamond (Eds.), *Ecology and evolution of communities* (pp. 16–50). Harvard University Press.

Loeffler-Henry, K., Kang, C., and Sherratt, T. N. 2023. Evolutionary transitions from camouflage to aposematism: Hidden signals play a pivotal role. *Science*, *379*(6637), 1136–1140.

Lotka, A. J. 1925. *Elements of physical biology.* Williams & Wilkins.

MacArthur, R. H. 1958. Population ecology of some warblers of northeastern coniferous forests. *Ecology*, *39*(4), 599–619.

Maran, T. 2017. *Mimicry and meaning: Structure and semiotics of biological mimicry*. Springer International.

Maturana, H. R., and Varela, F. J. 1980. *Autopoiesis and cognition. The realization of the living.* Reidel.

McGann, M., Di Paolo, E. A., Heras-Escribano, M., and Chemero, A. 2020. Editorial: Enaction and ecological psychology: Convergences and complementarities. *Frontiers of Psychology*, *11*(617898), 1–5.

Mensch, J. 2013. *Kant's organicism: Epigenesis and the development of critical philosophy*. University of Chicago Press.

Merleau-Ponty, M. 1962 [1945]. *Phenomenology of perception*. Routledge & Kegan Paul.

Nicholson, A. J. 1927. A new theory of mimicry in insects. *Australian Zoologist*, *5*, 10–104.

Nicholson, A. J., and Bailey, V. A. 1935. The balance of animal populations: Part 1. *Proceedings of the Zoological Society of London*, *1935*(3), 551–598.

Noble, R., and Noble, D. 2023. *Understanding living systems.* Cambridge University Press.

Pickering, J. 2007. Affordances are signs. *tripleC*, *5*(2), 64–74.

Portmann, A. 1952. *Animal forms and patterns: A study of the appearance of animals.* Faber & Faber.

Poulton, E. B. 1890. *The colours of animals: Their meaning and use* (2nd ed.). Kegan Paul, Trench, Trübner & Co.

Price, P. W., Bouton, C. E., Gross, P., McPheron, B. A, Thompson, J. N., and Weis, A. E. 1980. Interactions among three trophic levels: Influence of plants on interactions between herbivores. *Annual Review of Ecology and Systematics*, *11*, 41–65.

Resetarits, W. J., Jr., Pintar, M. R., Bohenek, J. R, and Breech, T. M. 2019. Patch size as a niche dimension: Aquatic insects behaviorally partition enemy-free space across gradients of patch size. *The American Naturalist*, *194*(6), 776–793.

Rojas, B., Nokelainen, O., and Valkonen, J. K. 2017. Aposematism. In T. K. Shackelford and V. A. Weekes-Shackelford (Eds.), *Encyclopedia of evolutionary psychological science* (5 pp.). Springer.

Ruxton, G. D., Allen, W. L., Sherratt, T. N., and Speed, M. P. 2018. *Avoiding attack: The evolutionary ecology of crypsis, aposematism, and mimicry* (2nd ed.). Oxford University Press.

Sanjuán, M., and Viejo, J. M. 2023. Life and mind: An introduction. In J. M. Viejo and M. Sanjuán (Eds.), *Life and mind: New directions in the philosophy of biology and cognitive sciences* (pp. 1–9). Springer.

Schlehaider, I. 2023. Apart from the experiences of subjects there is nothing, nothing, nothing, bare nothingness: Nature and subjectivity in Alfred North Whitehead. *Histories, 3*, 176–188.

Sheppard, P. M. 1961. Recent genetical work on polymorphic mimetic papilios. *Symposia of the Royal Entomological Society of London, 1*, 20–29.

Sheppard, P. M. 1975. *Natural selection and heredity* (4th ed.). Hutchinson & Co.

Silva, J. F., and Yrjönsuuri, M. (Eds.). 2014. *Active perception in the history of philosophy: From Plato to modern philosophy*. Springer.

Skelhorn, J., and Rowe, C. 2006. Prey palatability influences predator learning and memory. *Animal Behaviour, 71*(5), 1111–1118.

Summers, K., Speed, M. P., Blount, J. D., and Stuckert, A. M. M. 2015. Are aposematic signals honest? A review. *Journal of Evolutionary Biology, 28*(9), 1583–1599.

Sonesson, G. 2006. The meaning of meaning in biology and cognitive science: A semiotic reconstruction. *Sign Systems Studies, 34*(1), 135–213.

Thompson, E. 2007. *Mind in life: Biology, phenomenology, and the sciences of mind.* Harvard University Press.

Turner, J. R. G. 1983. "The hypothesis that explains mimetic resemblance explains evolution": The gradualist–saltationist schism. In M. Grene (Ed.), *Dimensions of Darwinism*, (pp. 129–169). Cambridge University Press.

Turner, J. R. G. 1984. Mimicry: The palatability spectrum and its consequences. *Symposia of the Royal Entomological Society of London, 11*, 141–161.

Uexküll, J. von 2010 [1934]. *A foray into the worlds of animals and humans. With a theory of meaning* (J. D. O'Neil, Trans; introduction by D. Sagan; afterword by G. Winthrop-Young). University of Minnesota Press.

Uexküll, T. von 1982. Introduction: Meaning and science in Jakob von Uexküll's concept of biology. *Semiotica, 42*(1), 1–24.

Valentine, L. E., Ramalho, C. E., Mata, L., Craig, M. D., Kennedy, P. L., and Hobbs, R. J. 2020. Novel resources: Opportunities for and risks to species conservation. *Frontiers in Ecology and the Environment, 18*(10), 558–566.

Vane-Wright, R. I. 1976. A unified classification of mimetic resemblances. *Biological Journal of the Linnean Society, 8*(1), 25–56.

Vane-Wright, R. I. 1981. Mimicry and its unknown ecological consequences. In P. L. Forey (Ed.), *The evolving biosphere* (pp. 157–168, 1 pl). British Museum (Natural History) & Cambridge University Press.

Vane-Wright, R. I. 2019. Agency, meaning and mimicry: A perspective from the Third Way. *Biosemiotics, 12*(1), 57–77.

Varela, F. J., Thompson, E., and Rosch, E. 1991. *The embodied mind: Cognitive science and human experience.* MIT Press.

Waddington, C. H. 1960. *The ethical animal.* Allen & Unwin.

Waddington, C. H. 1961. *The nature of life*. Allen & Unwin.

Walsh, D. M. 2015. *Organisms, agency, and evolution.* Cambridge University Press.

Walsh, D. M. 2018. Objectcy and agency. Towards a methodological vitalism. In D. J. Nicholson and J. Dupré (Eds.), *Everything flows. Towards a processual philosophy of biology* (pp. 167–185). Oxford University Press.

Whitehead, A. N. 1920. *The concept of nature*. Cambridge University Press.

Whitehead, A. N. 1978 [1929]. *Process and reality: An essay in cosmology*. The Free Press.

Wickler, W. 1965. Mimicry and the evolution of animal communication. *Nature, 208*, 519–521.

Wickler, W. 1968. *Mimicry in plants and animals* (R. D. Martin, Trans.). Weidenfeld & Nicolson.

Williamson, M. 1957. An elementary theory of interspecific competition. *Nature, 180*, 422–425.

11 Toward Expanded Reproduction: An Intraorganismal and Interorganismal Relational Account

Arantza Etxeberria and David Cortés-García

Reproduction is widely regarded as a key concept in biology and considered fundamental to life as we know it. However, it is often framed narrowly in terms of heredity (primarily genetic transmission) and epistemically confined to measuring differential fitness. This reductive perspective fails to account for the material and relational causal processes involved in organismal continuity and renewal. In a broad sense, reproduction is the process through which living entities produce new individuals of their kind. However, the phenomenon of biological renewal is expressed through various terms rooted in different theoretical traditions. These include generation, reproduction, replication, regeneration, and reconstitution, each offering different, though often overlapping, views of the process. The term "reproduction" emerged in biology in the mid-eighteenth century, replacing "generation" and shifting the view of life from isolated creation of entities to lineage-based continuity (Etxeberria, 2023; Jacob, 1970). Whereas reproduction encompasses organismal-level processes of generating offspring, later on, with the development of molecular biology, it was often overtaken by "replication," which focuses on the copying of genetic material (Wilkins and Bourrat, 2022). In contrast, recent philosophy of biology argues, from a developmental point of view, that reproduction includes the entire process throughout an organism's life cycle, not just gamete or zygote formation (Griesemer, 2016). This broader perspective extends to concepts like "regeneration," which used to refer to the restoration of lost or damaged organs and limbs, but now includes ecosystem recovery (Maienschein and MacCord, 2022). Lastly, "reconstitution" refers to phenomena of system recreation beyond lineage or genealogical continuity in holobionts and microbial symbioses; it emphasizes the dynamic interplay of relationships and processes that restore phenotypic variants through time (Veigl et al., 2022).

This plural conceptualization reflects two distinct and widely recognizable approaches developed during the twentieth century. One of them, rooted in embryology, views reproduction as a process-oriented physiological capacity expressed through organic development, and based on materiality. This approach aligns with ideas of regeneration and developmental dynamics. In contrast, the molecular genetic view regards reproduction as the transmission of essential elements, such as genes or genetic programs, that ensure continuity across generations. This perspective focuses on inheritance and replication, highlighting the informational aspects of life.

Historically, molecular biology framed reproduction primarily as replication, exemplified by Dawkins' replicators (Dawkins, 1976), focusing on the reliable transmission of information. Hull's philosophical distinction between replicators and interactors aimed to introduce developmental and environmental phenomena (Hull, 1980), while Griesemer's "reproducer," influenced by Developmental Systems Theory, claimed that any account of reproduction must incorporate material overlap, developmental causality, and environmental contributions, not just molecular copying (Griesemer, 2000, 2014). Subsequent philosophical developments elaborated on these issues: Godfrey-Smith's work on simple, collective, and scaffolded reproduction (2009) and O'Malley's focus on microbial and mutualistic reproduction (2016) sought to accommodate new biological findings, adjusting the replication paradigm accordingly. Research on collaborative and multispecies reproduction (Chiu and Gilbert, 2015) and on pregnancy as a complex biological relationship (Nuño de la Rosa et al., 2021) highlights the interdependent nature of reproductive processes as well as cooperative agency (Etxeberria, 2023; Nuño de la Rosa, 2023).

Bridging the gap between the embryological and the molecular traditions and the corresponding philosophical views of each of them requires the development of an expanded account of reproduction as argued by philosopher of biology Maureen O'Malley (2016). Our approach to *expanded reproduction* brings together and builds on organizational, collective, and distributed perspectives of how living systems actively construct and maintain reproductive processes beyond heredity. This reconceptualization of reproduction builds on separate words earlier theoretical and philosophical insights on the topic and on current biological knowledge, linking historical organismal perspectives with contemporary advances in extended evolutionary theory.

This chapter is divided into two sections. The first one examines the legacy of organicism and its morphogenetic and developmental conceptualizations of reproduction and contrasts these with informational views based on molecular transmission. We claim that organismal approaches, oriented toward organization, development, and interdependence, provided an alternative to dominant genetic and molecular frameworks since the twentieth century. This perspective challenges reductionist views and reshapes our understanding of reproductive processes. The second section elaborates on the concept of *expanded reproduction*, framing it as an organizational, relational, and collective process that operates in a distributed manner, encompassing both conflict and collaboration. This may reveal reproduction as a complex phenomenon that extends beyond individual organisms to capture its multifarious developmental, social, and ecological nature and its significance across organizationally complex, relationally rich, and distributed biological scales. Lastly, we introduce the concept of the *reproductive umwelt*, which helps to enhance the interorganismal perspective of how perception, interaction, and ecological embedding contribute to reproductive dynamics.

11.1 The Legacy of Organicism Concerning Reproduction

Organismal perspectives provide alternatives to gene-centered transmission accounts of reproduction by emphasizing biological organization, development, and interdependence. They highlight how reproduction recreates organization, integrating physiological, developmental, and ecological processes. Although organicism was a strong research program in the early twentieth century, it was marginalized with the rise of molecular biology. Since the early 2000s, however, it has regained renewed interest (Gilbert and Sarkar, 2000; Etxeberria and Umerez, 2006; Nicholson and Gawne, 2015; Triviño and Cerezo, 2024; Wolfe, 2024), and many current debates and challenges echo those of the early twentieth century.

The legacy of organismal approaches in shaping our understanding of reproduction is worth examining to assess its relevance for an expanded view of reproduction as a complex, multi-agent biological process, where agents are entities, from cells to organisms, that actively contribute to biological organization, relate to their surroundings, and create meaning. While organicism has traditionally focused on the organization and self-maintenance of living systems rather than on their reproductive capacities, this approach challenges the genetic view that tends to reduce reproduction to heredity, by stressing the developmental and interactive processes through which new entities arise. By revisiting historical and theoretical developments, from early embryology to contemporary organizational biology, we show that reproduction involves not only the transmission of genetic material but also the maintenance and reconstruction of biological organization across generations.

11.1.1 The Perspective of Embryology

In the late nineteenth century and early twentieth century, embryology embraced organismal philosophies as a middle ground between vitalism and mechanism, bridging materialism with property emergentism (Gilbert and Sarkar, 2000). A key principle of these perspectives was that parts gain meaning through their relation to the whole, a concept rooted in Kant's *Critique of Judgment* and which proved valuable for understanding embryological development. This perspective was adopted by embryologists and endured as embryology shifted from descriptive to experimental methods. Embryologists theorized reproduction as the reestablishment of complex organizational patterns, an account that contrasted sharply with that of transmission and copying underlying molecular genetics. They studied how new organisms emerge through the recreation of physiological and developmental organization, a process akin to how organisms regenerate lost parts. Key organismal concepts explored the formation of ordered systems and coordinated development, particularly in reproductive processes such as fertilization, embryonic induction, cell differentiation, gastrulation, and tissue interactions during morphogenesis or regeneration. This line of research emphasized how complex biological organization is constructed and maintained, rather than simply inherited.

The description of the *organizer* by Hilde Mangold and Hans Spemann in 1924 (Brandt, 2022; Bubnoff, 2024), alongside Ethel Browne's earlier work on induction in *Hydra* in 1909, was crucial in shaping developmental biology (Holstein, 2024). These studies revealed that specific embryonic regions can direct the fate of surrounding cells, showing that reproduction involves not just egg and sperm fusion but also complex developmental processes that actively shape the whole organism. Embryologists like Ross G. Harrison and Charles Manning Child further challenged gene-centric views, arguing that reproduction transmits and recreates complex organizational patterns, beyond genetic inheritance. Their experimental work with tissue cultures revealed the self-organizing capacities of living systems during development. Child's studies on *physiological gradients* and *metabolic fields* showed how reproduction and regeneration are regulated by metabolic activity and thereby ensure organizational continuity.

The dispute between Child and Thomas Hunt Morgan over regeneration and heredity reflected deeper tensions about reproduction. Whereas Morgan emphasized transmission genetics as governed by statistical laws, Child argued that reproduction depends on the establishment of new metabolic gradients (Blackstone, 2006; Gilbert et al., 1996). Against Weismannian and Mendelian views, Child concluded that the unit of heredity was not the gene but the developing organism as a whole (Esposito, 2017).

In sum, the organismal perspective adopted by early-twentieth-century embryologists reflects their focus on organization and physiological gradients. The works of Child and Harrison suggest that reproduction and regeneration share a common foundation, both functioning as processes that recreate biological organization at different scales. This perspective challenges reductionist genetic models by emphasizing organisms as dynamic self-organizing systems. Revisiting these early embryological frameworks provides crucial insights into reproduction as a relational, developmental, interorganismal, and environmentally embedded process.

11.1.2 The Theoretical Biogatherings and the Serbelloni Symposia on Theoretical Biology

A very influential framework drawing from organismal ideas was developed by the members of the Theoretical Biology Group (TBG) (Esposito, 2017; Haraway, 1976; Peterson, 2016). They formed an interdisciplinary network of biologists, philosophers, and mathematicians including figures such as Joseph Henry Woodger, Joseph Needham, Dorothy Needham, Dorothy Wrinch, and Conrad Hal Waddington, who began meeting regularly to develop alternative approaches to the mechanistic and reductionist views dominating biology with the common aim to unify biological theory around the concept of organization (Etxeberria and Umerez, 2006). Explicitly influenced by Whitehead's views on process and organismal interconnectedness, their aim was to develop a theoretical biology to advance in the problems of biology in the image of theoretical physics (Bertalanffy, 1933). While acknowledging the importance of experimental work and physical-chemical explanations, they argued that biological organization required its own theoretical framework. However, this effort of integrating biological knowledge and making biology an autonomous science was soon absorbed into the project of the Modern Synthesis, which centered biological theory on evolution, leading to the abandonment of organization as a central concept (Smocovitis, 1996).

Also, the field of molecular biology started to expand at the end of the 1930s through a convergence between genetics and biochemistry (Haraway, 1976; Kay, 2000; Morange, 2023). Following Pnina Abir-Am (1987), the origins of molecular biology may be traced to the research program of the Theoretical Biology Group. Their program, titled "mathematico-physico-chemical morphology," laid the foundation for what Warren Weaver would later term "molecular biology" (Weaver, 1970). It represented a fundamental shift in biological understanding, as it sought to explain life processes at the molecular level, which Lily Kay described as a "molecularization" in the understanding of living processes (Kay, 2000). Despite the fact that the origin of the molecular biology program can be traced back to organicist studies, both research programs developed in parallel, and the development of molecular biology contributed to shift the scientific focus away from questions of biological organization and development toward an intense exploration of genetics and hereditary mechanisms at the molecular level.

By the mid-twentieth century, as molecular biology gained dominance, some biologists and philosophers sought to explore broader theoretical conceptual foundations for biology. In this context, Waddington organized a series of four symposia held between 1966 and 1969 at Villa Serbelloni (Bellagio, Lake Como). Biologists, philosophers, physicists, and theoreticians were brought together with the shared goal of sketching what a theoretical biology might look like. These meetings represent the interplay and survival of organicist ideas on reproduction and biological organization, even as the views of molecular biology were becoming increasingly prevalent. The discussions sought to explore how different explanatory frameworks—ranging from genetics to systems theory—could be integrated into a broader understanding of biological organization, including reproduction. The fact that these debates persisted suggests that molecular biology had not entirely supplanted other approaches, and that questions about development, form, and systemic organization remained open theoretical challenges. This continuity reflects the influence of older, more holistic perspectives in biology.

A key tension at the symposia was how to reconcile holistic, process-based views of biological organization with the rapidly advancing framework of genetics and molecular biology, and to explore how these approaches could complement one another. While many contributors adopted the genetic perspective (Waddington 1968–1972), others, such as Brian Goodwin and René Thom, defended morphogenetic models that emphasized dynamical, holistic processes over the reduction of reproduction to gene transmission. This divide between morphogenetic and genetic approaches became a persistent source of debate, highlighting the broader challenge of integrating developmental and structural perspectives with molecular explanations of life. The implications of this split remain a subject of historical and philosophical inquiry, and the discussions surrounding it materialized in the publication of the corresponding four-volume series *Towards a Theoretical Biology* (1968–1972).

Three positions can be very broadly distinguished among the discussions in those volumes, which reflect distinct views on heredity and reproduction. One, based on developmental genetics, maintains a holistic and systemic view of how genes need to interact among themselves to counteract environmental threats. Another approach is more formal, focusing on the mechanistic and structural aspects of biology, often emphasizing the role of morphogenetic dynamics in shaping organisms. A third position highlights the importance of the complementarity between semiotic and material processes. We will now examine each of these approaches in turn.

First, the developmental genetics approach, importantly advanced by Waddington, reflects an effort to reconcile emerging molecular biology with broader biological principles in its account of reproduction and heredity (Peterson, 2016). While he adopted a genetics-based framework that emphasized heredity as transmission of information (Loison, 2022), his work also offered nuanced perspectives on the role of reproduction in biological understanding. Waddington regarded reproduction as crucial for comprehending *heredity*, conceived as information transmission, and *evolution*, understood as changes in hereditary information, which occur nonrandomly and respond to organismal principles. He explicitly challenged reductionist views that treated the "nature of living matter" as sufficient, arguing instead that biological organization requires incorporating reproduction, environmental embedding, heredity, and evolution. According to him, living systems are defined not only by their ability to synthesize specific structures from simpler molecules (metabolism) but, more importantly, by their capacity to reproduce themselves (Waddington, 1968a). In his view, reproduction serves as a means rather than an end, serving primarily as a mechanism for transmitting the specificity of proteins to ensure reliable functionality.

Waddington's use of information metaphors in his discussions about reproduction, heredity, and evolution reveals a particularly intriguing tension in his work. While he explicitly criticizes informational views of life, rejecting the idea that genes alone encode biological form, he nonetheless employed information metaphors to describe developmental and evolutionary processes (Waddington, 1968a). This duality places Waddington in an intermediate position: He adopted key elements of Weismann's barrier, recognizing the role of genes as carriers of hereditary information, but resisted the deterministic implications of such frameworks. This nuanced stance reflects Waddington as a thinker straddling two paradigms, engaging within the emerging frameworks of molecular biology and genetics, yet striving for a more comprehensive account of biological organization. His treatment of reproduction and heredity illustrates a deliberate effort to reconcile and integrate multiple conceptual approaches in biological thought.

A second major perspective discussed at the Serbelloni Symposia was biological structuralism, which aimed to uncover the universal principles underlying biological form and organization. Among its key proponents was Brian Goodwin, whose *relational structuralism* challenged the genetic-deterministic approach and sought to transform biology around the search for universal generative processes in development and reproduction, prioritizing structural constraints and general principles of organization over historical, accidental, or purely adaptive explanations (Goodwin, 1984).

Goodwin regarded the intergenerational continuity of cytoplasm between generations as a key factor in heredity, against the idea that the nucleus alone determines the characteristics of offspring. In his view, genes acted as modulators of developmental processes already constrained by preexisting "relational" rules immanent in the dynamics of developmental patterns, morphogenetic gradients, and field interactions, which together provided the framework within which reproduction unfolds. Thus, Goodwin's relational biology, centered on gradients and fields, critiqued Weismannist ideas of germ-soma separation and conceptualized reproduction as the propagation of structured, self-organizing systems that maintain their own stability and variation over time.

Another proponent of a structuralist approach to biological organization at the Symposia was René Thom. Best known for his work in catastrophe theory, Thom (1968) elaborated a distinctive perspective on biological organization and morphogenesis as governed by underlying mathematical structures, emphasizing topological constraints rather than genetic determinism. He challenged conventional views of biological individuality, arguing that organisms must be understood within broader relational configurations. According to him, "the essential unit entity in biology is not the isolated individual" (Thom, 1968, p. 32), but two fundamental relational aspects that are more decisive for biological organization. First, Thom pointed to lineages, which he describes as continuous configurations that link individuals across generations, framing reproduction as a structural continuity rather than a discrete genetic event. Second, he treated functional relations between individuals of different lineages, such as predation, mating, and other ecological interactions, as crucial to configure interlineage dynamics and the overall stability of biological organization. Thom's structuralist perspective suggests that reproduction involves the preservation and transformation of relational structures over time; his work later evolved into a more explicitly semiotic approach.

A third position at the Serbelloni Symposia emphasizes the complementarity between semiotic or informational and material or dynamic processes, particularly in relation to biological organization and reproduction. Developed by Howard H. Pattee, this approach considers that meaning and communication are intertwined with the physical and structural aspects of biological systems. Subsequent work in the field of biosemiotics has further sought to understand and make sense of how semiotic processes and meaning can be understood from a material biology perspective.

Pattee's insights on hierarchical systems and the formation of symbols draw upon concepts such as *constraint*, *record*, and *symbol* to explain the principle of *semantic closure*, the self-interpretation of symbols within the cell. His requirement of symbolic constraints to channel the system's material dynamics was grounded on how von Neumann required a minimum level or *threshold of complexity* to adapt to the environment and evolve, so that complexity did not degenerate throughout reproduction. Von Neumann had argued that to construct artificial automata that have a level of complexity comparable to that of life, structural or material conditions as well as informational conditions were required. Consequently, Pattee's complementarity principle requires that the physical and symbolic aspects of life are interdependent. This duality underpins Pattee's broader argument that living systems balance the dynamic flexibility of processes with the stability and control provided by symbols (see also Umerez, 2021).

Pattee's work relates well to Waddington's views that biological systems must be understood within their broader relational and environmental contexts. In his closing reflections on the first symposium, Waddington brings out biology's inherently multidisciplinary nature, stating that no biological unit, no matter how internally complex, can be fully understood without considering its relationship to the broader environment in which it exists (Waddington, 1968b, p. 218). Pattee develops this idea further, arguing that "'living' is unavoidably a *total* ecosystem property and not the property of an isolated collection of macromolecules" (Pattee, 1968, p. 219). Hence, biological organization is not merely an internal property of an organism but rather something that emerges through its interaction with the surroundings. This perspective offers valuable insights for understanding reproduction: Instead of focusing solely on what is transmitted from parent to offspring, we might better ask what ecological, social, and developmental conditions enable reproduction to succeed. In this sense, epigenesis, development, and biological reconstitution rely on interorganismal relationships that shape the viability of reproduction.

In sum, the three types of insights arising from these discussions highlight the tension between genetic and organismal (relational) accounts of reproduction within the organicist and organizational framework of biology.

11.1.3 The Autopoietic and Organizational Account

Later in the twentieth century, the theory of autopoiesis sought to advance further explanations of reproduction based on biological organization, beyond the molecular genetic conception that reproduction is the primary purpose of life (Maturana and Varela, 1987). Instead, it argues that reproduction arises as a consequence of autopoietic organization

(Cortés-García, 2026; Etxeberria, 2004). By prioritizing the autonomous organization of living systems, the autopoietic framework aligns with organismal views on development and physiology, although its links with the embryological and theoretical biology traditions remain limited, due to its roots in the cybernetic tradition.

In the autopoietic framework, reproduction is the division of a unit to form two units of the same type, not identical but sharing the same organization. This view departs from conventional models in two key ways. First, it subordinates reproduction to organization, treating it as a byproduct of already-formed systems rather than a mechanism that generates life. Second, it shifts focus from genetic transmission to organizational continuity. In this view, genes do not direct development, but act within a broader network of cellular relationships, without a privileged causal role. Cells, as self-producing systems, are not controlled by internal or external instructions but operate through autonomous organization.

In later advances in the organizational framework, Pattee's notion of constraint, as a mechanism for channeling dynamical processes in organized systems, was particularly influential. This approach aims at a synthesis of the concept of closure originating from autopoiesis theory with Pattee's notion of constraint. On this basis, it has advanced an explanation of biological organization as a closure of constraints: a self-sustaining network of interdependent processes that both generate and regulate the conditions for their own existence (Moreno and Mossio, 2015). Pattee's work thus provided a crucial theoretical foundation for understanding how material and informational dynamics coalesce to maintain the autonomy and functionality of living systems.

Closely related to organicist theories, the defense of extended inheritance has contributed to expanding reproduction beyond genetic transmission (Pontarotti, 2024). Extended inheritance includes epigenetic marks, socially transmitted behaviors, ecological niches, and symbiotic microorganisms, offering a richer understanding of how organisms sustain their continuity across generations through dynamic interactions, developmental interdependencies, and the regulation of material and symbolic flows.

Yet, in this framework, reproduction presents a paradox, as its role expands beyond individual organization. Reproductive processes primarily contribute to a collective domain of interactions rather than to the self-maintenance of a single system or organization. Defining this collective domain from a strictly individual perspective raises questions about how reproduction integrates into the broader dynamics of biological organization.

The novel organismal-relational framework of expanded reproduction conceives reproduction as an interactive and collaborative process rather than merely as genetic (or extended) transmission. Based on an interplay between material, informational, and environmental dimensions, this approach proposes a broader, expanded conception of reproduction in theoretical biology. The next section explores the concept of collective and expanded reproduction.

11.2 Expanded Reproduction

The expanded reproduction framework highlights the interorganismal, ecological, and semiotic dimensions of the reproductive process. From a relational perspective, reproduction emerges through complex interactions between organisms, shaped by environmental contexts, developmental dynamics, and ecological relationships, rather than by the execution of predetermined genetic instructions, a misleading notion critiqued by Noble (2016). The following section is structured around four key aspects for an expanded notion of reproduction: First, we explore how living organization and biological normativity shape reproduction (11.2.1). Second, we examine reproduction as both intraorganismally and interorganismally relational, revealing how it operates across levels of biological complexity (11.2.2). Third, we discuss the tension between conflict and cooperation in reproductive interactions, challenging competition-based models (11.2.3). Finally, we introduce the concept of the *Reproductive umwelt*, arguing that reproduction entails its own distinct sphere of meaning and perception (11.2.4).

11.2.1 Reproduction and Living Organization

Darwinian evolutionary traditions often see reproduction as the primary function and purpose of life, assuming that biological normativity—the set of principles that drive living processes—rests in the successful transmission of variation and the perpetuation of species (Etxeberria, 2023). As advanced in the previous section, this perspective is challenged by organizational approaches, which argue that the relational and systemic dimensions of biological organization precede heredity and evolution (Maturana and Varela, 1987). From this standpoint, normativity lies in the autonomous production and maintenance of an organized system, where interactions among parts construct and constitute the whole.

However, reproduction poses an important difficulty—namely, whether and how it contributes to individual organization. Organizational views, such as autopoiesis theory, address this by subordinating reproduction to self-maintenance, presenting it as a consequence rather than the ultimate goal of a living system's self-producing capacity (Bich and Skillings, 2024). Yet, this does not fully resolve the issue, since reproduction rarely occurs in isolation; it typically involves multiple systems, external structures, and interorganismal interactions. The challenge for organizational theories, then, is how to integrate reproduction into a framework that prioritizes autonomy, given its inherent reliance on relational and collective processes.

Reproduction is rarely an individual act; many examples in nature show that it rests on external scaffolding, collaboration, and physiological integration, as seen in pregnancy, cooperative breeding, and symbiotic relations. These dependencies call into question the notion of reproduction as self-contained within a single organism and instead suggest that biological normativity emerges through collective processes, thus creating tensions between individual reproductive agency (Nuño de la Rosa, 2023) and collective normative relations. While organizational approaches define biological normativity through part-whole relations at the organismal level, reproductive processes encourage us to extend this framework to incorporate interorganismal interactions (see 2.2). If reproduction requires cooperation, communication, or coordination, then normativity also emerges at a collective level, guiding behaviors and structures across populations or communities. Such an understanding redefines reproduction as a shared, structured process that shapes the continuity, stability, and evolution of biological systems. By recognizing it as a multi-agent process, we move beyond individual-centered frameworks.

An issue that has raised special interest within the organizational framework is whether reproductive traits and processes such as sex organs, gamete production, or embryo selection contribute to individual biological organization. It is widely considered that these traits fall outside the self-maintenance of the organism since they are not part of the closure typical of self-sustaining systems (Etxeberria, 2023; Nuño de la Rosa, 2023; Saborido et al., 2011). However, such an assumption overlooks how reproductive structures and functions shape organismal activities, development, and well-being. Indeed, the fact that reproduction is inherently interorganismal does not diminish the relevance of reproductive traits for organismal organization, even if their function is not immediately evident at an individual level. Some have argued that reproductive traits respond to second-order self-maintenance, attributing their function to ecological or lineage continuity (Saborido et al., 2011). However, this perspective remains incomplete. Reproductive normativity emerges at the intersection between intraorganismal and interorganismal dynamics, shaped by both individual self-maintenance and collective constraints and goals, even if those goals are not uniformly shared by all participants.

A theoretical and philosophical account of reproduction should address these overlapping dimensions of biological normativity, revealing why reproduction remains a conceptual challenge. Even as organisms adhere to norms of self-maintenance and autonomy, reproduction often requires that organisms engage in behaviors that extend beyond individual interests.

11.2.2 Reproduction as Intraorganismal and Interorganismal Relationality

Organismal views emphasize individual aspects of biological organization in reproductive processes and, consequently, often struggle with their inherently relational, multi-agential, and scaffolded character. Reproduction involves varied interactions among partners, generations, and even non-kin organisms like host–microbiome relations. We argue that recognizing reproduction as a relational and interorganismal process shifts attention from individual-level fitness and genetic transmission to broader networks of coordination, recognition, and shared biological norms, where reproduction appears as an emergent process embedded in organismal and ecological systems.

In a recent paper, we argue that reproductive characters are not merely shared features between organisms with common interests, as conflict-based models suggest (Cortés-García et al., 2024). Instead, many reproductive traits require both intraorganismal and interorganismal interactions. The former involve interactions among an organism's components for maintenance and reproduction, while the latter encompass relationships between organisms or their parts, as in fertilization, parent-offspring relations, and symbiotic codependences. This approach shifts reproductive biology from a focus on individual traits to a broader network of relational processes.

For these reasons, it is insufficient to understand reproduction merely as the production of copies and variations for selection. Reproductive capacity, as Evelyn Fox Keller argues, involves primitive operations that sustain metabolism, cell division, and other vital processes since the earliest cells (Keller, 1987), which are also foundational to the emergence of meaning-making in biological systems. These processes likely ensured cellular integrity and responded to environmental threats, gradually evolving to incorporate multiple agents and relational processes. However, reductionist views have hindered a deeper understanding of reproduction as a systemic process, by attributing all agency to the genome and the self-replicating ability of nucleic acids. Keller contends that this view overlooks the fact that genomes alone lack the capacities to build new organisms, as they operate only within a broader organizational and interactive context.

To study reproduction adequately, then, we must adopt a naturalistic approach that considers the physiological, material, and dynamic interdependencies involved among agents. Life is an inherently relational, and reproductive functions extend beyond the propagation of individual traits, transcending organismal boundaries. Pregnancy, for instance, exemplifies these interconnections, showcasing the complexity of life's continuity and evolution. Thus, we can understand reproduction as an organizational or systemic process, not just informational transmission (Pontarotti, 2024).

Consequently, reproduction cannot be reduced to notions of self-reproduction. Rather than multiplying an organism's kind or organization by its own means, it conveys a collaborative process of reconstruction, shaped by interactions with other organisms and external structures (Etxeberria, 2023). Evolutionary biology's fitness calculations often idealize reproduction as an individual value, but as frequently noted, organisms do not reproduce alone. Sexual reproduction is a particularly relevant case, for it requires interaction with sexual partners, making individual fitness dependent on the availability, fertility, and compatibility of others (Keller, 1987; Triviño and Nuño de la Rosa, 2016). Reproduction is thus a collective process, often scaffolded by both living and inanimate environmental structures and distributed among various agents.

This perspective echoes the insights of nineteenth-century biologist Karl Ernst von Baer, who framed the formation of a new organism not merely as the beginning of an individual existence but as part of a larger, ongoing organic process (Vienne, 2015). Von Baer's views underscore the relational nature of life, where organismal development and reproduction are inseparable from broader networks of biological interconnections.

Thus, such insights challenge the dominant gene-centered approaches that emerged in the 1970s during the "hardening" of the Modern Synthesis (Gould, 1983). This shift reinforced selection-driven and individualistic views of evolution, sidelining the role of organismal organization and relational dependencies (Cortés-García and Etxeberria, 2023). In contrast, an organismal approach attentive to biological organization and relationality shows that reproduction is not an isolated individual function but a collaborative process of reconstruction. Recognizing this relational and systemic dimension reshapes how we understand biological continuity, moving beyond the reductive view of reproduction as an independent act.

11.2.3 Conflict versus Collaboration in Reproductive Relations

While recent evolutionary biology increasingly acknowledges interorganismal relations in reproduction, these are often framed in terms of conflict and competition. The conflict-centered perspective sees evolutionary innovation emerging from competition, where organisms maximize fitness at others' expense. Under this view, reproductive relations, whether between parents and offspring or between sexes, are seen as antagonistic negotiations over limited resources. Pregnancy, for instance, is frequently described as a conflict-driven struggle between gestator and fetus.

In contrast, insights from organismal and developmental biology reveal reproduction as a collaborative process, rather than solely conflict driven. Examples include physiological integration during pregnancy, behavioral cooperation, and symbiotic relationships between species. Research on pregnancy demonstrates how gestator and fetus form a unified processual individual through physiological and immune integration that fosters successful birth (Nuño de la Rosa et al., 2021). These ideas about collaborative reproduction align with Pyotr Kropotkin's legacy, who was part of a tradition of naturalists who developed the idea that not only individuals but also life relations between individuals evolve. Similarly, Johannes Reinke's notion of *consortium* highlights the importance of biological associations and interdependencies in life processes. Building on these insights, the approach of expanded reproduction shifts the focus from individual fitness to interdependent biological processes.

While acknowledging that conflict exists in reproductive processes, this framework highlights how organisms often resolve emerging tensions by evolving novel mechanisms. For instance, Patricia Brennan's research on the evolution of bird genitalia (Brennan et al., 2017), though grounded in a conflict framework, reveals that adaptations in female genitalia not only counteract forced copulation but also enable greater control over fertilization, illustrating how reproductive traits evolve as active regulatory strategies rather than solely passive responses to conflict. Thus, recognizing both conflict and collaboration as evolutionary factors allows for a more nuanced understanding of reproduction, moving beyond simplistic competitive models toward an integrated framework of relational and adaptive reproductive dynamics.

11.2.4 Is There a *Reproductive Umwelt*?

While organism-environment interactions have been extensively studied, and the idea that agents are active participants in their own evolution has gained traction, reproduction has often been overlooked in these discussions. Reproductive agency extends beyond ecological and developmental roles to include reproductive goals, which shape behavior, communication, and interorganismal relationships. Recognizing this dimension demands an expanded understanding of *umwelt*, one that integrates reproduction as a fundamental layer of meaning in an organism's engagement with its world.

Jakob von Uexküll's concept of umwelt refers to the perceptual and meaningful world emerging as an organism engages with its environment. Traditionally, it has centered on survival to emphasize how organisms navigate their surroundings to secure resources and avoid threats (Uexküll, 1940). However, organisms also inhabit a world shaped by reproductive relations and goals. We propose that a distinct reproductive umwelt exists alongside the survival-oriented one, creating its own sphere of significance and meaning. This reproductive layer influences how organisms perceive and interact with their environment in ways specifically tied to reproduction, shaping behaviors and relationships beyond immediate survival needs. Survival and reproductive Umwelten are not completely distinct spheres, but deeply intertwined; the concept of reproductive umwelt highlights that reproduction is a dimension within the organism's meaningful world, differentially shaping its perception, behavior, and interactions.

The interorganismal perspective on reproduction can be meaningfully developed through the lens of Uexküll's umwelt theory, which is foundational in biosemiotics (Kull, 2001). The concept of the umwelt not only captures an organism's experienced and meaningful world but also represents an inherited system of recognition, shaping both organismal survival and reproduction. This inheritance of perceptual frameworks highlights that recognizing other organisms is a crucial factor in biological evolution, with significant implications for reproduction. This idea aligns with Paterson's concept of species recognition (1993), which rests on the mutual recognition of sexual partners, and also with recognition processes in pregnancy. However, the reproductive umwelt extends beyond mate recognition and selection, shaping how organisms perceive and interact with their environment, including other organisms. This network of inherited perceptual and relational capacities structures not only individual reproductive behaviors but also broader evolutionary trajectories (Deely, 2001; Kull, 2001). By highlighting the role of inherited perceptual and relational capacities, this perspective offers a rich, integrative framework between biosemiotic and organizational perspectives for understanding how these capacities evolve and persist. Reproduction is thus viewed as a dynamic, meaning-driven process embedded in the continuity of life.

Claus Emmeche (2004) expands on Uexküll's insights and elaborates on the qualitative aspects of the umwelt, arguing that an organism's perceptual world is not merely a passive reception of sensory *stimuli* but an active construction shaped by its needs, evolutionary history, and the interactions with its surroundings. The concept of the "qualitative umwelt" requires adopting the perspective of the organism itself rather than relying solely on that of an external observer. He further suggests that an organism's umwelt and its self-organization are deeply intertwined, each providing insights into the other, meaning that an organism's agency and experiential engagement with the world directly influence its developmental and evolutionary processes. Recognizing this qualitative aspect allows us to see how organisms actively shape their perceptual worlds rather than simply reacting to external conditions.

When linked to Uexküll's and Paterson's ideas, Emmeche's notion of the qualitative umwelt enriches an interorganismal perspective on reproduction. It reveals that relational capacities and reproductive processes are not just reactions to external *stimuli* but are rooted in the subjective, experiential worlds of organisms. This view underscores the active role organisms play in shaping the evolutionary and reproductive networks they inhabit, offering a crucial lens for understanding relational and reproductive dynamics across generations.

By integrating these perspectives, the reproductive umwelt emerges as a meaning-making dimension distinct from, yet partially overlapping with, the survival-oriented umwelt. While connected to survival, it encompasses unique goals and processes that shape reproductive interactions and evolutionary change.

11.3 Conclusion

In this chapter, we have examined expanded conceptualizations of reproduction and argued for the integration of insights from organicism, developmental biology, theoretical biology, biosemiotics, and organizational approaches to offer an expanded understanding of reproduction.

Contrary to conventional focus on transmission mechanisms, the expanded view of reproduction incorporates complex intraorganismal processes and interorganismal interactions. Through a historical review of organicism, we traced early perspectives of reproduction as an integrated process, drawing on insights from embryology and theoretical biology. They anticipated later critiques of gene-centric and molecular-based models, as seen in autopoietic and organizational accounts, which prioritize closure and self-production while treating reproduction as an outcome derived from those. In this light, the expanded view reveals reproduction as a fundamentally collective process, involving multiple agents and interactions that extend beyond individual organisms to shape developmental, social, and ecological dynamics.

To further develop this view, we introduce the idea of a *reproductive umwelt*, which illustrates the semiotic and perceptual dimensions of reproductive processes. Organisms experience, interpret, and respond to reproductive cues in their environment. Recognizing reproduction as both organizational and relational provides an expanded framework for understanding its role in biological continuity and change, as well as its impact on the emergence and maintenance of biological organization. Future research should further investigate how organizational and biosemiotic dimensions intersect in understanding reproductive behaviors, norms, and perceptual frameworks across different scales of biological organization.

Acknowledgments

AE and DCG acknowledge IAS-Research group funding IT1668–22 from Basque Government, grants PID2023–147251NB-I00 for project "OUTAGENCIES: Varieties of autonomous agency across living, humanimal and technical system" funded by MCIU/AEI/10.13039/501100011033 and FEDER/UE, and PID2021–127184NB-I00 for project "Metaphysics of Biology: Processes and Dispositions" of the Spanish Ministry of Science and Innovation. DCG has a predoctoral contract with the University of the Basque Country (PIF-2020).

References

Abir-Am, P. G. 1987. The biotheoretical gathering, trans-disciplinary authority and the incipient legitimation of molecular biology in the 1930s: New perspective on the historical sociology of science. *History of Science*, *25*(1), 1–70.

Bertalanffy, L. von 1933. *Modern theories of development: An introduction to theoretical biology* (J. H. Woodger, Trans.). Oxford University Press. [1962 reprint, Harper.]

Blackstone, N. W. 2006. Charles Manning Child (1869–1954): The past, present, and future of metabolic signaling. *Journal of Experimental Zoology Part B: Molecular and Developmental Evolution*, *306*(1), 1–7.

Bich, L., and Skillings, D. 2024. There are no intermediate stages: An organizational view on development. In M. Mossio (Ed.), *Organization in biology* (pp. 241–262). Springer.

Brandt, C. 2022. Vitalism, holism, and metaphorical dynamics of Hans Spemann's "organizer" in the interwar period. *Journal of the History of Biology*, *55*(2), 285–320.

Brennan, P. L. R., Gereg, I., Goodman, M., Feng, D., and Prum, R. O. 2017. Evidence of phenotypic plasticity of penis morphology and delayed reproductive maturation in response to male competition in waterfowl. *The Auk*, *134*(4), 882–893.

Bubnoff, A. von 2024. The Spemann-Mangold organizer discovery and society. *Cells and Development*, *178*, 203906.

Chiu, L., and Gilbert, S. F. 2015. The birth of the holobiont: Multi-species birthing through mutual scaffolding and niche construction. *Biosemiotics*, *8*(2), 191–210.

Cortés-García, D. 2026. Biological autonomy and reproduction. In X. E. Barandiaran and A. Etxeberria (Eds.), *Outonomy: Fleshing out autonomy beyond the individual.* Springer (55–64).

Cortés-García, D., and Etxeberria, A. 2023. Ontologies in evolutionary biology: The role of the organism in the two syntheses. In J. M. Viejo and M. Sanjuán (Eds.), *Life and mind: Interdisciplinary evolution research* (pp. 185–205). Springer.

Cortés-García, D., Etxeberria, A., and Nuño de la Rosa, L. 2024. The evolution of reproductive characters: An organismal-relational approach. *Biology & Philosophy*, *39*(5), 26.

Dawkins, R. 1976. *The selfish gene*. Oxford University Press.

Emmeche, C. 2004. Organicism and qualitative aspects of self-organization. *Revue internationale de philosophie*, *2*, 205–217.

Esposito, M. 2017. The organismal synthesis: Holistic science and developmental evolution in the English-speaking world, 1915–1954. In R. Delisle (Ed.), *The Darwinian tradition in context: Research programs in evolutionary biology* (pp. 219–241). Springer.

Etxeberria, A. 2023. Jacob's understanding of reproduction: Challenges from an organismic collaborative framework. *HOPOS: The Journal of the International Society for the History of Philosophy of Science*, *13*(2), 535–553.

Etxeberria, A. 2004. Autopoiesis and natural drift: Genetic information, reproduction, and evolution revisited. *Artificial Life*, 10(3), 347–360.

Etxeberria, A., and Umerez, J. 2006. Organización y organismo en la Biologıa Teórica ¿Vuelta al organicismo? *Ludus Vitalis*, *26*, 3–38.

Gilbert, S. F., Opitz, J. M., and Raff, R. A. 1996. Resynthesizing evolutionary and developmental biology. *Developmental Biology*, *173*(2), 357–372.

Gilbert, S., and Sarkar, S. 2000. Embracing complexity: Organicism for the 21st century. *Developmental Dynamics*, *219*, 1–9.

Godfrey-Smith, P. 2009. *Darwinian populations and natural selection*. Oxford University Press.

Goodwin, B. 1984. A relational or field theory of reproduction and its evolutionary implications. In M.-W. Ho and P. T. Saunders (Eds.), *Beyond neoDarwinism* (pp. 219–241). Academic Press.

Gould, S. J. 1983. The hardening of the Modern Synthesis. In M. Grene (Ed.), *Dimensions of Darwinism* (pp. 71–93). Cambridge University Press.

Griesemer, James. 2000. Reproduction and the reduction of genetics. In P. Beurton, R. Falk, and H. J. Rheinberger (Eds.), *The concept of the gene in development and evolution: Historical and epistemological perspectives* (pp. 240–285). Cambridge University Press.

Griesemer, J. 2014. Reproduction and scaffolded developmental processes: An integrated evolutionary perspective. In A. Minelli and T. Pradeu (Eds.), *Towards a theory of development* (pp. 183–202). Oxford University Press.

Griesemer, J. 2016. Reproduction in complex life cycles: Toward a developmental reaction norms perspective. *Philosophy of Science*, *83*(5), 803–815.

Haraway, D. J. 1976. *Crystals, fabrics and fields. Metaphors of organicism in twentieth-century developmental biology*. Yale University Press.

Holstein, T. W. 2024. The significance of Ethel Browne's research on hydra for the organizer concept. *Cells & Development*, *178*, 203907.

Hull, D. 1980. Individuality and selection. *Annual Review of Ecology and Systematics*, 11, 311–332.

Jacob, F. 1970. *The logic of life: A history of heredity*. Princeton University Press.

Kay, L. E. 2000. *Who wrote the book of life: A history of the genetic code*. Stanford University Press.

Keller, E. F. 1987. Reproduction and the central project of evolutionary biology. *Biology and Philosophy*, *2*, 383–396.

Kull, K. 2001. Jakob von Uexküll: An introduction. *Semiotica*, *134*, 1–59.

Loison, L. 2022. The environment: An ambiguous concept in Waddington's biology. *Studies in History and Philosophy of Science*, *91*, 181–190.

Maturana, H. R., and Varela, F. J. 1987. *The tree of knowledge: The biological roots of human understanding*. Shambhala Publications.

Maienschein, J., and MacCord, K. 2022. *What is regeneration?* University of Chicago Press.

Morange, M. 2023. *A history of biology*. Princeton University Press.

Moreno, A., and Mossio, M. 2015. *Biological autonomy: A philosophical and theoretical enquiry*. Springer.

Nicholson, D. J., and Gawne, R. 2015. Neither logical empiricism nor vitalism, but organicism: What the philosophy of biology was. *History and Philosophy of the Life Sciences*, *37*, 345–381.

Noble, D. 2016. *Dance to the tune of life: Biological relativity*. Cambridge University Press.

Nuño de la Rosa, L. 2023. Agency in reproduction. *Evolution & Development*, *25*(6), 418–429.

Nuño de la Rosa, L., Pavlicev, M., and Etxeberria, A. 2021. Pregnant females as historical individuals: An insight from the philosophy of evo-devo. *Frontiers in Psychology*, *11*, 572106.

O'Malley, M. A. 2016. Reproduction expanded: Multigenerational and multilineal units of evolution. *Philosophy of Science*, *83*(5), 835–847.

Paterson, H. E. H. 1993. *Evolution and the recognition concept of species: Collected writings* (S. F. McEvey, Ed.). Johns Hopkins University Press.

Pattee, H. H. 1968. Comments by H. H. Pattee to the basic ideas of biology. In C. H. Waddington (Ed.), *Towards a theoretical biology: Prolegomena* (pp. 219–220). Edinburgh University Press.

Peterson, E. 2016. *The life organic: The theoretical biology club and the roots of epigenetics*. University of Pittsburgh Press.

Pontarotti, G. 2024. Organization and inheritance in twenty-first-century evolutionary biology. In M. Mossio (Ed.), *Organization in biology* (pp. 219–239). Springer.

Saborido, C., Mossio, M., and Moreno, A. 2011. Biological organization and cross-generation functions. *The British Journal for the Philosophy of Science*, *62*(3), 583–606.

Smocovitis, V. B. 1996. *Unifying biology: The evolutionary synthesis and evolutionary biology*. Princeton University Press.

Thom, R. 1968. Comments by René Thom to the basic ideas of biology. In C. H. Waddington (Ed.), *Towards a theoretical biology: Prolegomena* (pp. 32–41). Edinburgh University Press.

Triviño, V., and Cerezo, M. 2024. Explorando las implicaciones metafísicas del organicismo: la fundamentalidad de los organismos. *Ludus Vitalis: Revista de Filosofía de las Ciencias de la Vida*, *30*(2), 1–27.

Triviño, V., and Nuño de la Rosa, L. 2016. A causal dispositional account of fitness. *History and Philosophy of the Life Sciences*, *38*(3), 1–18.

Uexküll, J. von 1982. The theory of meaning. *Semiotica*, *42*(1), 25–82.

Umerez, J. 2021. Control hierarchies: Pattee's approach to function and control as time-dependent constraints. In D. S. Brooks, J. DiFrisco, and W. C. Wimsatt (Eds.), *Levels of organization in the biological sciences* (pp. 279–296). The MIT Press.

Veigl, S. J., Suárez, J., and Stencel, A. 2022. Rethinking hereditary relations: The reconstitutor as the evolutionary unit of heredity. *Synthese*, *200*(5), 367.

Vienne, F. 2015. Seeking the constant in what is transient: Karl Ernst von Baer's vision of organic formation. *History and Philosophy of the Life Sciences*, *37*, 34–49.

Waddington, C. H. (Ed.). 1968–1972. *Towards a theoretical biology*. 4 vols. Edinburgh University Press.

Waddington, C. H. 1968a. The basic ideas of biology. In C. H. Waddington (Ed.), *Towards a theoretical biology: Prolegomena* (pp. 1–31). Edinburgh University Press.

Waddington, C. H. 1968b. Comment by C. H. Waddington on Note on topics for a second symposium. In C. H. Waddington (ed.), *Towards a theoretical biology: Prolegomena* (p. 218). Edinburgh University Press.

Weaver, W. 1970. Molecular biology: Origin of the term. *Science*, *170*(3958), 581–582.

Wilkins, John S., and Pierrick Bourrat. 2022. Replication and reproduction. In E. N. Zalta and U. Nodelman (Eds.), *The Stanford encyclopedia of philosophy*. https://plato.stanford.edu/archives/win2022/entries/replication/.

Wolfe, C. 2024. Varieties of organicism: A critical analysis. In M. Mossio (Ed.), *Organization in biology* (pp. 41–58). Springer.

12 Making Meaning Matter with Irruption Theory: Bridging Efficacy and Uncertainty by Satisfying the Participation Criterion

Tom Froese, Georgii Karelin, and Takashi Ikegami

12.1 Introduction

Biosemiotics holds that meaning plays an essential role in all living systems, from single cells to human language users. Organisms are sensitive to the meaning of their situation, including to their own and others' behavior. A well-studied example is chemotaxis, which enables bacteria to encounter glucose not as some neutral property of the physical environment but rather as an "object worth pursuing" (Kłóś and Płonka, 2021, p. 752). And this meaning of "attractant" for the bacterium is not reducible to the chemical properties of glucose, given that the general semiotic category of "attractant" is realizable by multiple kinds of objects. However, according to Pattee's (2001) assessment, there is an issue with such a fundamental distinction between matter and its meaning for the organism: This "epistemic cut" in the scientific world image has given rise to the illusion of a strict isolation of meaning from lawful physical dynamics. "The practice and study of semiotics do not appear to have any necessary relation whatsoever to physical laws. As Hoffmeyer and Emmeche (1991) emphasize, it is generally accepted that, '*No natural law restricts the possibility-space of a written (or spoken) text*,' or, in Kull's (1998) words, 'Semiotic interactions do not take place of physical necessity.'" (Pattee, 2001, p.17).

Pattee highlights that making a clear distinction between matter and meaning is not the same as the isolation of meaning from all material relations, and he therefore proposes that we need to take seriously the concept of a "physics of symbols." For example, although maximal compression of events by dynamical laws has yielded effective theories for the physics of nonliving phenomena, this strategy is only partially successful for phenomena in which meaning also plays a role: "We know from everyday decision-making that all our experiences are not completely compressible and that our life is not state-determined" (Pattee, 2001, p. 13).

We can go a step further and derive these information-theoretic and dynamical consequences from first principles, starting from the premise that meaning makes a difference, in its own right, in the physical universe. By building on Svozil's (2018) analysis of the physics of intentionality, we can identify clear limitations on the efficacy of meaning. When an agent moves in a way that is sensitive to meaning, and hence thereby performs an action that reaches across the epistemic cut, the resulting interface—whatever its precise nature—must "employ gaps in the intrinsic laws of the universe" such that it "excludes any kind of immanent predictability of the signals emanating from it" (Svozil, 2018, p. 158). In other words, from the perspective of scientific measurement, the efficacy of meaning on an agent's behavior would have to operate akin to a hidden variable injecting an "irreducible intrinsic indeterminism" (Svozil, 2018, p. 158) into its underlying dynamics. Accordingly, we conjecture that moments of increased efficacy of meaning are associated with less compressible dynamics, which could for instance be estimated as the entropy rate of the observed process (e.g., Mediano et al., 2023).

What this brief introductory analysis reveals is that an encounter between core concerns of biosemiotics, methodological progress in information theory, and theoretical advances in the cognitive science of intentional agency promises to be a mutually informative undertaking. Indeed, one of the most pressing frontiers of cognitive science is to provide a theory of how intentional agents, like ourselves and plausibly extending to all organisms, fit into the natural order described by the physical and life sciences (e.g. Azarian, 2022; Ball, 2023; Mitchell, 2023; Musser, 2023; Sapolsky, 2023). For example, long-standing questions regarding the possibility of teleology, which denotes the goal- or end-directedness of biological processes, become even more intricate when considering the intentional actions of humans (Noble and Ellis, 2022). Observers like ourselves have first-person experiences of purposeful action, in which there is a felt disposition to achieve an anticipated end alongside a corresponding sense of agency. Yet this familiar human form of teleological causality can be considered a special case of the general form of end-directed behavior with which theoretical biology has long struggled (García-Valdecasas and Deacon, 2024). It is therefore helpful to adopt a minimalist approach with the aim of systematically uncovering the necessary and sufficient conditions of teleological causality in organisms (García-Valdecasas, 2022). Without a workable scientific account of basic teleology—of how purpose can play a role in behavior generation—more specific claims of biosemiotics, including about communicative, representational, or other meaning-bearing aspects of living systems, will rest on shaky foundations. Meaning matters; the key challenge is to account for this mattering.

There are valuable attempts at addressing this challenge from the bottom up—for example, by elucidating the end-directedness of living systems in thermodynamic terms (e.g., Deacon, 2012; Swenson and Turvey, 1991; Tschacher and Haken, 2007). Yet, akin to the causal exclusion principle that is haunting accounts of mental causation and nonreductive physicalism more generally (Kim, 2005), these proposals may still suffer from a notable theoretical shortcoming: The assumed presence of end-directedness in these minimal systems is, empirically, indistinct from its absence.[1] For example, if we accept that ends supervene on biological processes in a physical world characterized by causal completeness at that scale, then appeals to end-directedness in behavior generation can at best be a heuristic tool. There would be no conceptual room for the efficacy of end-directedness as such. But if the presence of teleology is compatible with its absence, then this is also equivalent to treating teleology as physically ineffective, which is hard to square with the appearance of end-directedness across all biological scales (Ball, 2023). It would also ultimately be in tension with our own sense of agency, including as scientists pursuing ends by developing these very theories.

Accordingly, responding to the challenge requires developing a scientifically workable theory of the efficacy of end-directedness, which also manages to do justice to the first-person experience that the presence of end-directedness genuinely makes a difference to behavior. We can reformulate this challenge into an explicit criterion to assess how successful a theory of behavior generation is in securing the efficacy of end-directedness. For purposes of illustration, let us work with an example proposed by Moore in the context of the debate on mental causation: "Nonreductive physicalists endorse the principle of mental causation, according to which some events have mental causes: Sid climbs the hill because he wants to. Nonreductive physicalists also endorse the principle of physical causal completeness, according to which physical events have sufficient physical causes: Sid climbs the hill because a complex neural process in his brain triggered his climbing" (Moore, 2019, p. 479).

What is the role of end-directedness in Sid's behavior? A loosely related group of dynamical approaches would broadly claim that Sid's intention is either supervenient on—or even identical with—the organizational constraints that collective dynamics impose on neural or organismic activities (Deacon, 2012; Freeman, 1999; Juarrero, 1999; Kelso, 1995; Thompson and Varela, 2001). Traditionally, the focus has been on the brain and on mind-brain identity, but embodied versions where mental features are identified more broadly with organism–environment interaction dynamics are also conceivable (Myin and Zahnoun, 2018). Accordingly, if there were no evidence of appropriate organizational constraints on Sid's bodily processes, then this would indicate the absence of end-directedness in his behavior.

Yet this focus on organizational constraints raises the worry that dynamical approaches capture only formal properties (Vial and Cornejo, 2022). Specifically, if the organizational constraints were sufficient causes for changes in organismic activity, then what role would the ends themselves play in bringing about these changes? Put differently, if these constraints do all the actual work in getting Sid's behavior appropriately organized, his ends as such become superfluous. Or at least, the normative conditions specifically associated with ends, but not with mere constraints, would not make any difference—the efficacy would rest on the constraints alone. In other words, the characteristic properties that make ends distinctive from generic constraints might as well be nonexistent. The irony of such an approach to naturalizing teleology—for example, of aiming to accommodate a realist interpretation of intentions by recasting their role purely in terms of a nonintentional cause—is that it is self-undermining (Cea, 2023).

More generally, any theory of teleology's role in behavior generation that directly identifies that role with a concrete physiological process or cause is one step away from eliminativism. To put it differently, if a theory appeals to a particular already independently existent physiological factor as the role that end-directedness is supposed to play in the physiology of behavior generation, it can justifiably be asked whether that end-directedness as such does any work itself, or whether it is ultimately nothing but an ineffective epiphenomenon. For example, merely attributing a mental content to a particular order parameter of the associated neurophysiological dynamics does not help to explain how the corresponding behavior can be sensitive to its meaning; after all, the order parameter by itself is sufficient to describe the organization of neural dynamics, and whether it carries meaningful content is superfluous in this description.

Accordingly, we need a stricter criterion that places a two-fold demand on any explanation of end-directedness: It is not only the case that end-directedness of behavior must potentially be empirically distinguishable from its absence, but this distinguishability cannot be based on physiological aspects that already have a sufficient account in their own terms. It is in meeting the second demand that a theory can manage to avoid the exclusion of end-directedness, even under the strict assumption of a physical causal completeness, as posited by conservative naturalists (Kim, 2005; Moore, 2019). We capture this stronger demand for an irreducible role of end-directedness in behavior generation by proposing a Participation Criterion:

Participation Criterion: End-directedness of a behavior entails that, in principle, it is distinguishable by measurement from one without end-directedness.

Foreshadowing the proposal we will develop in more detail in section 12.4, we briefly note that one scientifically workable strategy for satisfying the Participation Criterion is to conceptualize the efficacy of end-directedness in behavior generation in terms of a spontaneous change in stochastic fluctuations. This kind of appeal to indeterminacy has the double advantage of specifically avoiding charges of causal exclusion due to overdetermination (Potter and Mitchell, 2022) while at the same time allowing for a novel possibility: that an additional factor dependent on end-directedness is making a measurable difference to the bodily process, albeit a factor that is itself not directly observable from that particular measurement perspective. For example, if an agent's behavior changes because of the meaning of a situation, and given that the meaning itself cannot be directly measured in objective terms nor reduced to something nonmeaningful, then the efficacy of meaning on behavior could only be captured indirectly as increased uncertainty (decreased compressibility) of the underlying dynamics.

It is important to preemptively address the reasonable concern that the Participation Criterion is too strong, such that it would be ruled out because of inconsistency with physics. Fortunately, however, there is sufficient causal slack in organisms' behavior. At least on some interpretations of quantum mechanics, when observers conduct a measurement, the result is not determined, not even by the previous state of the entire universe (Conway and Kochen, 2009). And this kind of indeterminism could conceivably be amplified across all scales of the organism: "In general, physics is non-linear and large effects of small changes are well known to happen. From this perspective, agency is simply a situation where scale separation does not hold: nothing puzzling here" (Rovelli, 2021). From this starting point, it may be less contentious to go one step further: We acknowledge that agency is dependent on such breakdowns of scale separation, while at the same time positing that the resulting uncertainty also opens the door to agency-dependent increases in stochastic fluctuations.

It is also noteworthy that just because a theory satisfies the Participation Criterion in principle, this does not mean that the difference can be easily measured in practice. The various physiological processes contributing to an organism's rate of entropy production (REP) across its multiscalar organization are difficult to disentangle. Accordingly, De Bari et al. (2023, p. 18) ask, "If one measures the total entropy production of an organism, what changes in REP are owed to the focal behavior (e.g. locomotion) and what to the other processes playing out at different scales (e.g. perception, motor control, metabolism)?" This is a fair methodological concern. Yet what is crucial for our argument is a conceptual point: If a theory were to assume that living systems are fully deterministic, then there would be no possibility of satisfying the Participation Criterion in the first place. Thus, somewhat paradoxically, the vast uncertainty inherent in biological processes provides a window of opportunity for theories that aim to explain how an agent's sensitivity to meaning makes a genuine difference to its behavior. For example, in Integrated Information Theory, a degree of indeterminism is required to ensure that an entity has intrinsic cause-effect power all the way to the most fundamental level, even though the large-scale behavior is largely predictable (Tononi et al., 2023).

In the next sections, we explore the Participation Criterion in its most basic form. We examine two of the most developed thermodynamic accounts of the origins of end-directedness—namely, the "law of maximum entropy production" by Swenson and colleagues as well as the "autogen" model by Deacon and colleagues. Each of the theories provides important insights, specifically regarding the roles of energy flow and of autonomous organization, in end-directed processes. Nevertheless, they fall short of satisfying the Participation Criterion, at least in their current formulations. We will then introduce the irruption theory of intentional agency as an example of how a theory of end-directedness could build on their insights while also satisfying the Participation Criterion. Our suggestions of how the efficacy of end-directedness manifests in terms of uncertainty that is measurable as information-theoretic entropy, and what is its broader role in physiological activity, remain speculative. Yet they tantalizingly point toward an expanded theoretical biology, in which matter, life, and mind are three distinct yet related domains of phenomena that play unique and complementary roles in the organism.

12.2 Matter: Energy Flow

One innovative theoretical perspective on the thermodynamics of the organism comes from ecological psychology (Swenson and Turvey, 1991). In contrast to systems-theoretic approaches that conceptualize the organism as a self-producing system in more abstract terms, such as Maturana and Varela's (1980) traditional autopoietic theory (cf. Ruiz-Mirazo and Moreno, 2004), ecological psychology's concept of an "autocatakinetic" (ACK) system is distinctive for its appeals to physics (Chemero, 2012). It is rooted in the thermodynamics of dissipative structures (Prigogine, 1997), often referred to as flow structures. Further, its hypothesis regarding the lawful origins of ACK systems is underpinned by the so-called "maximum entropy production principle" (Deacon, 2021), also sometimes called the "law of maximum entropy production" (LMEP), or, even more ambitiously, the "fourth law of thermodynamics" (Beretta, 2020; Morel and Fleck, 2006; Swenson, 2009, 2020). In a recent publication, Swenson (2023), a principal contributor to what he refers to as the "ACK-LMEP" paradigm, further raised the stakes by positing it as "a grand unified theory for the unification of physics, life, information and cognition (mind)" (Swenson, 2023, p. 1).

Yet despite ecological psychology's appeals to entropy production's lawlike nature, there are open questions about the epistemological and ontological status of the MEPP (Sánchez-Cañizares, 2023), including how universally applicable the MEPP is to nonliving and living systems (De Bari et al., 2023). There have been numerical works showing that in some multi-stable systems, the steady state with the highest entropy production is favored (Endres, 2017), but there are counterexamples (Bartlett and Virgo, 2016).

More importantly for our current purposes, there remains a specific concern about the adequacy of a straightforward application of the ACK-LMEP paradigm to the specific characteristics of the behavior of living beings (Barrett, 2020b; Froese et al., 2023). It seems that the ACK-LMEP paradigm does not (yet) have sufficient conceptual resources to distinguish the end-directedness of living, cognitive ACKs from the nonnormative processes of other, generic ACK systems—for example, a Bénard cell. Swenson (2020) has dismissed this concern as unfounded, but this dismissal may rest on a misunderstanding of the criticism (Barrett, 2020a). Indeed, Swenson's (2023) subsequent attempt to turn the ACK-LMEP paradigm into a "grand unified theory" has usefully brought this same concern to the forefront.

To unpack the concern in more detail, let us refer to a standard definition of ACK systems: "ACKs are flow structures, their identities constituted through flow, and defined as a system that maintains its 'self' as an entity constituted by and empirically traceable to a set of nonlinear (circularly causal) relations (constitutive relations) through the dissipation or breakdown of environmental potentials (resources) in the continuous coordinated motion of its components" (Swenson, 2023, p. 8).

In other words, ACKs are a specific class of dissipative structure, which includes both living and nonliving systems from cellular to planetary scales. The next step is to account for the lawful origin of ACK systems, which involves positing the LMEP as a general selection principle that provides an answer to the question of path selection:

> "Which paths out of available paths will a system take to get to equilibrium (maximize the entropy or minimize potentials)?" The second law, of course, is mute on the subject. It only says that in all natural processes the entropy increases. The answer to this question, and the one that solves the entire question of physical selection, the "why" of universal ordering, life and cognition is the law of maximum entropy production (LMEP) or the fourth law of thermodynamics . . . :
>
> (the world) a system will select the path or assembly of paths out of available paths that minimizes the potential or maximizes the entropy at the fastest possible rate given the constraints. (Swenson, 2023, p. 10)

Essentially, Swenson's argument is that out-of-equilibrium systems will spontaneously become more organized—for example, self-organizing into ACK systems—to the extent that this increase in order has the immediate consequence of an increase in entropy production due to increased efficiency of energy dissipation. However, we must proceed carefully in moving from nonliving to living systems. Contrary to the LMEP, it is not in the best interest of living ACK systems to always dissipate free energy at the fastest possible rate, especially given that this would entail approaching thermodynamic equilibrium with the environment at the fastest possible rate, which is equivalent to dying (Deacon and García-Valdecasas, 2023). In recognition of this problem, Swenson admits that a distinctive characteristic of living ACK systems is their capacity to resist the fastest *local* dissipation by redirecting dissipation toward spatiotemporally *distant* ends:

> This, the intentionality of living things, is life's central distinguishing feature. Living systems are epistemic (cognitive) systems that constitute their ACK over times and distances that are arbitrary with respect to local potentials using instead their "on board" potential . . . and *information* (in the semantic or meaningful sense) to seek out and access non-local potentials and access otherwise inaccessible dimensions of space–time. . . . The dramatic increase to otherwise inaccessible dissipative dimensions afforded by the origin and progressive ordering of life and its cognitive functioning answers the "why" question in the specific case. (Swenson, 2023, p. 12)

There is a lot to unpack in this paragraph, and several argumentative leaps require more careful deliberation.

To begin with, the laws of thermodynamics do not have foresight, and so nature can only "select" from among the paths that are locally available to it. In other words, the LMEP is spatiotemporally constrained to competing gradients in the here and now. A key unsolved issue in this regard is how to even determine the spatiotemporal scale or system boundary with respect to which maximum entropy production is defined (Sánchez-Cañizares, 2023; Virgo, 2010). Assuming that this fundamental issue can be solved for the case of a living system, an attractive idea is that local paths could be adaptively changed by investing stored up energy to create alternative potential energy gradients with better future prospects, which can then get "selected" because they dissipate potential in the fastest manner (Tschacher and Haken, 2007).

However, this pushes back the original problem to another unsolved problem—namely, the origins of stored energy potential. If the LMEP is assumed to be the driving principle behind the origins and progressive ordering of ACKs, then the sequence of thermodynamically allowed paths from a generic ACK system to the first living ACK systems must have been via paths of consistently increased rates of entropy production. Such a rate-dependent pathway from nonliving to living does not seem plausible. Life is distinctive, as Swenson acknowledges: "Living systems behave arbitrarily with respect to their local potentials" (Swenson, 2023, p. 12). An account of the origins of this arbitrariness and of the efficacy of meaning in living ACKs is still missing.

As an example of a rate-independent constraint on behavior, Swenson refers to the genetic system. However, the genetic code is sufficiently complex that it is unlikely to have arisen by chance, and hence, selection by evolutionary or proto-cellular processes would be required. And it is not permissible, at least not without invoking something akin to teleological backward causation, to appeal to future increases in energy dissipation to account for the selection of the present path. Hence, the locus of agency is directly identified with increasing entropy, as indicated by Swenson's insistence on the original notion of "striving" attributed to the Second Law: "'The universe,' Clausius . . . wrote (in an often misquoted phrase), 'strives (*strebt*) to increase its entropy to a maximum.'" (Swenson, 2023, p. 4). In accordance with this seemingly teleological interpretation of entropy production, there is an experimental research program in ecological psychology that attempts to ground the striving of organisms in the assumed end-directedness of dissipative structures (De Bari et al., 2023).

In sum, according to the ACK-LMEP paradigm, end-directedness is a property of all ACK systems, whether living or nonliving, because it is identified with universal entropic tendencies. The system's ultimate goal just *is* the maximization of entropy production, and it is the latter that is the basis for selection among state trajectories. Therefore, end-directedness as such no longer has any specific role of its own to play in behavior generation. This has the benefit of satisfying a strict naturalism, such as reductive physicalism, yet it does so by sacrificing realism about the efficacy of end-directedness. Seemingly by construction, the current formulation of the ACK-LMEP paradigm thereby fails to satisfy the Participation Criterion. This implies that its normative talk about striving toward goals and the role of meaning is open to being eliminated from its explanations of behavior generation altogether. From this perspective, it appears that a key difficulty for this paradigm—namely, to account for living systems' capacity to behave in ways that are arbitrary with respect to local potentials—is a consequence of its failure to make conceptual room for end-directedness to have its own efficacy.

12.3 Life: Individuation

What the ACK-LMEP paradigm needs to get clearer on is how it is possible for a dissipative structure to attain the behavioral flexibility of organisms. As a starting point, it needs to be able to answer the question, How is it possible for a dissipative structure to down-regulate its rate of energy dissipation, which would involve getting a degree of independence from the dissipation of local energy potentials? This question highlights a deep and unresolved tension between the assumed universal tendency of entropy rate maximization and the biological capacity for rate regulation. In the absence of this regulatory capacity, the LMEP becomes self-undermining: "This leaves us with a conundrum. In order to generate and maintain organization, living processes must take advantage of self-organizing processes, and yet they must also prevent these processes from depleting the very gradients that drive them. So, how can life both use self-organization at the same time that it prevents or holds off its terminal tendencies?" (Deacon and García-Valdecasas, 2023, p. 8).

The dissipative structures that are investigated by ecological psychology cannot (yet) address this question. However, another line of theoretical research into the thermodynamics of end-directedness that has positioned itself as providing an answer is Deacon's "autogen" model (Deacon, 2012, 2021). An autogen consists of two interdependent processes—namely, reciprocal catalysis and self-assembly: "Each of these self-organizing terminal processes—reciprocal catalysis and capsid shell self-assembly—generates the boundary conditions that the other requires, but in addition prevents the other from reaching an irreversible terminal state. As a result, the synergistic coupling of both processes will develop toward a target state that, although relatively inert, preserves the potential for both self-organizing capacities to recur when conditions are right. This targeted disposition is teleological (i.e. future-oriented)" (Deacon and García-Valdecasas, 2023, pp. 10–11).

An advantage of the autogen model is that, by reciprocally counteracting the tendency of physical processes to run down, the problem of the whole structure's tendency for maximum dissipation of local energy potentials has been avoided. However, the solution raises a different concern (Froese, 2021): An autogen has a disposition to become inert, unless it is externally forced to react. In other words, a one-sided tendency was averted only at the cost of replacing it with another one-sided tendency—namely, the minimization of dissipation of energy potentials until dissipation ceases altogether: complete stasis. We went from one extreme tendency to another—from maximum flow to no flow—both of which are tendencies that by themselves fail to capture the flexible behavior of living systems. As Deacon (2023) points out, this lack of a capacity to initiate behavior is by design, as it helps to simplify the autogen model. For example, there is no need to assume that an autogen has the capacity to accumulate and store potential energy, and hence, questions about the origins of this capacity can be deferred.

Still, the autogen model notably sets the bar higher for end-directedness compared to the ACK-LMEP paradigm. The latter identifies end-directedness with the self-organized increase of entropy production in a preexisting physical system, such as an electrical dissipative structure consisting of metal beads in a fluid (De Bari et al., 2023). The autogen model is situated in the more complex domain of chemistry, in which an enclosed system self-organizes out of specific interdependent processes of catalysis and crystallization. This process of individuation is taken to be a "distinctive end-directed dynamic" (García-Valdecasas and Deacon, 2024, p. 75), but it is just a heuristic that plays no role in the dynamics of the autogen model. Nothing but chemical interactions are at work in the model. Indeed, Deacon is explicit about not assigning efficacy to end-directedness as such: Teleology is part of the broader class of what he calls "absential" phenomena, whose absence from direct observation is a property that he argues facilitates their naturalization in terms of constraints (Deacon, 2012). Absential phenomena do not even have physical efficacy when considered as constraints alone. As Deacon and Cashman (2016) clarify, doing work requires both contextual constraints and energy release: "The 'efficacy' confusion is also related to this misidentification of absence with non-being. Defining the concept of *constraint* in terms of absent degrees of freedom makes it tempting to think of absences *doing* things. But absences themselves don't do work, nor do they resist work. And yet there is no work without absence. The absent degrees of freedom are only part of the story, necessary but not sufficient. Physical work requires the release of energy in a *constrained* context" (Deacon and Cashman, 2016, pp. 419–420).

In sum, the teleological causality of the autogen model is a "physically embodied disposition" with a "material existence that can be preserved or lost" (García-Valdecasas and Deacon, 2024, p. 75). We can conclude that, like the ACK-LMEP paradigm, the autogen model by construction does not satisfy the Participation Criterion: Given that the chemical dynamics of this model can be completely specified in terms of the physics of constrained energy release, the presumed presence of teleology as such makes no difference compared to its absence.

Still, the autogen model has provided us with useful clues about what to look for as we move from chemical systems to living systems: We need an account of how a dissipative structure could gain the capacity to flexibly inhibit its own tendencies. Ideally, this capacity for inhibition should enable the structure to free its processes from always being driven by local energy gradients, thereby permitting it to become responsive to nonlocal energy potentials and hence ultimately making available new forms of behavioral complexity.

In addition, we can build on Deacon's "absential" approach and go a step further: Appeals to the presence of teleology are only permissible as a "hypothesis of last resort," to paraphrase Sagan (Sagan et al., 1993). Positing an efficacious role for end-directedness in behavior generation is only justifiable for those measurable differences for which an immediate physiological cause is absent or at least has not been observed.

We therefore could not agree with Deacon's claim that "a good model should include no unknown or undescribed processes" and "include no opaque (black box) properties" (Deacon, 2021, p. 541). In contrast, we believe that taking both end-directedness and its "absential" nature seriously highlights the need of broadening the scope of admissible natural phenomena to those that are characterized by intrinsic uncertainty. In general, science has no problem dealing with uncertainty, which can, for example, be quantified using information-theoretic entropy. Changes in the relative amount of entropy can then be used to track changes in relative efficacy of end-directedness. Uncertainty does not undermine agency but enables it to properly unfold. Following Kauffman's work on the evolution of the biosphere, we could say that a condition of possibility of our meaningful participation in the world is the inherent uncertainty of nature (Kauffman and Roli, 2021).

12.4 Mind: Irruption

Let us take stock. We act freely in accordance with our goals. Yet we do not have firsthand access to precisely how our goals are transformed into the appropriate physiological basis of our behavior. At the same time, when we scientifically investigate the physiological basis of behavior generation, we cannot directly measure anything like end-directedness playing a role—there is purely physiological activity. We have therefore argued that explaining end-directedness in terms of lower-level dynamics is simply not possible—because goals do not exist on that lower level in the first place. This limit on the intelligibility of end-directedness is a severe challenge.

The recently proposed irruption theory takes this in-principle limit at face value (Froese, 2023; Froese and Karelin, 2023). As Deacon rightly highlighted, end-directedness is "absential"—it does not show up as such in our observations of the physiological basis of behavior. At the same time, the efficacy of end-directedness cannot be completely absent at that scale, either. Instead—and this is crucial—we need to start working with the fact that while both end-directed and physiological aspects are involved in behavior generation, only the latter are directly accessible via measurement. We are therefore led to posit the following research hypothesis: "End-directedness of a behavior is associated with measurable changes at the scale of physiology that cannot be fully predicted purely from that physiological basis alone. " If so, then we need to operationalize the unpredictable changes in the physiological basis that are specifically resulting from end-directedness, which are referred to in the theory as "irruptions." These measurable consequence of meaningful participation in the basis of behavior generation—irruptions—can be modeled as variable stochastic perturbations or noise introduced into the living system by a "black box," which stands for the efficacy of end-directedness that cannot be measured as such. To be fair, this is a highly unusual way of conceiving of the efficacy of end-directedness, and so it is worth unpacking irruption theory in more detail as a set of smaller axioms and related theses, which in themselves are less controversial. Irruption theory starts by accepting that an agent's motivations as such, including being directed at future ends, is efficacious:

> **Axiom 1: Motivational efficacy**. An agent's motivations, as such, make a difference to the material basis of the agent's behavior. (Froese, 2023, p. 9)

Irruption theory accepts that the difference that is made in this way to the physiological basis is not traceable to their agent-level source. As Deacon (2012) highlighted, when observing and measuring the material record, motivations are "absential" phenomena:

> **Axiom 2: Incomplete materiality**. It is impossible to measure how motivations, *as such,* make a difference to the material basis of behavior. (Froese, 2023, p. 9)

This sets up an apparent tension between the behavioral efficacy of agent-level motivations and their absence in the physiological basis. However, instead of rejecting one of these two axioms, irruption theory introduces a third axiom that makes all three axioms mutually consistent:

> **Axiom 3: Underdetermined materiality**. An agent's behavior is underdetermined by its material basis. (Froese, 2023, p. 10)

Now comes the novel theoretical move with which the Participation Criterion is satisfied: The relative level of indeterminacy of the physiological basis of behavior is dependent on the presence of end-directedness, due to irruptions making a difference. In other words, there are end-directed-dependent changes at the scale of physiology in terms of stochastic variability that would be absent otherwise. However, it remains to be spelled out how these irruptions relate to the generation of appropriately end-direct behavior. For this purpose, the theory proposes three theses (Froese, 2023, p. 11), which we adapt to the case of end-directedness:

Irruption Thesis: A living system is organized as an *incomplete system* such that it is open to the efficacy of end-directedness via increased physiological underdetermination.

Scalability Thesis: A living system is organized as a *poised system* such that it amplifies microscopic irruptions to macroscopic fluctuations that impact end-directed behavior.

Attunement Thesis: A living system is organized as an *attuned system* such that it responds to scaled up irruptions in an end-directed manner.

The core idea of the *Irruption Thesis* has already been introduced; here, it is operationalized as taking an active role in modulating its specific condition of possibility—namely, the relative level of underdetermination of the physiological basis of behavior. The *Scalability Thesis* assumes that the window of opportunity for irruptions is most likely located at the smallest scales, but given the "strange loop" self-referential organization of the brain and body (Hofstadter, 2007; Varela, 1984), an alternative possibility is that irruptions occur at the system-level scale.

The *Attunement Thesis* ensures that irruptions give rise to appropriate behavior because the space of possibilities that they spontaneously open is then closed down in a self-organized manner in accordance with internal and external constraints. Much existing work in embodied cognition and "4E cognition" more generally slots in nicely here, such as attunement in the context of metastable dynamics of brain and behavior (Bruineberg et al., 2021; Tognoli and Kelso, 2014).

Regarding the *Irruption Thesis*, a key issue is how to measure the interference in physiological processes due to end-directedness and how to model the efficacy of this interference. An attractive possibility is to focus on the information-theoretic concept of entropy: Given that entropy is a measure of uncertainty in a system, then irruptions could be measured in terms of a temporary increase in entropy rate. For example, this fits well with a growing literature showing an association between cognition and broken detailed balance in brain dynamics (e.g., Lynn et al., 2021). Relatedly, there is a tradition in artificial life that demonstrates how the basis for adaptive behavior can be simulated by stochastic breaks in system dynamics (e.g., Ikegami and Suzuki, 2008). Irruption theory's contribution to this research is to provide an explanation for these kinds of practical successes—that is, for why the onset of end-directedness can be measured and modeled by bursts of unpredictable state changes. It is because making a cognitive effort and striving toward some end is indeed efficacious in making a physiological difference, but this difference that is made cannot be directly controlled by the striving, nor can the original source of this difference be measured as such. In this regard, the difference introduced into the system due to irruption shows up for measurement as increased uncertainty, and the source of this uncertainty cannot be traced back to physiological or smaller-scale factors.

An application of the Irruption Thesis to the thermodynamic scale could be promising but remains speculative (Froese and Karelin, 2023). Here, we can offer only a brief sketch. In the context of an ACK-LMEP or autogen model, increased end-directedness in a system's processes could be equivalent to increased noise levels. At first sight, this efficacy of end-directedness as a disordering factor might seem counterproductive, but it depends on the context. As we saw, the ACK-LMEP paradigm was missing a mechanism for the inhibition of tendencies toward the maximum rate of energy dissipation, for which the autogen model overcompensated by introducing a tendency toward the minimum rate of energy dissipation. Irruptions could provide a minimal living system with the capacity for end-directed regulation of the rate at which energy is dissipated. For example, stochastic perturbations could degrade energy sources or decrease efficiency of work-constraint cycles, both of which will slow processes down.

This appeal to thermodynamic inhibition as the primary consequence of end-directedness is consistent with the primordial goal of life—namely self-preservation as the "mother-value of all values" (Jonas, 1992). At the origins of life, one essential goal was preventing the system to cross its metabolic boundary of viability, and hence, inhibition of thermodynamic tendencies would have been an adaptive response. A more flexible regulation could then be achieved by a simple mechanism of rein control (Harvey, 2004). Moreover, inhibition continues to be the default mechanism of regulation for more complex forms of life (Jost, 2021). Yet in the context of these evolved living systems, the end-directed-dependent presence of irruptions will also have correspondingly more complex consequences, even if their immediate impact remains the same—a contribution to stochastic fluctuations. Starting with the realization of the ubiquity of 1/f noise in natural systems (Bak, 1996), it has been increasingly recognized that noise plays an essential role in the adaptive workings of the brain (e.g. Mitchell, 2023; Northoff, 2018), and in the organism more generally (e.g. Ball, 2023; Longo and Montévil, 2014; Roy and Majumdar, 2022).

Irruption theory could therefore be elaborated to contribute to more thermodynamic grounding of enactive accounts of adaptivity (Di Paolo, 2018). For example, sufficiently large irruptions could also serve to "reset" the living system's state more generally by temporarily flattening the attractor landscape, thereby broadening its exploration of state space, which in conjunction with a basic associative memory can facilitate self-optimization of constraints via a mechanism akin to generalization (Froese et al., 2023). This comes close to Mitchell's recent argument in support of a two-stage model of action selection, which he elaborates in his systematic defense of agency and free will: "Importantly, in this model, it's not that *individual* random events at the quantum level decide what the organism does or generate new ideas. It's that the general randomness and thermal fluctuations cause a kind of variability in neural networks that can jostle them out of the ruts of habit and into potentially novel states" (Mitchell, 2023, p. 189).

However, as Schurger and colleagues point out, "theorists who want to identify the source of action as the agent will have to tell a story that somehow makes a case for the noisy trigger being part of or attributable to the agent" (Schurger et al., 2021, p. 566). Irruption provides a potential source for this "active modulation of randomness" (Mitchell, 2023, p. 188), which would make a measurable difference compared to the absence of end-directedness.

To be fair, the explanation provided by irruption theory is somewhat unusual in form, as it posits what could be described as a *dynamic exclusion principle*: Within the operations of a living system, the end-directed efficacy of meaning appears to preclude the simultaneous causal efficacy of physiology, and vice versa. This conceptual principle allows living systems to be distinguished from other natural systems in which causal efficacy tends to align with a system's most fundamental measurable properties. In this regard, the explanatory form of irruption theory is structurally reminiscent of Pauli's exclusion principle in particle physics, which states that two identical fermions (such as electrons) cannot occupy the same quantum state simultaneously. Pauli's mathematical principle allows fermions to be distinguished from bosons, which can occupy the same quantum state simultaneously. Future work could aim to develop this conceptual analogy more fully—for example, by exploring the possibility of grounding this dynamical exclusion principle in a mathematical formalization.

12.5 Conclusion

We have analyzed three theories of end-directed behavior, and we have found essential ingredients in each of them. The ACK-LMEP paradigm has demonstrated that we can get self-organized energy flow from physics alone, while the autogen model highlights the role of codependent processes in a self-individuating chemical system, and irruption theory introduces the possibility of end-directed efficacy in terms of active regulation of levels of uncertainty. Taken together, the ACK-LMEP paradigm, the autogen model, and irruption theory highlight the complementary roles of (1) energy flow maintenance, (2) systemic constraint construction, and (3) state constraint destruction, respectively. All three roles are seemingly necessary to explain the end-directed behavior of living systems.

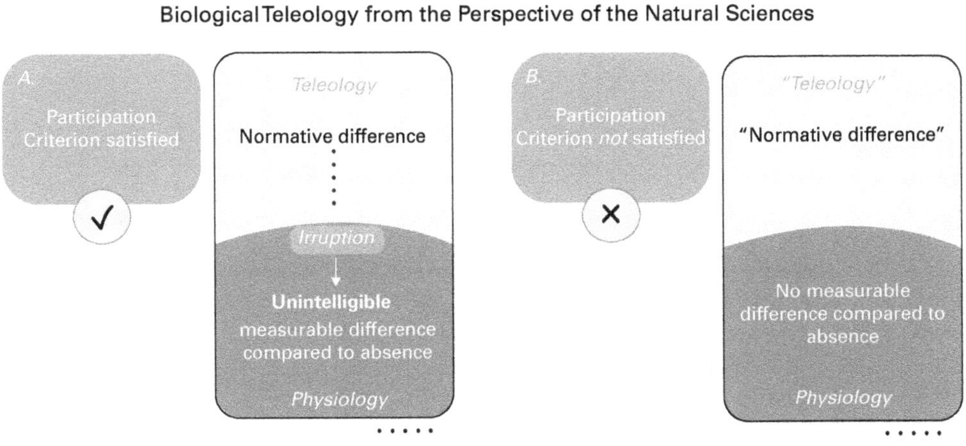

Figure 12.1
Illustration of the proposed Participation Criterion for theories of end-directedness. Panel A: Irruption theory states that end-directedness (teleology) makes a measurable difference to living matter (physiology). However, end-directness itself is not measurable from a physiological perspective, and hence, the difference that it makes to the processes underlying behavior generation is not intelligible (irruption). Panel B: According to theories in which end-directedness is directly identified with some measurable physiological or physical cause or constraint, there can be no effects specifically due to end-directedness as such, and therefore no measurable difference compared to its absence.

Arguably, this unique form of complexity is not accidental, but an essential and irreducible aspect of our own ambiguous being in the world. As long noted by phenomenologically minded thinkers, "our bodies are both subjects open to the things surrounding us, and themselves such things" (van Buuren, 2018, p. 34). Accordingly, the task of an expanded theoretical biology is to describe the end-directedness of behavior as an efficacious relationship between two distinct domains of phenomena, teleology and physiology, but in such a way that neither domain can be directly identified with the other. This approach requires taking seriously the possibility of irreducible cross-domain consequences, from the scale of minimal living systems to that of the human mind (Nicolescu, 2012; Wagemann, 2011). As illustrated in figure 12.1, once it is accepted that end-directedness is irreducible to physiology and yet is efficacious in behavior generation, it follows that it makes a measurable difference that is not intelligible from the perspective of physiology. Measurement can only grasp this difference as an unpredictable deviation from the expected physiological inertia of the body. On the positive side, we can leverage the concept of uncertainty to make room for a participatory role of end-directedness. From simple inhibition to symbolic negation, we expect that irruption can become a useful concept to account for the difference between meaningless movement and meaningful behavior.

Acknowledgments

This research was initiated while TI was visiting the Okinawa Institute of Science and Technology Graduate University through the Theoretical Sciences Visiting Program. We thank Roger Pullin and Alexander Hölken for detailed comments on earlier drafts. We are grateful to Alina Hernandez Porrello for help with the design of the figure.

Note

1. TF is grateful to Stephen Esser for this precise formulation of the fundamental problem.

References

Azarian, B. 2022. *The romance of reality: How the universe organizes itself to create life, consciousness, and cosmic complexity*. BenBella Books.

Bak, P. 1996. *How nature works: The science of self-organized criticality*. Copernicus.

Ball, P. 2023. *How life works: A user's guide to the new biology*. The University of Chicago Press.

Barrett, N. 2020a. Extremal properties and self-preserving behavior. *Adaptive Behavior*, *28*(2), 113–118.

Barrett, N. 2020b. On the nature and origins of cognition as a form of motivated activity. *Adaptive Behavior*, *28*(2), 89–103.

Bartlett, S., and Virgo, N. 2016. Maximum entropy production is not a steady state attractor for 2D fluid convection. *Entropy*, *18*(12), 431.

Beretta, G. P. 2020. The fourth law of thermodynamics: Steepest entropy ascent. *Philosophical Transactions of the Royal Society A: Mathematical, Physical & Engineering Sciences*, *378*, 20190168.

Bruineberg, J., Seifert, L., Rietveld, E., and Kiverstein, J. 2021. Metastable attunement and real-life skilled behavior. *Synthese*, *199*, 12819–12842.

Cea, I. 2023. On motivating irruptions: The need for a multilevel approach at the interface between life and min. *Adaptive Behavior*, *32*(1).

Chemero, A. 2012. Modeling self-organization with nonwellfounded set theory. *Ecological Psychology*, *24*, 46–59.

Conway, J. H., and Kochen, S. 2009. The strong free will theorem. *Notices of the American Mathematical Society, 56*(2), 226–232.

De Bari, B., Dixon, J., Kondepudi, D., and Vaidya, A. 2023. Thermodynamics, organisms and behaviour. *Philosophical Transactions of the Royal Society A: Mathematical, Physical & Engineering Sciences, 381*(20220278).

Deacon, T. W. 2012. *Incomplete nature: How mind emerged from matter*. W. W. Norton & Company.

Deacon, T. W. 2021. How molecules became signs. *Biosemiotics, 14*, 537–559.

Deacon, T. W. 2023. Minimal properties of a natural semiotic system: Response to commentaries on "How molecules became signs." *Biosemiotics, 16*, 1–13.

Deacon, T. W., and Cashman, T. 2016. Steps to a metaphysics of incompleteness. *Theology and Science, 14*(4), 401–429.

Deacon, T. W., and García-Valdecasas, M. 2023. A thermodynamic basis for teleological causality. *Philosophical Transactions of the Royal Society A: Mathematical, Physical & Engineering Sciences, 381*(20220282).

Di Paolo, E. A. 2018. The enactive conception of life. In A. Newen, S. Gallagher, and L. De Bruin (Eds.), *The Oxford handbook of 4E cognition* (pp. 71–94). Oxford University Press.

Endres, R. G. 2017. Entropy production selects nonequilibrium states in multistable systems. *Scientific Reports, 7*, 14437.

Freeman, W. J. 1999. *How brains make up their minds*. Weidenfeld & Nicolson.

Froese, T. 2021. To understand the origin of life we must first understand the role of normativity. *Biosemiotics, 14*, 657–663.

Froese, T. 2023. Irruption theory: A novel conceptualization of the enactive account of motivated activity. *Entropy, 25*(5), 748.

Froese, T., and Karelin, G. 2023. The enactive account of motivated activity and the hard problem of efficacy (HPE): Artificial life meets the physics of life. In H. Iizuka, K. Suzuki, R. Uno, L. Damiano, N. Spychala, M. Aguilera, E. Izquierdo, R. Suzuki, and M. Baltieri (Eds.), *Proceedings of the Artificial Life Conference 2023 (ALIFE 2023)*. The MIT Press.

Froese, T., Weber, N., Shpurov, I., and Ikegami, T. 2023. From autopoiesis to self-optimization: Toward an enactive model of biological regulation. *BioSystems, 230*(104959).

García-Valdecasas, M. 2022. On the naturalisation of teleology: Self-organisation, autopoiesis and teleodynamics. *Adaptive Behavior, 30*(2), 103–117.

García-Valdecasas, M., and Deacon, T. W. 2024. Origins of biological teleology: How constraints represent. *Synthese, 204*(75).

Harvey, I. 2004. Homeostasis and rein control: From Daisyworld to active perception. In J. Pollack, M. A. Bedau, P. Husbands, T. Ikegami, and R. A. Watson (Eds.), *Artificial life IX: Proceedings of the ninth international conference on the simulation and synthesis of living systems* (pp. 309–314). The MIT Press.

Hofstadter, D. 2007. *I am a strange loop*. Basic Books.

Hoffmeyer, J., and Emmeche, C. 1991. Code-duality and the semiotics of nature. In *Biosemiotics: Information, Codes and Signs in Living Systems* (pp. 34–35).

Ikegami, T., and Suzuki, K. 2008. From homeostatic to homeodynamic self. *BioSystems, 91*(2), 388–400.

Jonas, H. 1992. The burden and blessing of mortality. *The Hastings Center Report, 22*(1), 34–40.

Jost, J. 2021. Biology, geometry and information. *Theory in Biosciences, 141*(2), 65–71.

Juarrero, A. 1999. *Dynamics in action: Intentional behavior as a complex system*. The MIT Press.

Kauffman, S., and Roli, A. 2021. The world is not a theorem. *Entropy, 23*(11), 1467.

Kelso, J. A. S. 1995. *Dynamic patterns: The self-organization of brain and behavior*. The MIT Press.

Kim, J. 2005. *Physicalism, or something near enough*. Princeton University Press.

Kłóś, A., and Płonka, P. M. 2021. The semiotic approach to bacterial chemotaxis. *Biosemiotics, 14*, 743–766.

Longo, G., and Montévil, M. 2014. *Perspectives on organisms: Biological time, symmetries and singularities*. Springer.

Lynn, C. W., Cornblath, E. J., Papadopoulos, L., Bertolero, M. A., and Bassett, D. S. 2021. Broken detailed balance and entropy production in the human brain. *Proceedings of the National Academy of Sciences, 118*(47), e2109889118.

Maturana, H. R., and Varela, F. J. 1980. *Autopoiesis and cognition: The realization of the living*. Kluwer Academic.

Mediano, P. A. M., Rosas, F. E., Luppi, A. I., Noreika, V., Seth, A. K., Carhart-Harris, R. L., . . . Bor, D. 2023. *Spectrally and temporally resolved estimation of neural signal diversity*. bioRxiv.

Mitchell, K. J. 2023. *Free agents: How evolution gave us free will*. Princeton University Press.

Moore, D. 2019. Causal exclusion and physical causal completeness. *Dialectica, 73*(4), 479–505.

Morel, R. E., and Fleck, G. 2006. A fourth law of thermodynamics. *Chemistry, 15*(4), 305–310.

Musser, G. 2023. *Putting ourselves back into the equation: Why physicists are studying human consciousness and AI to unravel the mysteries of the universe*. Farrar, Straus and Giroux.

Myin, E., and Zahnoun, F. 2018. Reincarnating the identity theory. *Frontiers in Psychology, 9*(2044).

Nicolescu, B. 2012. Transdisciplinarity: The hidden third, between the subject and the object. *Human and Social Studies, 1*(1), 13–28.

Noble, D., and Ellis, G. 2022. Biological relativity revisited: The pre-eminent role of values. *Theoretical Biology Forum, 115*(1–2), 45–69.

Northoff, G. 2018. *The spontaneous Brain: From the mind-body to the world-brain problem.* The MIT Press.

Pattee, H. H. 2001. The physics of symbols: Bridging the epistemic cut. *BioSystems, 60*, 5–21.

Potter, H. D., and Mitchell, K. J. 2022. Naturalising agent causation. *Entropy, 24*(4).

Prigogine, I. 1997. *The end of certainty: Time, chaos, and the new laws of nature*. The Free Press.

Rovelli, C. 2021. Agency in physics. In C. Calosi, P. Graziani, D. Pietrini, and G. Tarozzi (Eds.), *Experience, abstraction and the scientific image of the world* (pp. 25–40). FrancoAngeli.

Roy, S., and Majumdar, S. 2022. *Noise and randomness in living system*. Springer.

Ruiz-Mirazo, K., and Moreno, A. 2004. Basic autonomy as a fundamental step in the synthesis of life. *Artificial Life, 10*(3), 235–259.

Sagan, C., Thompson, W. R., Carlson, R., Gurnett, D., and Hord, C. 1993. A search for life on Earth from the Galileo spacecraft. *Nature, 365*(6448), 715–721.

Sánchez-Cañizares, J. 2023. Is the maximum entropy production just a heuristic principle? Metaphysics on natural determination. *Synthese, 201*, 121.

Sapolsky, R. M. 2023. *Determined: A science of life without free will*. Penguin Press.

Schurger, A., Hu, P. B., Pak, J., and Roskies, A. L. 2021. What is the readiness potential? *Trends in Cognitive Sciences, 25*(7), 558–570.

Svozil, K. 2018. *Physical (a)causality: Determinism, randomness and uncaused events*. Springer.

Swenson, R. 2009. The fourth law of thermodynamics or the law of maximum entropy production (LMEP). *Chemistry, 18*(5), 333–339.

Swenson, R. 2020. The fourth law of thermodynamics (LMEP) and cognition from first principles: Commentary on Barrett's "On the nature and origins of cognition as a form of motivated activity." *Adaptive Behavior, 28*(2), 105–107.

Swenson, R. 2023. A grand unified theory for the unification of physics, life, information and cognition (mind). *Philosophical Transactions of the Royal Society A: Mathematical, Physical & Engineering Sciences, 381*(20220277).

Swenson, R., and Turvey, M. T. 1991. Thermodynamic reasons for perception-action cycles. *Ecological Psychology, 3*(4), 317–348.

Thompson, E., and Varela, F. J. 2001. Radical embodiment: Neural dynamics and consciousness. *Trends in Cognitive Sciences, 5*(10), 418–425.

Tognoli, E., and Kelso, J. A. S. 2014. The metastable brain. *Neuron, 81*, 35–48.

Tononi, G., Albantakis, L., Boly, M., Cirelli, C., and Koch, C. 2023. *Only what exists can cause: An intrinsic view of free will.* arXiv: 2206.02069v3.

Tschacher, W., and Haken, H. 2007. Intentionality in non-equilibrium systems? The functional aspects of self-organized pattern formation. *New Ideas in Psychology, 25*(1), 1–15.

van Buuren, J. 2018. *Body and reality: An examination of the relationships between the body proper, physical reality, and the phenomenal world starting from Plessner and Merleau-Ponty*. Bielefeld: transcript.

Varela, F. J. 1984. The creative circle: Sketches on the natural history of circularity. In P. Watzlawick (Ed.), *The invented reality* (pp. 309–324). W. W. Norton and Company.

Vial, I., and Cornejo, C. 2022. Not complex enough for complexity: Some intricacies of interpersonal synergies theory. *New Ideas in Psychology, 64*(100914).

Virgo, N. 2010. From maximum entropy to maximum entropy production: A new approach. *Entropy, 12*, 107–126.

Wagemann, J. 2011. The structure-phenomenological concept of brain-consciousness correlation. *Mind & Matter, 9*(2), 185–204.

IV ESSAYS

13 Bioinformatics Meets Biosemiotics and Signal Peptides Emerge as Dicisigns

Henrik Nielsen, Mariana Vitti Rodrigues, and Claus Emmeche

13.1 Introduction

In both the field of bioinformatics (developing methods and software tools for understanding biological data) and the field of biosemiotics (investigating biological systems' generation and use of signs, interpretation, and meaning), *pattern recognition* seems to be a fundamental aspect, both of the basic biochemical processes underlying living beings and of the methods used by the researchers to learn about these processes. Both fields are interdisciplinary and deal with biological information at various levels, especially the molecular level, as one of the major insights of biological research has been an in-depth understanding of the link between the phenomenon of biochemical *specificity* of the cell's enzymes and other macromolecules and the *sequence information* of bases in DNA and RNA and amino acids in proteins (Judson, 1979). Specificity and pattern recognition are keys to functionally organize a variety of different components within a system, whether they are molecules in a cell or agents in a city. In his foundational studies of biosemiotics, Jesper Hoffmeyer compared the complexity of the cell's intracellular traffic to a city and noted that the "semiotics of this intracellular traffic is one of the most fascinating topics of modern biochemistry" (2008, p. 145).

Most secretory proteins (having their final location outside the cell) and some transmembrane proteins, both in eukaryotes and prokaryotes, begin with a sequence of amino acids termed the signal peptide (SP). The SP is cleaved off by an enzyme, signal peptidase, while the protein is being transported across the membrane and is thus not part of the mature protein. The existence of SPs was originally hypothesized in 1971 (Blobel and Sabatini, 1971) and experimentally shown a few years later (Blobel and Dobberstein, 1975a, b). This work formed the basis for the Nobel prize that Günter Blobel was awarded in 1999 "for the discovery that proteins have intrinsic signals that govern their transport and localization in the cell."

Seen from the bioinformatics angle, SPs were among the first properties of proteins to be targets for prediction, with the first prediction method being published as early as 1983 (von Heijne, 1983). Since then, many methods for predicting SPs from amino acid sequences were invented, some of them based on simple statistical approaches and others based on machine learning (Nielsen et al., 2019). The most well-known of these methods is SignalP, which was first made available as a web server in 1995 and published two years later (Nielsen et al., 1997). SignalP has gone through many developments and is now in its sixth major version (Teufel et al., 2022)—a historical account of the evolution of SignalP is found in Nielsen et al. (2024). It has always been based on some flavor of artificial neural networks, combined with hidden Markov models in versions 2 and 3, but the newest version (6.0) is particularly interesting from a biosemiotic perspective since it is based on protein language models.

13.1.1 On Language Models

Language models are a set of Artificial Intelligence techniques within the Natural Language Processing (NLP) field that aim to perform human language tasks. Language models can be developed to classify, translate, predict, and generate text. In general terms, language models are usually trained on the *Masked Language Modeling* task, where some percentage of the words are hidden from the computer model, which has to predict the missing words from their context. For example, in the phrase "My dog is _____," the blank part can be completed with words such as "sleeping," "eating," and "barking." A trained language model would assign high probability to such words, intermediate probability to a semantically nonsensical word such as "reading," and low probability to a grammatically wrong word such as "horse."

The point of the language model is that it can be *pretrained* on a massive amount of text data taken from books, webpages, and encyclopedias, thereby gaining a general "understanding" of which words typically appear together in context.

This can be exploited by using the internal numerical representations (the *embeddings*) from the language model in so-called *downstream tasks*, where the model is used to, for example, classify or translate texts. Importantly, the large body of text used to pretrain the model does not need to be labeled (with classes or translations), while the downstream task is typically done on a much smaller labeled dataset.

Nowadays, LLMs have become popular with the release in late 2022 of ChatGPT, a chatbot created by using Generative Pre-trained Transformers (GPT-3.5 and GPT-4). The success of large language models extrapolated the realm of natural language processing and was seen as a useful tool to analyze protein function based on their primary structure.

13.1.2 On the Language Metaphor: Extending to Protein Language Models

Since the mapping of the genetic code in the 1960s, the language metaphor for basic biochemical processes (e.g., the genetic "code"; "translation"; "transcription"; developmental "program"; "signal" transduction) has been widely used not just among molecular biologists but also in interdisciplinary discussions among biologists, philosophers, and linguists, and in the popular media (Emmeche and Hoffmeyer, 1991; Keller, 2002). The contemporary algorithmic use of LLMs to predict biological features has taken this metaphor to surprisingly new levels of scientific inquiry.

The success of language models in NLP inspired their usage in the study of proteins where the primary structure of a protein is treated as a sentence and the amino acids are seen as words. Protein language models (pLMs) are trained by randomly masking an amino acid and, by considering the context (the amino acids that surround the blank space), predicting the probabilities of each of the twenty possible amino acids in the blank space. Following this metaphor, large pretrained pLMs, such as ProtBERT (Elnaggar et al., 2022) and ESM-2 (Lin et al., 2023), were developed, enabling tasks such as calculating the *probability* of a protein, *generating* new (biochemically likely) proteins, and creating context-dependent *representations* of amino acids for downstream prediction of structure, function, location, and post-translational modifications. Just like in NLP, the strength of the pLMs is that they can be pretrained on a large corpus of unannotated sequences (hundreds of millions of available protein sequences derived from genomic and metagenomic studies), while the downstream tasks are trained on much smaller annotated datasets.

The strength of the language metaphor is evident from some of the titles of papers describing pLMs: "Modeling aspects of the language of life through transfer-learning protein sequences" (Heinzinger et al., 2019) and "ProtTrans: Toward understanding the language of life through self-supervised learning" (Elnaggar et al., 2022). To which extent is the language metaphor suitable to explain processes of life? What does it mean to say that one can use a pLM to *understand* the language of life? Let us reflect upon these questions by considering the case of signal peptides.

13.1.3 On Signal Peptides

The production of a secretory protein in a eukaryotic cell starts like any other protein: The mRNA is produced in the nucleus and transported to the cell's cytosol where it interacts with a free ribosome (Cooper, 2000). The part that is produced first, the N-terminal part, contains an SP consisting of a small number (typically fifteen to forty) of amino acids, containing a region with hydrophobic side chains. This sequence is recognized by a cytosolic protein-RNA complex called SRP (Signal Recognition Particle), which halts the translation and aids in the transport of the mRNA-ribosome complex to an SRP receptor found on the membrane of the endoplasmic reticulum (ER). When it arrives at the ER, the signal sequence is transferred to the translocon, which is a protein-conducting channel in the ER membrane that allows the newly synthesized polypeptide to be translocated to the ER lumen. The dissociation of SRP from the ribosome restores the translation of the secretory protein. The signal sequence is removed by a signal peptidase, and the translation continues while the produced chain moves through the translocon (cotranslational translocation) (see figure 13.1).

Prokaryotes also have SPs, but there are several differences in the process (Briggs and Gierasch, 1986). First, prokaryotes lack ER, so translocation happens across the plasma membrane. Second, only a minority of prokaryotic SPs depend on SRP; the rest are recognized by the chaperone SecB and pushed through the translocon by the ATPase SecA. The translocation by SecA is not co-translational but happens after the protein has been synthesized. Conventionally, we think of prokaryotes as being simpler than eukaryotes, but regarding SP action, the opposite is the case. Not only do prokaryotes have two ways of targeting proteins to the membrane (SRP and SecA+B), but they also have two types of translocon (Sec and Tat) and three types of signal peptidases (I, II, and III).

As described above, SPs can be predicted from sequence data by several bioinformatics tools, including SignalP. However, SignalP 6.0 (Teufel et al., 2022)[1] was the first tool to be able to detect all known types of SPs. This was made possible by the use of pretrained protein language models that were able to generalize from very few (less than one hundred) known examples of certain rare SP types that occur in prokaryotes.

13.1.4 On Dicisigns

According to Peircean semiotics, a dicisign can be described as a double sign, composed of an icon and an index, structured by a syntax. A *sign*, for Peirce, is a *relation* structured by

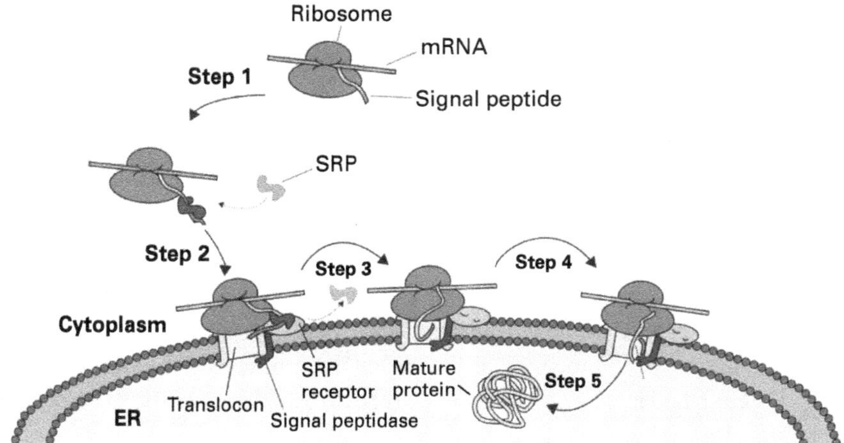

Figure 13.1
The function of signal peptides in protein synthesis in a eukaryotic cell. Step 1: The SRP (signal recognition particle) binds an emerging signal peptide (the first part of the nascent protein chain), and protein synthesis is paused. Step 2: The ribosome-SRP complex binds to the SRP receptor at the ER membrane. Step 3: The SRP is released, and protein synthesis resumes, while the nascent protein is threaded through the translocon. Step 4: The signal peptidase cleaves off the signal peptide. Step 5: The mature protein is released to the ER lumen.

three elements: (i) the sign-vehicle or representamen, (ii) the object of the sign, and (iii) the interpretant of the sign. In the triad, the *sign-vehicle* can be anything that represents its object by embodying its form according to its own abilities (different kinds of signs can embody different forms available in their object). The *object of the sign* directly determines the sign-vehicle that instantiates its form and indirectly determines its interpretant by means of the forms conveyed by the sign-vehicle. The *interpretant of the sign* is characterized as the possible effects of the sign-vehicle, as determined by its object, on the system that interprets the sign (the interpreter or system surrounding the sign process). The interplay among these three elements allows a sign, in a broad sense, to establish a semiotic chain through the process of *semiosis*—that is, the action performed by signs that generates interpretants.

The characterization of a sign involves three logical functions, meaning that what constitutes a sign-vehicle, object, or interpretant depends on their logical place in the semiotic chain. As an example, imagine you receive a gift from a friend. This gift can be understood as a sign (vehicle) of the appreciation your friend has toward you (object of the sign) and, in turn, generates a warm feeling of affection in you (interpretant of the sign). This feeling might prompt you to hug your friend, this hug being a sign-vehicle whose object (affection—that is, the interpretant of the previous sign, also called a more developed sign) might generate feelings of reciprocity and appreciation in your friend (another interpretant), and so on.

Without going into detail about Peirce's theory of signs, it is worth mentioning that Peirce classified signs according to what the sign is in itself (qualisign, sinsign, legisign), according to how it refers to its object (icon, index, symbol), and according to its relation to the interpretant (rheme, dicisign, argument). Our analysis focuses on the dicisign aspect of the process: A *dicisign* is a sign that says or predicates something about some object and, also, separately indicates the object it represents; thus, it conveys information about its object by means of description and reference. Therefore, better than other sign categories, the dicisign suits an account of an SP as part of a semiotic process. The concept of dicisign (or natural proposition, cf. Stjernfelt, 2014) is to be understood as a sign type with a proposition-like structure, not limited to human semiosis. Remember that the traditional view of a proposition ("*s* is *p*," "Sue is proud") is rooted in and restricted to human language, but in a more general sense, according to semiotics, we find this structure also in nonhuman extralinguistic domains. The multimodality of dicisigns, which allows an analysis beyond human language, leads to a suggestion that *language models may learn to "read" natural signs* by emulating the processual aspect of interpretation and communication of forms.

In the process of semiosis, a sign can be characterized as a *medium* that communicates the available forms of the object through the generation of interpretants. In this context, Peirce (EP2, 544) states that "a Sign may be defined as a Medium for the communication of a Form . . . as a medium, the Sign is essentially in a triadic relation, to its Object which determines it, and to its Interpretant which it determines . . . That which is communicated from the Object through the Sign to the Interpretant is a Form. . . . The Being of a Form consists in the truth of a conditional proposition. Under given circumstances, something would be true."

This aspect is important when we analyze SPs as a type of natural proposition because it helps a better understanding of the *pattern recognition process* when the form embodied by an object, such as an enzyme-substrate interaction or SP-SRP interaction in biochemistry, is recognized by other forms or sign-interpretants, guiding future processes. The Peircean concept of form can be interpreted as regularity, or habit, which allows a given system to interpret the functionality of a sign as indicative of a particular class of entities, events, facts, and processes (El-Hani et al., 2009, p. 93). As the form of a door handle invites an interpretive system to grasp and move it up or down, so the sign allows an interpreter to be guided by the potential effects or dispositions that the form *affords* on the system. Thus, the concept of form emphasizes the *dispositional character* of the sign: The *form* embodied in the sign-vehicle represents a disposition that guides the interpretation of its object, generating meaning, or by directing a course of action, under certain circumstances. Think here of the common meaning of disposition as a way in which something is placed or arranged, especially in relation to other things, enabling certain processes, just like the functional groups of organic molecules—for example, a ketone, phenyl, or aldehyde—enable specific classes of chemical reactions.

As hinted at above, a dicisign is composed of an icon and an index being structured by a syntax (figure 13.2). Both icon and index are signs that represent a relationship between a sign-vehicle and its object. An icon embodies the form of its object (existent or not) according to its own representational capacity (Peirce, CP 2.247). Within the structure of a dicisign, an icon functions as the descriptive part of a proposition by embodying the characteristics of the object it represents by its own representative abilities. For example, an unsaturated icon is a term without a subject: ____ is red.

An index, in turn, is a sign that "refers to the Object that it denotes by virtue of being really affected by that Object," such as a footprint on the sand or by virtue of appearing in correlation with that Object (Peirce, CP 2.248). In a dicisign, an index functions as the referential part of the dicisign by pointing out the object it represents: this chair____. Stjernfelt (2014, p. 62) explains that "indices[. . .]would then play the role of connecting certain selected icons to reality, granting them existence and thus ensuring their survival over others."

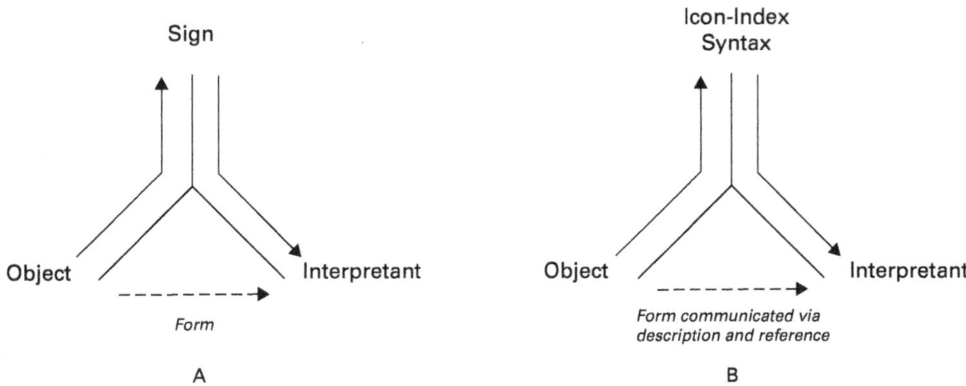

Figure 13.2
(A) Diagram of the general structure of the sign and its correlates, mediating the form of the object to the interpretant. (B) Diagram of a dicisign's triadic structure. The representamen part of the dicisign has a double structure, pointing out the dicisign's object (its indexical part), and describing that object in some way (its iconic part), thus being structured by a simple sort of syntax (juxtaposition between icon-index).

A portrait with a legend illustrates the dicisign structure: The portrait *describes* the object by iconic resemblance, and the legend *indicates* the object by indexical reference (Peirce, EP2, 307). Both the description and the reference to the object must be topologically connected—that is, the legend has to be close to the painting to be able to, by conventional rules, indicate who is described by the painting. According to Stjernfelt (2022), this topological connection consolidates the syntax of a multimodal proposition by the co-localization of an icon and an index forming a structure in a "spatio-temporal neighborhood of some sort." The interpretant of the dicisign "represents a real existential relation, or genuine Secondness, as subsisting between the Dicisign and the Dicisign's real object" (Peirce, CP 2.310). Moreover, "the interpretant is not merely the predicate, but the claim, made possible by the syntax, that the predicate actually holds about the existing object" (Stjernfelt, 2014, p. 68).

Once the Dicisign is structured by the Icon-Index co-localization, it generates an interpretant of existence that asserts that the sign-object relationship holds (is true). It means that the interpretant *claims* that the iconic description of the object is indexically connected to its object. In other words, the interpretant claims that the proposition mirrors the fact it represents by means of description and reference. The referential aspect of proposition-like signs gives rise to the possibility for assigning truth and falsehood to it, meaning that a fundamental characteristic of dicisigns is that they can be *false*. False propositions occur due to a mismatch between Icon and Index co-localization, when the syntax that "glues" description and reference together does not hold. However, how can a natural proposition be false? If we remember that a sign is *anything* that stands in the place of *something else*, dicisigns—understood as propositions that *represent* factual state-of-affairs—might be considered false when misrepresenting the fact the dicisign claims to exist by means of icon-index juxtaposition. For instance, in evolutionary biology, mimicry—the evolved resemblance of one species (e.g., a yellow-black striped hoverfly) to another (a dangerous wasp) to deceive a third species (an insect-eating bird) into confusing them (to avoid the harmless hoverfly)—is an adaptive, yet lying, use of dicisigns (such as "yellow-black striped insects___" (an index sign) and "___ are dangerous" (an icon sign)) making such natural propositions false (cf. Kleisner and Maran, 2019).

Another characteristic of dicisigns is that they can be considered in the imperative mode when they compel another sign to assume or change the other sign's role in the semiotic chain (cf. West, 2015 and 2018 for further application of the imperative function of dicisigns). A suggestive example of an *imperative dicisign* is a label with the address of the receiver of a letter. A letter in an envelope to Mrs. P. in Greenville needs her address on the front, in addition to stamps. It may include the zip code of her area. This address is an imperative dicisign that the postal system interprets as saying "send S to P" where S is the letter and P predicates a destination where Mrs. P. should receive it. The zip code increases the likelihood that the letter goes to the right Greenville of the circa 30 towns by that name in the US.

In sum, the sign-vehicle of the dicisign comprises its double structure (an icon and an index); the interpretant is the claim that the object is in fact such and such. In the next section, we explore how we can see that a biochemical dicisign claims something about some other thing or process in terms of what in biology would correspond to truth-value—namely, adaptive value for survival and reproduction (falsity would be maladaptive, dysfunctional processes).

13.2 Analyzing Signal Peptides as Dicisigns

Can the signal function of an SP be analyzed with tools from Peircean semiotics? Our *hypothesis* is that the notion of a dicisign, as introduced above, more so than other sign categories, provides a plausible explanation of how the SP functions as part of a semiotic process. Dicisigns or natural propositions convey information, which seems to be central to the SP phenomenon: The function of an SP mirrors the function of a dicisign. This aligns with the intuitive idea that SP "says something" (provides actionable information) about a real aspect of the world: "This protein__" (the index sign) is joined with the icon sign "____ should be translocated into the ER" (assuming we are talking about the eukaryotic situation, which will be used as the example in this section). In other words, SPs are a particular example of how biological (i.e., nonmental and nonlinguistic) dicisigns function. However, it is not at all an easy matter to analyze SPs in the light of their dicisign properties, which is why we are detailing three possible interpretations concerning what to attribute as the description—that is, the iconic aspect, and the reference or index, in the SP or translocation process. These three analyses are related to three different aspects of the information mediated by the dicisign, all contributing to a biosemiotic interpretation of signal peptide translocation processes. Thus, we see at least three ways of analyzing SPs as dicisigns (in part due to the fact that a general theory, like that of dicisigns, is underdetermined by concrete particular instances of what the theory is describing).

13.2.1 Analysis 1

- **Object**: Protein translocation of *this* protein across the ER membrane
- **Dicisign (icon–index syntax)**: SP–covalent bond between the SP and the mature part of the protein
- **Interpretant**: SP-SRP binding

In this analysis (see figure 13.3A), a dicisign functions by means of the colocalization of the SP-part and the "mature protein"-part (MP) of the long "immature" preprotein (i.e., the covalent bond between the two parts). The "proper significant effect" or interpretant action is the process consisting of a binding between the SP and the SRP, and this triggers further interpretant actions, such as the placement of the ribosome-SP-SRP-receptor complex at the ER membrane that completes the translocation process. Thus, in this analysis, the SP embodies a specific form (an icon of its binding to the SRP) and the covalent bond points to the mature protein (indexical function of the dicisign), and this informs the ongoing processing of the SP-SRP-MP complex. Using a zip code analogy, the dicisign would be the glue (syntax) that attaches the address label to this particular letter, saying "connect this letter envelope to this postal area."

13.2.2 Analysis 2

- **Object**: ER (as a destination)
- **Dicisign (icon–index syntax)**: {SP-SRP}–SRP-receptor
- **Interpretant**: Protein translocation across the ER membrane

While in Analysis 1, the dicisign had the meaning of binding (SP-SRP) as a part of the total process, here, the focus is on the overall process (protein translocation), which becomes the very meaning of the dicisign, now understood as the big complex consisting in the {SP-SRP}–SRP-receptor. The SP-SRP complex represents the iconic aspect of the dicisign, having a shape that embodies a form that is recognized by the SRP-receptor. Shape, biochemically speaking, could be understood semiotically as functioning as a sign-vehicle of coded processes (e.g., the key-and-lock system of enzymes, proteins, and transcription factors in the cell, or the code in language that connects sounds, words, and meanings). When the SP-SRP complex is recognized by the SRP-receptor, it forms a dicisign by indicating its object: the ER as a destination. Using the zip code analogy, the zip number (this big complex) corresponds to a sign indicating a functional location of a protein (the ER lumen). Note that what was the object in Analysis 1 is now the interpretant.

13.2.3 Analysis 3

- **Object**: SP cleavage
- **Dicisign (icon–index syntax)**: SP–translocon
- **Interpretant**: Signal peptidase catalytic function

The SP bound to the translocon functions as a substrate for the signal peptidase. The position of the SP in the translocon is important; without binding to the translocon, the SP is not a substrate. So the SP describes, by its cleavage site motif, exactly where the cleavage should take place, while the translocon indicates the signal peptidase.

13.2.4 General Comments on the Analyses

Effectuating an SP as an imperative dicisign of course depends on a whole context of physicochemical and semiotic processes (like postmen able to decode the meaning of zip codes), just like securing the functional transport of a protein across the right membrane demands

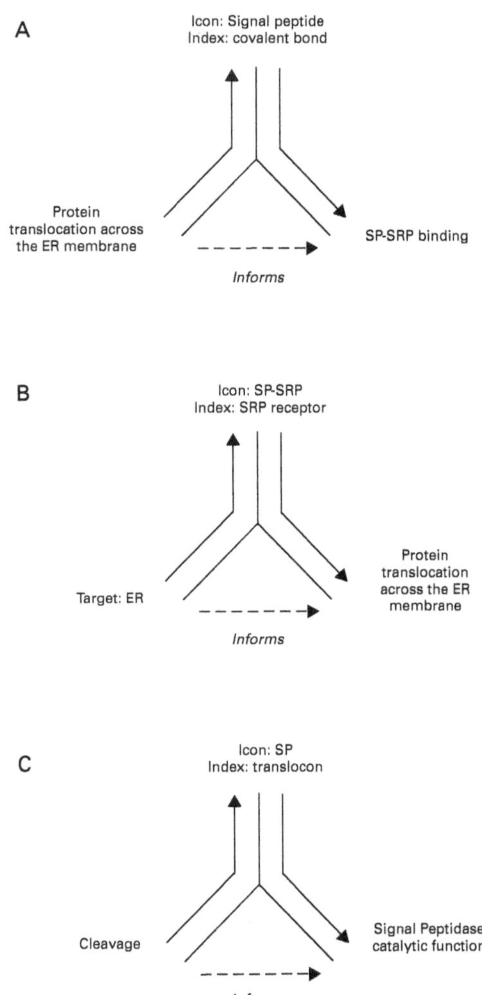

Figure 13.3
Represents three possible interpretations of signal peptide as a semiotic phenomenon. Panel A presents Analysis 1, panel B presents Analysis 2, and panel C presents Analysis 3.

a complex of microprocesses (chemical, but also protosemiotic to the extent they can be described as involving molecular recognition as a form of pattern recognition) involving not only the SP but also the SRP, the right location at the membrane of the SRP-receptor, and the binding of that receptor to the SP-SRP-ribosome complex (see figure 13.1).

As in all semiotic systems, there are several levels of analysis possible, such as the material building blocks, their dynamic functionalities, and the significance in terms of meaning related to the survival and reproduction of the cell or organism. An advantage of analyzing SP function as an instance of semiosis is that the "mechanistic" (biochemical) details of the workings of the systems are relatively well known, though some details are missing. The semiotic analysis does not question the need to understand the biochemical details, but it supplies this with a functional language focused on general patterns of biologically meaningful transformations within the cell. Just like chemical reactions ultimately rely on the micro-workings of quantum physical processes, quantum physics cannot capture the generalities of molecular reaction patterns of chemical functional groups (cf. Chang, 2015). Compare "code" in the concept of the genetic code, which actually *is* a code (also in the semiotic sense), in the same way "signal" in the concept of an SP is not just a biochemical but also a genuine semiotic term being an instance of a sign.

The existence of covalent chemical bonds between the SP part and the rest of the protein affords an instance of colocalization. Also weak interactions—for example, hydrogen bonds, hydrophobicity, and electrostatic interactions—which play a crucial role in general molecular recognition (and binding), can afford the same (semiotic) function. Two macromolecules interacting through weak bonds may cause conformational changes that again may open new sites of potential molecular recognition processes, thus affording new steps in the semiosic[2] interpretation involved in SP-mediated translocation. One general drawback on such an analysis may be the phenomenon of underdetermination, shown above, which leaves several options open for applying the semiotic tools in the analysis. We showed three analyses but could probably have shown even more—possibly one for each step in figure 13.1 (where our three analyses roughly correspond to steps 1, 3, and 4). An answer here may be that the different analyses are all contributing to an overall understanding of the concatenation and juxtaposition of interpretation processes involved in cellular metabolism. The whole biochemical "machinery" of SP-mediated protein translocation is complex, involving many steps, and both the biochemical and biosemiotic analyses may focus on its different parts.

One aspect of dicisigns mentioned in section 13.1.4 is that they can be false. But what is a false SP? A true SP signals the translocation of a protein that *should* be translocated because it has its proper biological function outside the cell (or somewhere in the secretory pathway of the cell). A false SP, by comparison, would be an SP coupled to a protein that should not—from the point of view of the organism—be translocated, or a mutated SP that does not bind properly to other components and therefore fails to get its protein translocated. In both cases, a protein becomes mislocalized. Mislocalization of proteins is not a merely theoretical possibility in the semiotic analysis, but a real phenomenon that seems to occur in a large number of human genetic disorders (Lacoste et al., 2024).

13.3 On the General Relation between Molecular Biology, Bioinformatics, and Biosemiotics

Molecular biology, as an experimental laboratory-based field of research, uses the vocabulary of chemistry and biochemistry to describe cellular substances (including peptides, nucleic acids, lipids, carbohydrates), their conversion via metabolic processes, and the different physical, chemical, and biological factors influencing these processes. It is not an autonomous field because it depends on lab-based research in chemistry and biology, interacting with fields like molecular kinetics and genetics, and it is generating sequence data on DNA, RNA, and proteins. These data are often further analyzed in the cross-disciplinary field of *bioinformatics* (drawing on computer science approaches to information theory, mathematical modeling, and machine learning) aiming to find general principles and patterns in the data that can shed light on basic biological processes, including the information-dependent intracellular processes of metabolism and transport. Thus, as a field, bioinformatics may help generate new hypotheses about molecular regularities in the cell that may be explored and tested experimentally in the lab. *Biosemiotics*, also a cross-disciplinary field, integrates semiotic and biological approaches via a focus on living systems from the perspective of sign interpretation and the emergence of meaning in cells and organisms, providing deeper analysis and understanding of functional and holistic features therein. We hope that the present case of SP as an instance of dicisign interpretation can help exemplify one of the connections between (molecular) biology, bioinformatics, and biosemiotics.

In summary, bioinformatic tools are used to analyze the massive datasets generated by techniques like DNA sequencing. By analyzing these datasets, bioinformatics helps uncover patterns, relationships, and functions of biomolecules. This information feeds back into molecular biology research. Biosemiotics explores the significance of information flows, thus adding another layer of analysis by looking at the information encoded within biomolecules and how they communicate with each other, hoping to comprehend major patterns of communication, both in the biological domain and beyond.

What is contributed by the biosemiotic analysis is more than an analogy; it is a statement that the semiosis is a real aspect of nature and not just a convenient shorthand for complicated chemical processes. It is easy to put up an analogy such as "an SP is like a zip code," but that refers to a human artifact (the zip code), which would not exist without billions of years of biological evolution involving a plethora of biological phenomena, including SPs. Thus, the analogy is circular. Biosemiotics offers a noncircular way of talking about terms like "function," "information," "code," "signal," and "cue." As is stated by Kull et al. (2009, p 170), "biosemiotics has the scientific task of (1) grounding such terms in a physicobiological context, (2) defining and interrelating such terms with the constant aim of avoiding the anthropomorphisms that threaten when they are left with implicit definitions only, and (3) so to make biology theoretically complete."

In this scenario, one might ask, how can a semiotic analysis shed light on the limits of computer (language) models used for gaining knowledge about signal peptides? Although the analyses above do not answer this question, *all* analyses (including semiotic and methodological) of the limits of LLMs as they are used in research have to deal with the relation between *encoding* and *decoding* (where "encoding" includes collecting data, building a model, and training it on datasets, and "decoding" includes critically interpreting the output answers and validating them in the context of a field's state-of-the-art knowledge about the relevant system's causal processes). This is what Robert Rosen called *the modeling relation* (Rosen, 1991, pp. 57–64), and it crucially involves representations, inferences, and interpretations (and thus, from our point of view, is intrinsically semiotic).

As we have seen, the semiotic approach allows for using the dicisign concept in several ways to analyze the "meaningful" (functional, adaptive) realization of protein synthesis and translocation, depending on which parts of the process one is focusing upon. Apparently, in the case of language models, there is no predefined range of correlations between a word (or amino acid) and its neighbors. What does it take, in terms of computation and of real biochemical processes, to "predict" or "identify" an SP? The cell—as a network of physical processes that are at once biochemical and semiotic—identifies an SP by the mentioned weak interactions and processes involving also covalent bonds, and local properties (say, of an amino acid) may have distant effects (on the same or other proteins), like in spoken language, where the meaning of a long sentence may change if we add, by its very beginning, for example, "it is misleading to say that . . ." LLMs are far from perfect in capturing more global interrelationships in the meanings of written texts (and "hallucinations" can be seen as indicative of this imperfection, possibly indicating false dicisigns, with a mismatch between description and reference).

13.4 Discussion and Limitations

It may seem surprising that the large generative language models, developed within the realm of human language (text prediction), with some success could be extrapolated to the realm of SPs and protein function, considering the vast differences in complexity of these two semiotic systems. However, as pointed out by Srinivasan,[3] the analogy between protein "language" and natural language is limited.

An important aspect that differentiates a bioinformatic analysis from a biosemiotic one is the role of meaning-making and communication processes in the object of study. The argument for seeing SPs as dicisigns pertains to phenomena that lie above the threshold that distinguishes biosemiotic entities from biochemical reactions: the possibility to make choices (Kull, 2018). Thus, a potential objection to our analyses of SPs as dicisigns could be that the phenomenon also works in vitro, in a test tube without living cells present. After all, the first experimental demonstration of SP action occurred in a cell-free system: "functional rough microsomes" derived from murine myeloma cells (Blobel and Dobberstein, 1975b). A microsome is a biochemical artifact, a vesicle derived from fragmentation of the endoplasmic reticulum. Our answer to this would be that the SP in a microsome system ceases to be a dicisign since its truth value—the biological function and hence the effect of survival and/or reproduction for the cell or organism—has been removed. In vitro, it is impossible to discern a true SP/protein pair from a false one: An SP coupled to a "wrong" protein—that is, a protein that should not (from an organismal viewpoint) be translocated—would function just as well as a biologically "right" SP/protein pair. For the organism, however, secreting a wrong protein could lead to pathological consequences.

Thus, seeing SP as dicisigns implies two things: First, dicisigns are formed only in a given state of information—that is, in a given context that allows it to be interpreted by something else and, thus, guide further semiosis. SPs, to work as dicisigns, have to be functioning within a cell environment. Thus, in scenarios of in vitro experimentation of SPs, they do not act as dicisigns: Even though they can be part of the translocation process they represent, there will be no biological function involved. Second, seeing SPs as dicisigns is relevant to the phenomenon of misplaced proteins due to SP malfunction, a misguided syntax that falsely claims that "this protein belongs to this place."

13.5 Conclusion

Considering the limitations of our analysis mentioned above, we conclude that SPs can be seen as natural dicisigns that are crucial for the distinctive distribution of proteins inside cells and the selective transport of some of them to the extracellular environment. The fact that dicisigns possess an inherent logical syntax (with a subject-predicate structure) indicates that cellular metabolism embodies a complex bio-logic that is both biochemical in its embodiment and biosemiotic in its processual organization.

The pLMs we have described are not literally but only in a metaphorical sense "learning to read the language of life." The very notion of SP-guided protein synthesis and translocation as embodying a "language of life" is a misleading metaphor, given that language in the strict sense is specific to human oral and written communication, so a better and much broader term than language is "semiotic system." In the case discussed here, the models have been trained on large sets of data and have (in a statistical and nonsubjective sense) "learned" to predict the existence of signal peptides based on only non-annotated sequence information. This instance of pattern recognition and prediction pertains to a purely "textual" domain of data of sequences that for us as researchers *represent* real biological molecules (including DNA and polypeptides) but are not themselves part of any biological system—just like an output from ChatGPT can be taken (for us) to *represent* an answer to a question posed in human language. It is us, as semiotic agents, that can "translate" the output strings into meaningful words and sentences, and be impressed by those instances where the statistically based outputs are congruent to plausible and (for us) reasonable answers, and disappointed or puzzled when SignalP 6.0 predicts some SPs that are unlikely and never seen in nature.

Given these caveats, it would be better to say that pLMs can be used to *help humans read* some aspects of the *semiotics* of life.

Acknowledgments

Thanks to the FAPESP (project n. 2020/03134-1) SP research grant, and to the participants of the XXIII Gatherings of Biosemiotics 2023, and to members of the bioinformatics section at the Department of Health Technology, Technical University of Denmark. We would also like to thank Ellen Vallentin Asmus for drawing figure 13.1.

Notes

1. See also Nielsen, 2022, Behind the paper: "SignalP 6.0 predicts all five types of signal peptides using protein language models," in *Bioengineering and Biotechnology*, https://communities.springernature.com/posts/signalp-6-0-predicts-all-five-types-of-signal-peptides-using-protein-language-models

2. The term *semiosic* pertains to the very processes of sign action (semiosis), while *semiotic* pertains to semiotic theories and concepts.

3. Srinivasan, 2024. A review on protein language models, Apoorva Srinivasan's Blog, https://www.apoorva-srinivasan.com/plms/.

References

Blobel, G., and Dobberstein, B. 1975a. Transfer of proteins across membranes. I. Presence of proteolytically processed and unprocessed nascent immunoglobulin light chains on membrane-bound ribosomes of murine myeloma. *The Journal of Cell Biology*, *67*, 835–851.

Blobel, G., and Dobberstein, B. 1975b. Transfer of proteins across membranes. II. Reconstitution of functional rough microsomes from heterologous components. *The Journal of Cell Biology*, *67*, 852–862.

Blobel, G., and Sabatini, D. D. 1971. Ribosome-membrane interaction in eukaryotic cells. In L. A. Manson (Ed.), *Biomembranes* (pp. 193–195). Plenum Press.

Briggs, M. S., and Gierasch, L. M. 1986. Molecular mechanisms of protein secretion: The role of the signal sequence. In C. B. Anfinsen, J. T. Edsall, and F. M. Richards (Eds.), *Advances in protein chemistry* (pp. 109–180). Academic Press.

Chang, H. 2015. Reductionism and the relation between chemistry and physics. In T. Arabatzis et al. (Eds.), *Relocating the history of science* (pp. 193–209). Springer.

Cooper, G. M. 2000. *The cell: A molecular approach* (2nd ed.). Sinauer Associates.

El-Hani, C. N., Queiroz, J., and Emmeche, C. 2009. *Genes, information, and semiosis*. Tartu University Press.

Elnaggar, A., Heinzinger, M., Dallago, C., Rehawi, G., Wang, Y., Jones, L., Gibbs, T., Feher, T., Angerer, C., Steinegger, M., Bhowmik, D., and Rost, B. 2022. ProtTrans: Toward understanding the language of life through self-supervised learning. *IEEE Transactions on Pattern Analysis and Machine Intelligence*, *44*, 7112–7127.

Emmeche, C., and Hoffmeyer, J. 1991. From language to nature: The semiotic metaphor in biology. *Semiotica*, *84*(1–2), 1–42.

Heinzinger, M., Elnaggar, A., Wang, Y., Dallago, C., Nechaev, D., Matthes, F., and Rost, B. 2019. Modeling aspects of the language of life through transfer-learning protein sequences. *BMC Bioinformatics*, *20*, 723.

Hoffmeyer, J. 2008. *Biosemiotics: An examination into the signs of life and the life of signs*. University of Scranton Press.

Judson, H. F. 1979. *The eighth day of creation: Makers of the revolution in biology*. Simon and Schuster.

Keller, E. F. 2002. *Making sense of life: Explaining biological development with models, metaphors, and machines*. Harvard University Press.

Kleisner, K., and Maran, T. 2019. Introduction to signs and communication in mimicry. *Biosemiotics*, *12*, 1–6.

Kull, K. 2018. Choosing and learning: Semiosis means choice. *Sign Systems Studies*, *46*(4), 452–466.

Kull, K., Deacon, T., Emmeche, C., Hoffmeyer, J., and Stjernfelt, F. 2009. Theses on biosemiotics: Prolegomena to a theoretical biology. *Biological Theory*, *4*(2), 167–173.

Lacoste, J., Haghighi, M., Haider, S., Reno, C., Lin, Z.-Y., Segal, D., Qian, W. W., Xiong, X., Teelucksingh, T., Miglietta, E., Shafqat-Abbasi, H., Ryder, P. V., Senft, R., Cimini, B. A., Murray, R. R., Nyirakanani, C., Hao, T., McClain, G. G., Roth, F. P. . . . Taipale, M. 2024. Pervasive mislocalization of pathogenic coding variants underlying human disorders. *Cell*, *187*(23), 6725–6741.

Lin, Z., Akin, H., Rao, R., Hie, B., Zhu, Z., Lu, W., Smetanin, N., Verkuil, R., Kabeli, O., Shmueli, Y., Dos Santos Costa, A., Fazel-Zarandi, M., Sercu, T., Candido, S., and Rives, A. 2023. Evolutionary-scale prediction of atomic-level protein structure with a language model. *Science*, *379*, 1123–1130.

Nielsen, H., Engelbrecht, J., Brunak, S., and von Heijne, G. 1997. Identification of prokaryotic and eukaryotic signal peptides and prediction of their cleavage sites. *Protein Engineering, Design and Selection*, *10*, 1–6.

Nielsen, H., Tsirigos, K. D., Brunak, S., and von Heijne, G. 2019. A brief history of protein sorting prediction. *Protein Journal*, *38*, 200–216.

Nielsen, H., Teufel, F., Brunak, S., and von Heijne, G. 2024. SignalP: The evolution of a web server. In F. Lisacek (Ed.), *Protein bioinformatics* (pp. 331–367). Springer.

Peirce, C. S. 1931–1958. *The collected papers of Charles Sanders Peirce* (vols. I–VI ed. by C. Hartshorne and P. Weiss, vols. VII–VIII ed. by A. W. Burks). Harvard University Press. [Quoted as CP, followed by the volume and paragraph]

Peirce, C. S. 1998. *The essential Peirce. Vol. 2* (1893–1913) (ed. by the Peirce Edition Project). Indiana University Press. [Quoted as EP2, followed by page number]

Rosen, R. 1991. *Life itself*. Columbia University Press.

Stjernfelt, F. 2014. *Natural propositions: The actuality of Peirce's doctrine of dicisigns*. Docent Press.

Stjernfelt, F. 2022. Co-localization as the syntax of multimodal propositions: An amazing Peircean idea and some implications for the semiotics of truth. In F. Stjernfelt, *Sheets, diagrams, and realism in Peirce* (pp. 70–109). De Gruyter.

Teufel, F., Almagro Armenteros, J. J., Johansen, A. R., Gíslason, M. H., Pihl, S. I., Tsirigos, K. D., Winther, O., Brunak, S., von Heijne, G., and Nielsen, H. 2022. SignalP 6.0 predicts all five types of signal peptides using protein language models. *Nature Biotechnology*, *40*, 1023–1025.

von Heijne, G. 1983. Patterns of amino acids near signal-sequence cleavage sites. *European Journal of Biochemistry*, *133*, 17–21.

West, D. 2015. Recommendations as imperative propositions in the operation of abductive reasoning: Peirce and beyond. In M. Bergman and J. Queiroz (Eds.), *The Commens encyclopedia: The digital encyclopedia of Peirce Studies "New Edition. Pub. 150526-0100a."*

West, D. 2018. The work of Peirce's Dicisign in representationalizing early deictic events. *Semiotica*, *225*, 19–38.

14 Meaning in Cells: A Relational Biosemiotic Approach

Federico Vega

14.1 Introduction

What is an organism? What distinguishes a living being from inanimate matter? Are the laws of physics and the sciences derived from them sufficient to answer these questions?

Theoretical biologist Robert Rosen answers the last question in the negative and argues for the need to complement the accepted biophysical laws with a principle that governs the organization of the system. He explains organisms as natural systems composed of *components*—where a component is a part of the system to which a function can be attributed—and analyzes them based on Aristotelian causes. He concludes that an organism is a natural system closed to efficient causation and proposes the (M, R) system—a system composed of three biological functions: metabolism, repair, and replication—as a minimal model of organismic being. Finally, to apply his relational biology to the work of biologists, he raises the need to realize this formal system on physicochemical entities and processes.

Conversely, logician and philosopher Charles Sanders Peirce posits that in order to study nature, a type of action that cannot be characterized as ordinary physical action must be considered: semiotic action. A semiotic action implies the existence of a semiotic relation, which is defined as a triadic relation (one not reducible to dyadic relations) between three subjects—sign, object, and interpretant—in which the object determines that the sign has the efficacy to determine the interpretant (an effect upon an agent), such that the interpretant leads the agent from the sign to the object.[1]

This chapter presents and applies to the cell a theoretical framework for the definition and study of life that integrates Rosen's relational biology and Peirce's theory of signs. Section 14.2 puts forward this *relational biosemiotics* and explains the possibility of realizing, the actual realization, and the habit of realizing semiotic actions that correspond to the interpretation of signs that produce meanings in natural systems. It is stated that an organism is a natural system that realizes *semiotic* closure to efficient causation, and a semiotic (M, R) system is proposed as a model.

On this theoretical basis, the problem of realizing the model is addressed in sections 14.3 and 14.4. First, a basic cell is presented that realizes semiotic closure to efficient causation, and then the model is extended to add to the system the capacity to self-regulate so as to adapt to changes in its environment.

In section 14.5, the analysis of the semiotic relations introduced is further elaborated. The results are summarized in section 14.6.

14.2 Notes on Relational Biosemiotics

This section recapitulates and complements the basic concepts of a relational biosemiotics introduced in Vega (2021, 2023, 2024), as developed from Rosen's relational biology and Peirce's semiosis.[2]

14.2.1 Components, Semiotic Relations, and the (M, R) System

Rosen discusses organisms as natural systems composed of components, where a component is a part of the system to which a function can be assigned and is the basic unit of its organization. He models each component with a mapping between two sets, f: A → B, and analyzes it in terms of Aristotelian causes. The material cause is the input; the efficient cause is the entity[3] that carries out the action that produces the flow from the input to the output; the formal cause is the way in which the component's action develops; and the final cause is the function that the output of the component performs in the system.

If, following Peirce, we admit the presence of semiotic action in a natural system, then the sign, the object, and the interpretant must necessarily be natural entities. Moreover, if we describe organisms as systems composed of components, we should conclude that the three subjects of the semiotic relation are, respectively, the material cause, the output, and the efficient cause of the component that carries out the semiotic action—that is, an action in which it can be stated not only that the efficient cause leads from the material cause to the output in the way indicated by the formal cause (as in any component) but also that the output (the object) determines that the input (the sign) determines the efficient cause (the interpretant).[4]

In figure 14.1, the mapping on the left represents a component in which the efficient cause is f, the material cause is an a ∈ A, the output is a b ∈ B, and the formal cause is the structure of the mapping, which can be read ". . . entails that . . . entails . . ."[5] The two mappings on the right exemplify a semiotic relation defined by the action of the component I: S → O and the rule O: S → I, which indicates that the output (the object) determines that the input (the sign) determines the efficient cause (the interpretant).

Rosen claims that a natural system is an organism if and only if it realizes closure to efficient causation—that is, if all its components have efficient causes that are generated within the system and effects that contribute to the production of other efficient causes. Furthermore, it is this closure that converts the parts of a natural system into components of an organism and assigns them their functions within the system—that is, their final causes.

The model he proposes for organisms is the replicative (M, R) system, which is composed of three biological functions: metabolism, repair, and replication. The first two correspond to sets of components: Metabolism incorporates entities from the environment and produces outputs that serve as inputs to other components or leave the system; repair preserves metabolism by synthesizing its efficient causes from its outputs.

However, replication is a mathematical operator, built on the metabolic and repair functions, which completes system closure. It should be noted that replication does not produce

Figure 14.1
Left: a component as a mapping from set A to set B. Right: a semiotic relation as the union of a component I: S → O and the determination of its efficient cause. Continuous line arrows illustrate the effect of the component (solid-headed) and the flow from input to output (hollow-headed). Dashed line arrows symbolize the determination of the interpretant.

additional entities but rather constrains the interrelations between the existing ones; it defines rules that allow the following semiotic account of relational biology.[6]

Until the late 1980s, Rosen analyzed organisms as anticipatory systems, where an anticipatory system is defined as "a system containing a predictive model of itself and/or of its environment, which allows it to change state at an instant in accord with the model's predictions pertaining to a later instant" (Rosen, 2012, p. 313).

However, *Life Itself* (Rosen, 1991) represents a qualitative change in Rosen's thinking. As Pattee (2007, p. 2274) points out, "In *Life Itself* Rosen no longer characterized life by its internal predictive models that allow adaptive evolution. Instead he develops a timeless relational concept of organisms based on natural and inferential "entailments" derived from Aristotle's causal categories."

Rosen (2006, p. 11) explains that he realized that "(M, R) systems have an inherent anticipatory aspect, built into their very organization." He argues as follows: In mathematics, the analog to anticipation is impredicativity (self-reference). Closure to efficient causation establishes an impredicative loop; therefore, an (M, R) system includes anticipatory behavior.

Following Rosen, it seems appropriate to affirm that the natural system that realizes the (M, R) system interprets what this model indicates about what might occur in the future from the state of the system and the environment in the present, and acts accordingly to preserve closure to efficient causation.

Nevertheless, the definition of a semiotic relation in a natural system such as that presented above allows for a more suitable account of semiosis in organisms.[7] As an alternative to the replicative (M, R) system, I posit a semiotic (M, R) system—a system in which Rosen's functions of metabolism and replication are included, but closure to efficient causation is completed by semiotic relations established in certain components in which the output determines that the material cause determines the efficient cause. As in the case of the replication function, the determination of the efficient cause (only) adds constraints to the model. Figure 14.2 includes diagrams of the replicative and the semiotic (M, R) systems.

14.2.2 Interpretants, Objects, and Signs

We will delve now into the three subjects of the semiotic relation. Peirce identifies three types of interpretants—immediate, dynamic, and final—and two types of objects—immediate and dynamic. The *immediate interpretant* is an abstraction, a potentiality in which the interpretability of the sign consists, a response to the general features of the object and to the syntax of the semiotic relation. The *immediate object* is the object as it is supposed to be, a partial replica of the real object that includes some of its features. The *dynamic interpretant*

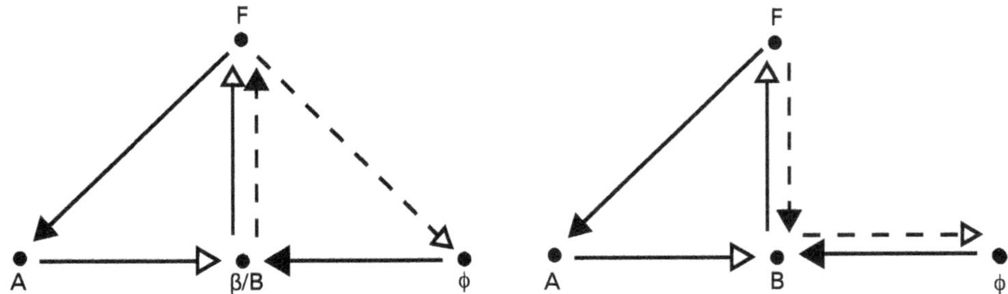

Figure 14.2
Left: the replicative (M, R) system, modified from Rosen (1991, p. 251, figure 10C.6). F and Φ denote the sets of components that make up metabolism and repair, A and B the sets of inputs and outputs of metabolism, and β an operator that corresponds to the replication function. Right: the semiotic (M, R) system obtained by replacing the replication function with the determination of the efficient causes of repair. This second model is a modification of one that Rosen (1991, p. 238) postulated for complex systems. Dashed lines are used to represent mappings that do not produce additional entities but rather constrain the interrelations between the existing ones.

is the effect on the agent that is actually involved in an interpretive process. It leads from the sign to the dynamic object, which is the object as it really is—that is, as it is known at the end of the inquiry. The *dynamic object* is the result to which the interpretive process leads. The *final interpretant* is again a potentiality: the ideal interpretant of the sign for the interpretive purpose. In each interpretive process, the dynamic interpretant actualizes the immediate interpretant and contributes to the definition of the final interpretant.

Simultaneously, Peirce classifies signs into icons, indexes, and symbols, depending on the type of relationship established between the sign and the object. Icons are characterized by being mediated by a similarity between the sign and the object, indexes by a correlation between both—a physical connection or a spatial or temporal association—and symbols correspond to formal or conventional associations.

Considering the above, the following can be stated about the three subjects of a semiotic relation established in a component of an organism. The immediate interpretant is the potential efficient cause—that is, a set of features of an entity that make it suitable to act on the material cause and obtain an entity that has the features of the output. The immediate object is a set of features of the output that make it suitable for determining that the input determines the immediate interpretant. The dynamic interpretant and the dynamic object are two existing entities that fulfill those features in an action of the component. The final interpretant is a habit by which entities of the efficient cause type tend to produce entities of the output type from the material cause.

Finally, the identification of the type of sign will have to be based on the analysis of the features that define the relationship between the entities that are the material cause and the output of the component.

14.2.3 Interpretation, Agent, Purpose, and Meaning

Short (2007) proposes a definition of interpretation built on the naturalized account of purpose, where a purpose is a *general type* of outcome or effect that explains the existence of something because it is selected for having effects of that general type—that is, a type of outcome that explains why there are outcomes of that type—and it is *explanatory* because it is selected-for as a *means*. He states, "An interpretant interprets a representamen[8] as a sign of an object if and only if (a) the interpretant is or is a feature of a response to the representamen for a purpose, (b) the interpretant is based on a relation, actual or past or apparent or supposed, of the representamen to the object or of things of the type of the representamen to things of the type of the object, and (c) obtaining the object has some positive bearing on the appropriateness of the interpretant to the purpose" (Short, 2007, p. 158).[9]

It should be highlighted that, in an interpretation (an actual semiotic action), the interpretant, the representamen, and the object correspond respectively to the dynamic interpretant, the sign, and the dynamic object. However, the expressions "a feature of" and "things of the type of" also introduce the following into the definition: the immediate interpretant, the immediate object, and the characteristics of the sign—and so the origin and evolution of the interpretive action.[10]

Certainly, Short argues that the concepts of sign, interpretation, and meaning are intentional concepts and that the intentional action is limited to animals. Therefore, to apply his definition to the organism, it is necessary to naturalize the concept of intentionality. Starting from a proposal by Stjernfelt (quoted in Favareau and Gare, 2017) in which he links *intention* and *the cyclic organization of metabolism*, I have argued elsewhere (Vega, 2021) that it is the closure to efficient causation that makes it meaningful to talk about intention, since the directionality of intention is governed by system closure. Thus, it can be stated that "in the action of a component in an organism, the efficient cause interprets an input as a sign of an output if and only if (a) the efficient cause is or is a feature of a response to the input for a purpose, (b) the efficient cause is based on a relation, actual or past or apparent or supposed, of the input to the output or of things of the type of the input to things of the type of the output, and (c) obtaining the output has some positive bearing on the appropriateness of the efficient cause to the purpose."

It should be noted that the statement "the efficient cause interprets" does not mean that the agent who performs the interpretation is the efficient cause. Neither the efficient cause (the dynamic interpretant) nor the input (the sign) nor the output (the dynamic object) act according to a purpose. In the interpretive process in an organism, the agent who acts purposefully is the natural system that realizes closure to efficient causation, and the purpose is the set of features of the output that explain its existence because it is produced for having the effect of contributing to system closure, both by performing a function and by constraining the semiotic action.

An interpretive process implies the realization of two closures: one global, which defines the organism, and another local, which defines the semiotic relation. Interpretation is executed locally, but only because the system realizes closure to efficient causation: the interpreting agent is the organism.

All components have final causes—that is, produce outputs that perform functions that contribute to closure to efficient causation. However, only those components in which the output determines that the material cause determines the efficient cause carry out semiotic actions that correspond to interpretations that produce meanings. Therefore, in an organism, meaning is a special kind of final cause; it is the final cause of a component in which the product obtained constrains its own production in the manner indicated by the semiotic relation. And it is not only a final cause; it is also the set of local rules that complete system closure, which in turn assigns functions to all components and defines and maintains the system.

14.3 A Basic Cell That Realizes a Semiotic (M, R) System

This section presents and analyzes a cell model that realizes semiotic closure to efficient causation.[11] First, the semiotic (M, R) system is constructed, and then the identified semiotic relations are further explored.

14.3.1 Metabolism and Repair

The model of the metabolic system that I propose for the simplest cell consists of four components, as represented in figure 14.3.

Mapping 1 summarizes the union of two processes.[12] In the first, nutrient transporters (tr_N) bound to the cell membrane introduce entities from the environment ($Input_{C\&A}$) into the system. These entities are the substrate of a second process, in which the catabolic and anabolic enzymes of intermediary metabolism ($enz_{C\&A}$) catalyze the production of the building blocks of the cell interior and membrane—nucleotides (nc), amino acids (aa), and fatty acids (fa)—as well as entities that serve as energy sources (ESource). Mappings 2 and 3 model the formation of the membrane. First, lipid synthesis enzymes (enz_{LIP}) catalyze the formation of

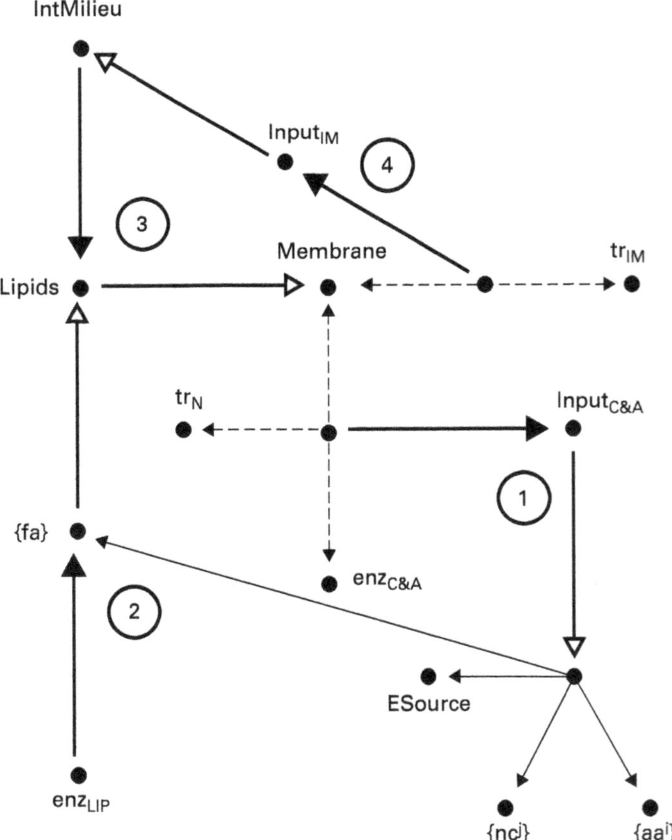

Figure 14.3
The metabolic system. In addition to the notation rules in figure 14.1, thin arrows represent the simultaneous contribution of several elements (dashed lines) and alternative options (continuous lines). Superscripts refer to specific elements. Curly brackets denote sets.

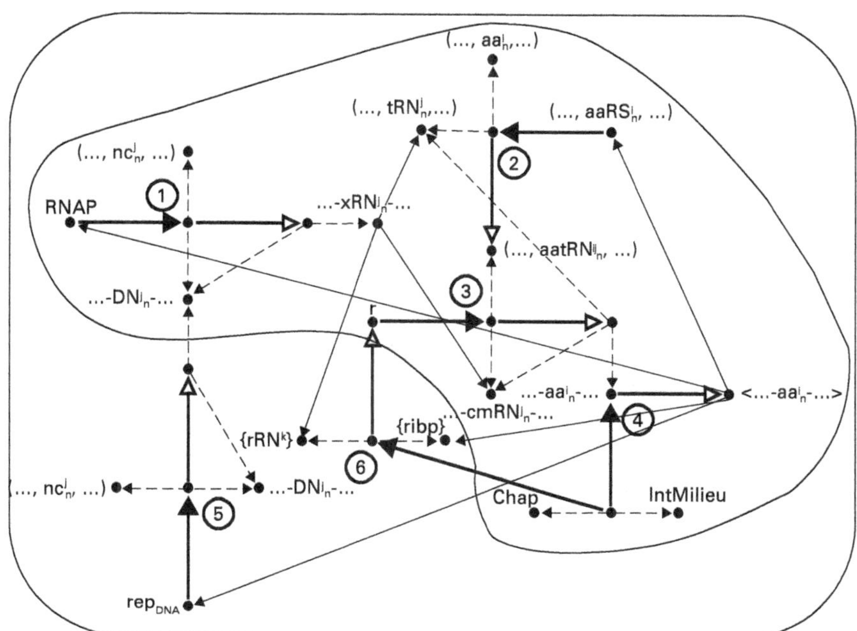

Figure 14.4
The repair system. In addition to the notation rules in the figures above, subscripts refer to positions in a series. That is, for each element X, (. . . , X^i_n, . . .) and . . . -X^i_n- . . . stand respectively for ordered sequences and chains of Xs, and < . . . -aa^i_n- . . .> denotes a protein—that is, a folded chain of amino acids.

lipids from fatty acids. Then, the intracellular milieu assists their assembly. Finally, mapping 4 shows that transport proteins (tr_{IM}) and the membrane maintain the intracellular milieu using additional inputs from the environment ($Input_{IM}$)—for example, electrolytes.

It is now necessary to add a system that replaces the efficient causes of metabolism that degrade. I propose a repair system that consists of the six components in figure 14.4. The first four, which correspond to protein synthesis, can be grouped into three parts: transcription (mapping 1), translation (mappings 2 and 3), and folding (mapping 4). The other two components replace the DNA chains that are part of the transcription input (mapping 5) and the ribosome (mapping 6).

Consider protein synthesis. First, an enzyme RNA polymerase (RNAP) catalyzes the production of a messenger, transfer, or ribosomal RNA (mRNA, tRNA, or rRNA, summarized as xRNA) from a nucleotide sequence, using a DNA chain as a template. Second, an aminoacyl tRNA synthetase (aaRS) binds an amino acid and a tRNA and produces an amino acyl tRNA (aatRNA). Third, a ribosome is the efficient cause of a process in which the material cause consists of a chain of mRNA codons (cmRNA)[13] and an aatRNA sequence, and the output is formed by the incoming cmRNA chain, a tRNA sequence, and an amino acid chain. Finally, the folding that forms the protein is assisted by chaperones and the intracellular milieu, as shown in mapping 4.

Additionally, DNA repair, represented in mapping 5, is a process in which DNA repair proteins (rep_{DNA}) produce a copy of a DNA chain; and ribosome repair—mapping 6—is carried out by processes catalyzed by chaperones and the intracellular milieu,[14] which have ribosomal proteins (ribp) and rRNAs as input and a ribosome as output.

By putting together and connecting the mappings in figures 14.3 and 14.4, a model of cell metabolism and repair is obtained. Next, we discuss whether the system obtained realizes closure to efficient causation.

14.3.2 A Semiotic Closure to Efficient Causation

Examining the components of metabolism and repair one by one, we can immediately verify that all the entities that participate in their efficient causes are produced in the processes described. However, it should be noted that each aminoacyl tRNA synthetase that is part of the efficient cause of one of the components involved in the synthesis of a protein is, in turn, a protein. Therefore, to complete closure to efficient causation, some rules must be fulfilled that relate each protein with those involved in its synthesis.

In relational biology, the response to this question would be the construction of a replication function to obtain a cell model that realizes the replicative (M, R) system.[15] As an alternative, the identification of the *semiotic* relations that complete a semiotic (M, R) system is studied below.

Here, we analyze protein synthesis as a global process, integrating the components of transcription, translation, and folding into one. The action of this component, shown in the continuous line mapping in figure 14.5, can be explained as follows. RNA polymerases, a pair formed by a ribosome and a sequence of aminoacyl tRNA synthetases, and a pair formed by chaperones and the intracellular milieu, acting together on a DNA chain, nucleotide sequences and an amino acid sequence, produce a protein, the input DNA chain, a chain of messenger RNA codons, and a sequence of transfer RNAs.

In this case, it can be affirmed that the output determines that the input determines the efficient cause, as represented in the dashed line mapping in figure 14.5. This is so because only a specific input (. . . -DN^j_n- . . . , (. . . , nc^j_n, . . .), (. . . , aa^i_n, . . .)) can be the base to produce a specific output (. . . -DN^j_n- . . . , . . . -$cmRNA^j_n$- . . . , (. . . , $tRNA^j_n$, . . .) <. . . —aa^i_n— . . .>), and the aspects of the input that are suitable to represent the output determine the efficient cause (RNAP, (r, (. . . , $aaRS^i_n$, . . .)), (chap, IntMilieu)).[16]

By joining the mappings in figures 14.3 and 14.4 and the dashed line mapping in figure 14.5, we obtain a model of a cell that realizes semiotic closure to efficient causation—that is, a semiotic (M, R) system.

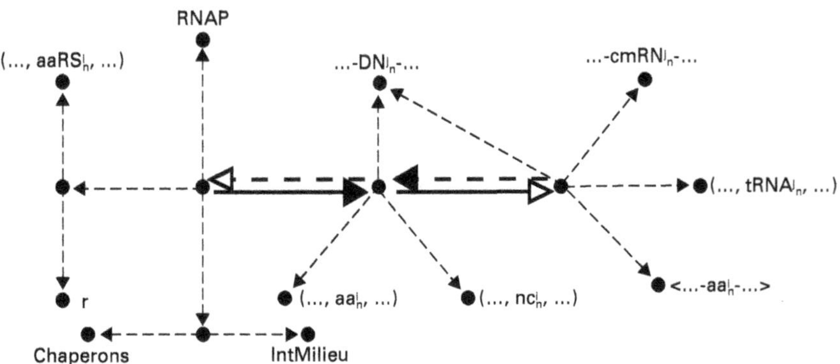

Figure 14.5
The semiotic relation established in protein synthesis is the union of the action of the component that summarizes the process (in continuous lines) and the determination of its efficient cause (in dashed lines).

It is now appropriate to review whether the system analyzed meets the requirements that an organism is supposed to satisfy. The construction of the basic cell directly responds to some of them. The proposed model is a system that is subject to the laws derived from physics, thermodynamically open and far from equilibrium, which builds itself, including its membrane, and maintains itself. It is also a system from which both reproduction—through cell growth and division—and the ability to participate in open-ended evolution—based on DNA mutations and changes in other entities and processes involved in protein synthesis—can be explained. However, this basic cell does not explicitly present the capacity to self-regulate to adapt to changes in its environment. We address this issue below.

14.4 A Cell That Auto-Regulates to Adapt to the Environment

The repair function of the postulated basic cell assumes the availability of inputs to be incorporated into the system. Next, we consider what would have to be added to the model to provide it with the capacity to self-regulate in order to adapt to changes in the environment. To do so, we discuss the regulation of transport and metabolism of lactose in the bacterium *E. coli*, discovered by Jacob and Monod (1961), in which two consecutive DNA portions are involved: the LacI and the Lac operon. The first consists of a promoter, the LacI gene, and a terminator, and the second includes a promoter, an operator, the LacZ, LacY, and LacA genes, and a terminator. The regulatory system in the absence of glucose is represented in figure 14.6, as part of an (M, R) system.

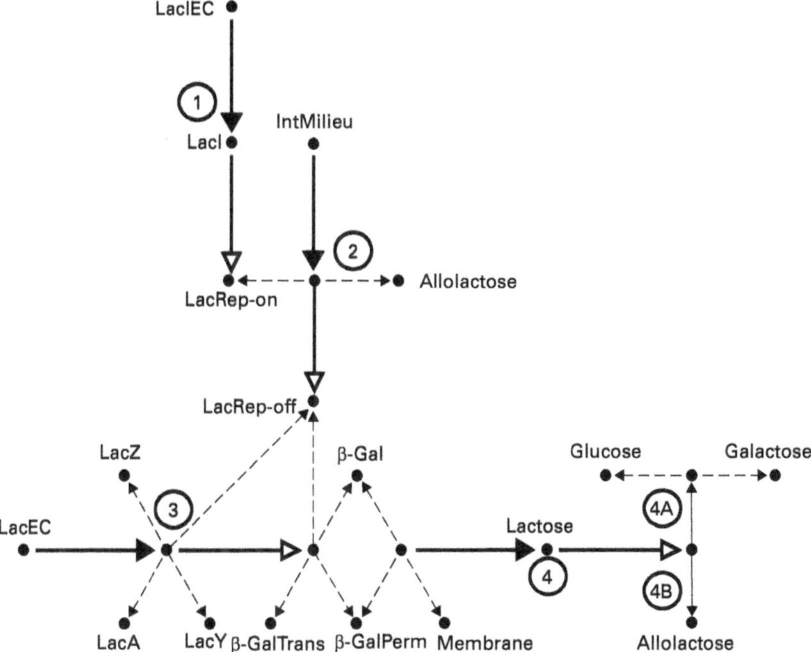

Figure 14.6
Regulation of lactose transport and metabolism includes one metabolism component (mapping 4) and three repair components (mappings 1 to 3). LacIEC, LacEC and IntMilieu denote the efficient causes of components one and two. Note that one of the alternative outputs of metabolism, the allolactose, is also part of the input of component 3.

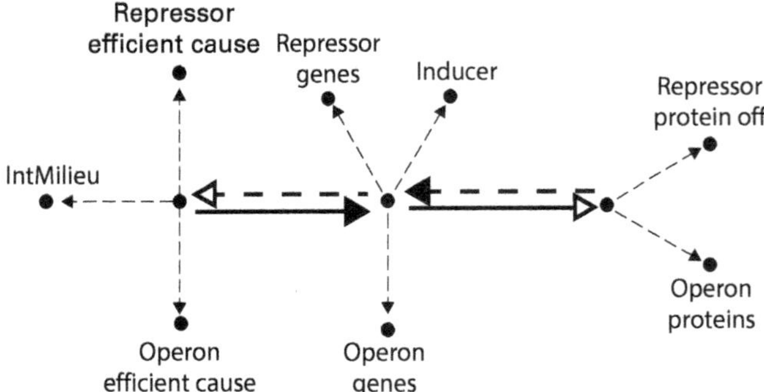

Figure 14.7
The semiotic relation established in the regulatory system.

The metabolic component, which carries out its action when there are lactose molecules in the environment, includes two consecutive processes—summarized in mapping 4 (similar to mapping 1 in figure 14.3)—and can produce two alternative outputs. First, a membrane-bound β-galactoside permease (transport protein) introduces a lactose molecule into the cell, and then β-galactosidase (an enzyme) usually catalyzes the separation of lactose into glucose and galactose (output 4A)—which are sources of energy for subsequent processes—or occasionally converts the lactose into an allolactose (output 4B).

The repair of metabolism consists of three components, represented in mappings one to three. First, an RNAP binds to the LacI promoter and initiates the synthesis of the repressor protein of the Lac operon, which binds to the Lac operator and blocks the start of transcription of the operon genes. At this point, we will say that the repressor is active (LacRep-on). Second, the intracellular milieu assists the binding of an allolactose, which acts as an inducer, to the repressor and separates it from the operator—that is, deactivates it (LacRep-off)—thereby allowing transcription to begin. Third, the joint action of the RNAP and the rest of entities that make up the efficient cause of protein synthesis, acting on the LacZ, LacY, and LacA genes, synthesize the proteins β-galactosidase, β-galactoside permease, and β-galactoside transacetylase.

If lactose is depleted from the environment, the production of allolactose ceases, which blocks the expression of the operon genes and thus the division of the lactose molecules present in the cell into pairs of glucose and galactose.

It should be noted that the four processes described are particular cases of the components described in section 14.3.1, and that by integrating components 1 to 3, we obtain a component that participates in the semiotic relation represented in figure 14.7.

As represented in figure 14.7, the efficient cause formed by the intracellular milieu and the efficient causes of the synthesis of both a repressor and the proteins corresponding to the operon, acting on a material cause (consisting of an inducer and the genes corresponding to the repressor and the operon), produce as output a deactivated repressor and the proteins corresponding to the operon. Crucially, the output determines that the input determines the efficient cause of the component, thus defining the semiotic relation.

By adding the mappings in figure 14.7 to those of the basic cell model, we obtain an extended semiotic (M, R) system for a cell that self-regulates to adapt to the presence or absence of a specific type of molecule in the environment.

14.5 A Closer Look at Semiotic Relations in the Cell

Next, we analyze in detail the semiotic relations that have been introduced. To do so, it is convenient to begin by reviewing the synthesis of aminoacyl tRNAs.

14.5.1 The Aminoacyl tRNA

It is immediately apparent that in each of the components 1 to 4 in figure 14.4, the output determines the input, since that specific input is the only one that can produce that output. That is, the output determines the entities that form part of the input of the component as well as their integration in pairs. However, only in the synthesis of aminoacyl tRNAs can it also be affirmed that the output determines that the input determines the efficient cause, as argued below.

The synthesis of an aatRNA can be summarized in two steps. First, a specific aminoacyl tRNA synthetase binds to it a specific amino acid—there are twenty aaRSs and twenty amino acids—and an appropriate transfer RNA—each aaRS can bind one or more specific tRNAs. Second, the aaRS facilitates the union of the amino acid and the tRNA that produces the aatRNA and then releases it. Thus, it can be stated that a specific aatRNA determines that the pair formed by the corresponding amino acid and tRNA determines a specific aaRS.

As a result, a semiotic relation is established in which the immediate interpretant corresponds to the features of an aaRS that make it suitable to produce an entity having the features of an aatRNA from an amino acid and a tRNA. The dynamic interpretant and the dynamic object are, respectively, a specific aaRS and the aatRNA produced by it; the final interpretant is the habit by which an entity having the features of an aaRS tends to catalyze the production of an entity having the features of an aatRNA from the amino acid and tRNA. Moreover, the relationship between the input and the output of the component is built on the complementarities that favor the binding of an amino acid and a tRNA to an aaRS, a spatial correlation that brings the incoming molecules together, and the complementarity inherent to the covalent bond that produces the aatRNA.

It should also be noted that each aatRNA establishes an association between the anticodon of the tRNA[17] and the amino acid, which is *not* mediated by similarities or correlations between them, but must necessarily be based on *the suitability of that association for the purpose of the system*—that is, to the realization of closure to efficient causation. Considering only the physicochemical processes involved, nothing would have prevented the formation of other aatRNAs, different from those present in current cells, and therefore other anticodon-amino acid associations. Possibly, in natural systems prior to them, entities simpler than aminoacyl tRNA synthetases could have catalyzed various alternatives of nucleic acid and amino acid unions. Later, some of these unions would have become more frequent due to the contribution of their anticodon-amino acid associations to system persistence. This would also have favored the gradual adaptation of the catalysts that enhanced the processes that were being selected. Remember that each realization of a dynamic interpretant updates the immediate interpretant of subsequent interpretive processes and contributes to the definition of the final interpretant.

It can therefore be stated that the synthesis of each aatRNA is a symbolic semiotic action based on indexical and iconic relations, as represented in figure 14.8.

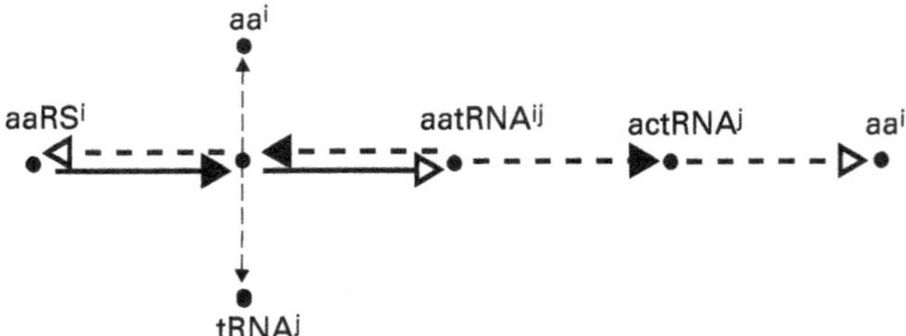

Figure 14.8
The symbolic semiotic relation established in the synthesis of an amino acyl tRNA. The two mappings on the left represent the definition of a semiotic relation: An aaRS produces an aatRNA from an amino acid and a tRNA, and the aatRNA determines that the pair (aa, tRNA) determines the aaRS. The mapping on the right indicates that, in the aatRNA produced, an arbitrary relationship is established between the anticodon of the transfer RNA and the amino acid, which assigns to the semiotic relation a symbolic (and not just iconic or indexical) character.

The synthesis of an aatRNA and the determination of its efficient cause constitute a semiotic relation in which it can be affirmed that an aaRS interprets a pair formed by an amino acid and a tRNA as a sign of an aatRNA,[18] and the meaning obtained in this interpretation is the joint contribution of the aatRNA and the rules that constrain the process, to system closure. Moreover, it can be stated that—through the aatRNA—the aaRS also interprets a tRNA anticodon as a sign of an amino acid, and the corresponding meaning is the contribution of the anticodon-amino acid association to system closure.

14.5.2 Protein Synthesis

To apply the concepts of (immediate, dynamic, and final) interpretant, (immediate and dynamic) object, and type of sign to protein synthesis, it is sufficient to take into account the following, as shown in figure 14.5:

- The efficient cause (the interpretant) consists of an RNAP, a pair formed by a ribosome and an aaRS sequence, and a pair formed by chaperones and the intracellular milieu: (RNAP, (r, (. . . , aaRS^i_n, . . .)), (chap, IntMilieu)).
- The material cause (the sign) is formed by a DNA chain and two sequences, one of nucleotides and one of amino acids: (. . . -DN^j_n- . . . , (. . . , nc^j_n, . . .), (. . . , aa^i_n, . . .)).
- The output (object) consists of the input DNA chain, a chain of mRNA codons, a tRNA sequence and a protein: (. . . -DN^j_n— . . . , . . . -cmRNA^j_n- . . . , (. . . , tRNA^j_n, . . .), < . . . -aa^i_n- . . .>).

However, to scrutinize the semiotic relation, it is convenient to separately review components 1 to 4 in figure 14.4. First, transcription is based on analogies (iconic relationships) between pairs of nucleotides, and on the correlation (indexical relationships) between the order these molecules occupy in the sequence and the DNA and RNA chains. Second, as explained above, the production of each aatRNA is based on an indexical relationship and several iconic ones between the molecules that bind together, and on a symbolic association between a tRNA anticodon and an amino acid. Moreover, the inclusion in the diagram of a sequence of processes of this type implies additional correlations between the entities participating in them. Third, the action of the ribosome is based on analogies and correlations between the elements that make up sequences of aatRNAs and mRNA codons, and an amino acid chain. Finally, the folding of the amino acid chain is based on structural properties of the molecules involved—that is, also on analogies and correlations.

The semiotic relation established in protein synthesis is built on these four components and the determination of the efficient cause of the synthesis of aatRNAs. Thus, the global process inherits from them the features that define the immediate and dynamic object, and the immediate, dynamic, and final interpretant that explain the possibility of realizing, the actual realization, and the habit of realizing, the semiotic action that produce the protein—as well as the features of the relationship between the material cause and the output that identifies the sign as a symbol built on indices and icons.[19]

It is the closure to efficient causation that converts the mappings in figure 14.5 into the semiotic relation corresponding to protein synthesis, and the mappings in figure 14.8 into the semiotic relation of the synthesis of an amino acyl tRNA. But additionally, it is the determination of the interpretant in the formation of aatRNAs that leads to the determination of the interpretant in protein synthesis, which in turn completes system closure.

Finally, it can be stated that a ribosome and an aaRS sequence (along with an RNAP and the union of chaperons and the intracellular milieu) interpret a DNA chain (and a nucleotide sequence and an amino acid sequence) as a sign of a protein (and the DNA chain, a tRNA sequence, and a cmRNA chain). Furthermore, the meaning produced in the synthesis of a protein is the joint contribution of the protein (along with the DNA chain, a tRNA sequence, and a cmRNA chain) and the rules that constrain the process, to system closure.

14.5.3 Regulation

It is now worth highlighting the role played by the repressor and the inducer. The features that make these molecules suitable to participate in the efficient and material causes and the output of the component represented in figure 14.7 contribute to the formation of the immediate interpretant and object, and thus to the definition of the semiotic relation. The dynamic interpretant and object, and the sign include the inducer and repressor that realize these features. The final interpretant is the habit of carrying out the regulation.

Similarly to what was explained above for the formation of aatRNAs, and then extended to protein synthesis, possibly, in natural systems prior to present-day self-regulated cells, some precursors of current repressors and inducers would have implemented simpler (and less efficient) solutions for blocking and unblocking the activation of regulatory processes, which would have given rise to semiotic relations. Then, each realization of a dynamic interpretant would have updated the immediate interpretant of subsequent interpretive actions and contributed to the definition of the current final interpretant—that is, to the development in the system of the present self-regulation habit.

Finally, it can be stated that the joint action of the efficient causes of the repressor and the operon, and the intracellular milieu results in the interpretation of an inducer and the genes corresponding to the repressor and the operon, as a sign of a deactivated repressor and the proteins corresponding to the operon. Furthermore, the meaning produced in regulation is the joint contribution of the output and the rules that constrain the process, to the enlargement of system closure.

14.6 Conclusion

The proposed theoretical framework can be summarized in the following statements. First, closure to efficient causation converts a natural system composed of parts, which carry out processes in which efficient and material causes and outputs can be identified, into an organism made up of components, to which final causes—that is, functions—can also be attributed. Second, in some components, the union of two closures—the global one that defines the organism and a local one that associates the action of the component and the determination of its efficient cause—defines a semiotic relation in which the efficient cause is the interpretant, the material cause is the sign, and the output is the object. In this case, the action of the component is a semiotic action that corresponds to the interpretation of a sign, and the meaning produced is both the final cause of the component and the rules that define the local closure that complete the systemic closure. Third, in an organism, it is the semiotic relations that complete closure to efficient causation. As a conclusion, an organism is defined as a natural system that realizes semiotic closure to efficient causation, and here, a semiotic (M, R) system is posited as a model.

To apply the proposed integrated account of Rosen's relational biology and Peirce's semiosis to organisms, we have begun here with a basic cell, a system that includes two types of semiotic relations concerning the syntheses of amino acyl tRNAs and proteins—two processes in which the output determines that the material cause determines the efficient cause. In the first, an amino acyl tRNA synthetase catalyzes the union of an amino acid and a transfer RNA and thereby establishes an arbitrary association between a tRNA anticodon and an amino acid. All this completes a symbolic semiotic relation that is based on iconic and indexical relations between the molecules that participate in the process. In the second, DNA transcription, aatRNA synthesis, the action of the ribosome, and the folding of an amino acid chain establish in protein synthesis a symbolic semiotic relation, built on iconic, indexical, and symbolic relations.

Then, to adapt to the environment, the cell model needs to include a third semiotic relation corresponding to the regulation of metabolism. The resulting cell has the capabilities that we attribute to organisms: self-construction, self-organization, self-maintenance, interaction, anticipation to internal and external changes, reproduction, and participation in a process of open-ended evolution.

It is worth highlighting the relationship established between the genetic code, protein synthesis, and closure to efficient causation. The transcription of a DNA codon into an mRNA codon, the complementarity of the mRNA codon and a tRNA anticodon, and the association of the latter with an amino acid together define a rule of the genetic code. The semiotic action corresponding to protein synthesis relies on (and, through system closure, contributes to defining) those rules, and also others that constrain the interrelations between all the entities involved in the entire process.

As for interpretation and meaning, and without forgetting that the agent that interprets is the organism, it can be affirmed that, in the synthesis of a protein, the joint action of a ribosome and an aaRS sequence (along with an RNAP and the union of chaperons and the intracellular milieu) results in the interpretation of a DNA chain (and a nucleotide sequence and an amino acid sequence) as a sign of a protein (and the DNA chain, a tRNA sequence, and a cmRNA chain), and the meaning produced in the synthesis of a protein is the joint contribution of the output and the rules that constrain the process to cellular closure.

Similar considerations about semiotic action, interpretation, and meaning apply to regulation.

There are two final considerations to extend the application of relational biosemiotics. First, the analysis of more complex cells and multicellular organisms, as well as their developmental processes and participation in open-ended evolution, should be based on the identification and analysis of additional semiotic actions—that is, processes in which the product obtained constrains its own production in the manner indicated in the definition of a semiotic relation.

Second, to study the interactions between organisms and with the environment, it should be noted that each organism incorporates inputs—some of which it interprets as signs—and releases outputs—some of which can be interpreted as signs by other organisms. The set of possible interactions of an organism constitutes its umwelt and provides it with the ability to compete and collaborate with other organisms, and contribute to niche construction.

Notes

1. It should be highlighted that *determines* here does not mean *produces* but rather *settles, fixes, tends towards, and/or selectively constrains.*
2. Primarily Rosen, 1972, 1991, 2012; Peirce, 1931–1936, 1998, 2009; and Short, 2002, 2007.
3. In this text, the term *entity* refers to one or more objects, processes, or a mixture of both.
4. Recall footnote 1.
5. In more complex mappings, such as those shown below, formal cause also includes the union of the elements that form part of the efficient and material causes and the output. Moreover, it should be noted that, in a natural system, efficient and formal causes are not separable; they lead together from the input to the output. For simplicity, references to the formal cause are usually omitted.
6. Kull (Pattee and Kull, 2009) affirms that Rosen practiced biosemiotics, probably without being conscious of it or, at least, without using that term. Pattee (Pattee and Kull, 2009) argues that relational biology depends on semiotic relations rather than on material ones.
7. Louie (2019) explains that relational biology states that anticipation is a necessary (but not sufficient) condition for life. However, Nadin (2012, p. 1) argues that "only in acknowledging the anticipatory condition of the living can grounding for semiotics be found." The proposed relational biosemiotics claims that an organism is a natural system that realizes semiotic closure to efficient causation, and thus, life and semiosis are coextensive.
8. Peirce uses the word *sign* to refer to both the semiotic relation and to one of the elements that form part of it. He also refers to the latter as *representamen* or *sign vehicle*. Except for references to Short's definition, in this text, *sign* is used instead of *representamen*.
9. Letters R, X, O, and P in the original text have been replaced by the concepts they stand for.
10. Furthermore, the expression "current or past or apparent or supposed," which completes the definition, can serve as the basis for analyses that exceed the scope of this chapter, such as the relationship between interpretation and memory, or the possibility of error that underlies each interpretive process.
11. Based on the works cited in section 14.2 and Hofmeyr's (2019) survey of the functional organization of cell biochemistry.
12. In a relational model, the level of disaggregation of the components must be adequate for the analysis that is carried out. In this case, separating the component into two would not contribute anything relevant to the argument being presented.
13. A codon is a group of three consecutive nucleotides.
14. The intracellular milieu is in turn repaired in metabolism.
15. A cell that realizes the replicative (M, R) systems is introduced in Vega (2023).

16. By determining the aaRS sequence that is part of the efficient cause.

17. A tRNA anticodon is the complementary of an mRNA codon.

18. Remember that, according to the definition of interpretation, the aaRS is the interpretant, and the agent that performs interpretation is the cell.

19. Specifically, the symbolicity of protein synthesis derives from that of the formation of aatRNAs.

References

Favareau, D., and Gare, A. 2017. The Biosemiotic Glossary Project: Intentionality. *Biosemiotics*, *10*, 413–459.

Hofmeyr, J. H. S. 2019. Basic biological anticipation. In R. Poli (Ed.), *Handbook of anticipation: Theoretical and applied aspects of the use of future in decision making* (pp. 219–233). Springer.

Jacob, F., and Monod, J. 1961. Genetic regulatory mechanisms in the synthesis of proteins. *Journal of Molecular Biology*, *3*(3), 318–356.

Louie, A. H. 2019. Relational biology. In R. Poli (Ed.), *Handbook of anticipation: Theoretical and applied aspects of the use of future in decision making* (pp. 191–218). Springer.

Nadin, M. 2012. Reassessing the foundations of semiotics: Preliminaries. *International Journal of Signs and Semiotic Systems*, *2*(1), 1–31.

Pattee, H. H. 2007. Laws, constraints and the modeling relation—history and interpretations. *Chemistry and Biodiversity*, *4*, 2272–2295.

Pattee, H. H., and Kull, K. 2009. A biosemiotic conversation: Between physics and semiotics. *Sign Systems Studies*, *37*(1–2), 311–331.

Peirce. C. S. 1931–1936. *The collected papers, vol. 1–6* (C. Hartshorne and P. Weiss, Eds.). Harvard University Press.

Peirce, C. S. 1998. *The essential Peirce, vol. 2* (Peirce Edition Project, Ed.). Indiana University Press.

Peirce, C. S. 2009. The logic of signs: Selections from the writings of Charles Sanders Peirce. In D. Favareau (Ed.), *Essential readings in biosemiotics* (pp. 121–148). Springer.

Rosen, R. 1972. Some relational cell models: The metabolism-repair systems. In R. Rosen (Ed.), *Foundations of mathematical biology* (pp. 217–253). Academic Press.

Rosen, R. 1991. *Life itself: A comprehensive enquiry into the nature, origin and fabrication of life*. Columbia University Press.

Rosen, R. 2006. Autobiographical reminiscences of Robert Rosen. *Axiomathes*, *16*(1–2), 1–23.

Rosen, R. 2012. *Anticipatory systems. Philosophical, mathematical, and methodological foundations* (2nd ed.). Springer. (1st edn, 1985, Pergamon Press, Ontario)

Short, T. L. 2002. Darwin's concept of final cause: Neither new nor trivial. *Biology & Philosophy*, *17*(3), 323–340.

Short, T. L. 2007. *Peirce's theory of signs*. Cambridge University Press.

Vega, F. 2021. An integrated account of Rosen's relational biology and Peirce's semiosis. Part I: Components and signs, final cause and interpretation. *Biosemiotics*, *14*, 697–716.

Vega, F. 2023. The cell as a realization of the (M, R) system. *Biosystems*, *225*, 104846.

Vega, F. 2024. The cell as a semiotic system that realizes closure to efficient causation: The semiotic (M, R) system. *Biosystems*, *240*, 105226.

15 Organic Intelligence: The Role of Repetition and Mathematical Groups in Natural Discernment

Daniel C. Mayer-Foulkes

15.1 Overview

Organisms make sense of the world. This is the starting point of biosemiotics. They achieve this by distinguishing (separating, marking) and by *intelligere* (choosing, selecting among). This chapter aims to demonstrate that intelligere must frame all inquiry into aliveness, focusing on locomotion and movement as foundational enactive intelligere that underpins all sense-making.

Discernment, synonymous with intelligere, encompasses both discriminating significant differences and penetrating into the essence of a subject. The term refers to the process of organizing sensed experiences as well as the outcomes of that process.

To illustrate the primacy of intelligere or discernment, it is shown that a single schema can explain how organisms manage locomotion in their habitats, manipulate objects, and handle themselves as objects. This schema is mathematical groups. As the foundation of abstract algebra, mathematical groups are applied across a vast range of fields, highlighting the nature of intelligibility and its unfolding. Sense-making is shown to develop through layers and networks of this form, which is inherently linked to the nature of aliveness.

15.2 Lulu the Woolly Monkey

The grandmother monkey Lulu (*Lagothrix lagothricha*) lived to the ripe old age of thirty-two. As the mother of Danny, the group leader, she remained remarkably composed when confronted by aggressive young males redirecting their aggression toward her. Lulu displayed exceptional evasion skills, deftly swinging around the territory and pausing at strategic locations to gauge her pursuers' next moves. She often held onto ropes for her own use, delaying her escape until the last moment when a male dashed at her along a perimeter beam. If he swung on a rope to cut her off, she anticipated his trajectory and adjusted hers accordingly, either retreating to a previous spot, aiming for a more distant location, or remaining still. Sometimes, she swung in wide arcs on a center rope to keep her options open, leaving the juvenile males stumped at the opposite end of an enclosure.

These observations were made by the author during a decade spent living and working at The Monkey Sanctuary in Cornwall, UK, founded by Leonard Williams, who authored several books, including *Challenge to Survival: A Philosophy of Evolution*, forwarded by C. H. Waddington. Lulu's skills offer valuable insights into sense-making. How can we articulate its underlying logic?

15.3 Sense-Making

Organisms make sense of their world through schemas that organize webs of relations. A useful distinction in Spanish illustrates this: The habitat of Lulu and her cohort *has sense* ("*tiene sentido*"), pointing to its intrinsic properties. In this case, the monkey territory is designed by humans to offer suitable affordances for monkeys (Gibson, 1979). Spacious indoor and outdoor areas are equipped with elaborate beams and ropes, which Lulu and the rest of the colony grow up *making sense* of ("*haciendo sentido*"). Over the years, she has mastered a vast repertoire of interrelated maneuvers, developing both mental and physical competencies. As Hoffmeyer (2008) notes, "meaning is not simply a property of an object; it is created in the interaction." Lulu's wild environment would possess a different intrinsic sense, coevolved with monkeys. In both contexts, mastery is achieved through ongoing active engagement.

Sense-making begins before direct experience and transcends it. An inner drive seeks and organizes experiences. Baby monkeys reach for whatever is at hand while riding on their mothers' backs, exemplifying intelligere. They actively select among various available objects, demonstrating early discernment. This preconceptual organizing ability forms the deep foundation for understanding, highlighting the ongoing unfolding underlying Lulu's skills.

Discernment encompasses the capacities that support an organism's skills, the skills themselves, and the means to implement them, whether acquired over generations by a species or through individual developmental processes. It is an embodied or enactive notion, as these skills involve bodies and their organization. A key question arises: Does discernment adhere to a common set of principles? This chapter affirms that it does. While there may be numerous schemas, this discussion focuses on mathematical groups (or "symmetries"). This schema is simple and intuitive, which is significant as it highlights our own deep enactive familiarity with it. It applies across a wide range of contexts, from elementary examples to modeling in various sciences and highly abstract instances in mathematics.

15.4 Insights and Experience

The central claim, echoing Bernard Lonergan, is that "experience itself calls for intelligibility" (Lonergan, 1992, p. 337). No experience, regardless of how rudimentary, is devoid of discernment. For humans, this means that experience mediated by insights, demands verification. The same holds true for monkeys, from a baby monkey's initial hesitant reach for a rope to the advanced skills exhibited by Lulu. An underlying architecture supports the incremental process by which experiences are patterned and verified through repetition. Lulu interprets and assembles her environment not only in the moment but through an ongoing compilation of learned patterns. Every movement along a beam, every swing, is intentional and part of a scheme continuously refined through loops of learning. Examination of her skill reveals that mathematical groups provide a schema to frame her sense-making, a universal form likely operative in all organisms.

The ability to account for diverse experiences, and their unfolding, through this single universal form supports the notion of intelligere at the heart of biosemiotics.

15.5 Mathematical Groups

Group theory is fundamental to abstract algebra. A mathematical group is defined as a set of elements that, when combined through a binary operation, satisfy four properties: closure, associativity, identity, and invertibility. This concept is best understood through examples. Consider how the set of real numbers under addition (designated as "R+") fulfills the properties of a group:

1. *Closure*: If a and b are real numbers, and members of R+, then $a + b$ is also a member of R+. Thus, when using the operation of addition, the sum of two real numbers is another real number.
2. *Associativity*: The way a sequence of additions is parsed does not affect their sum. For any real numbers a, b, and c, it holds that $a + (b + c) = (a + b) + c$.[1]
3. *Identity*: There exists an identity element for this group, specifically 0, because for any real number a, the equation $a + 0 = a$ holds true. Thus, any number plus 0 equals that same number.
4. *Inverse*: Each element has an associated inverse that cancels its effect under addition. For every real number a, there is number $-a$ such that $a + -a = 0$. This complements Identity: the addition of any number and its negative equals zero.

Since R+ fulfills these four criteria, it is a mathematical group.

The proposition, then, is that this scheme or form underlies how organisms interact with their surroundings. It follows that it may be shown how this form obtains in Lulu's skill. If the four criteria obtain, it suggests that this form is at work in how Lulu organizes her experience and, likewise, how other organisms also organize theirs.[2]

15.6 Elements of Movement

Let us begin at an elementary level. Displacement from one point to another can be described as a combination of translations and rotations, both of which can be conceptualized in terms of groups. Keep in mind Lulu's swings on ropes and runs along beams, to which we will return.

Starting with translations: In a city, one might walk a few blocks, and then another few blocks; the result is measured in blocks (Closure). Regardless of how these are parsed into subsets, the total number of blocks walked remains consistent (Associativity). One may choose to remain in place (Identity), and if one walks several blocks in one direction and back the same distance, one returns to the starting point (Invertibility). In sum, translation is intelligible in terms of mathematical groups.

Now, consider turns. Think of a toy clock. The set of hours fulfills the properties of a group under addition because a circle is a line whose ends meet. When turning the hands of the clock, regardless of their position, they indicate an hour that belongs to the set (Closure). Turn the hands forward, then again, and then again; the parsing of these turns into subsets does not affect the result (Associativity). You can refrain from turning the clock hands (Identity) or turn them back to their original position (Inverse). Just as the clock divides a circle into twelve hours, it can also be divided into 360 degrees to specify angles. Therefore, any set of segments that divides a circle constitutes a mathematical group.

In short, displacement—and thus navigation—in terms of translations and rotations can be understood through mathematical groups. This concept is highly intuitive; we know it instinctively, as do countless other creatures. Movement entails results defined in terms of movement (Closure); move, then move again, and then again—how this is parsed is irrelevant (Associativity); remaining still changes nothing (Identity); and moving to a point and back demonstrates (Invertibility).[3]

This form, applicable to displacement, also extends to object manipulation. For instance, place a book in front of you ready to read. If the only allowed operation is a 180-degree turn, these positions form a set of four elements. You can transition from any position to any other through a sequence of 180-degree turns. Verify for yourself that this set fulfills the conditions of a group. This concept is intuitive for any creature that handles objects.

Furthermore, groups can transform or become subsets of larger groups. For instance, allow for 90-degree flips, turns, and rotations of a book, transforming the previous group and making it a subset of this new one. Displacement in a city with buildings introduces a vertical axis, as does a habitat that includes trees. The branching structure of trees creates intriguing possibilities; a monkey high in a tree can access the ends of other branches without retracing its steps, much like skywalks between tall buildings. To grasp this topology, imagine crumpling a paper map of a city into a ball. Despite the new bends, routes between locations remain unchanged, with new shortcuts available, while displacement stays intelligible in terms of lines and turns, and thus in terms of groups.

Beyond displacement and object manipulation, the form of groups also applies to our self-handling as objects. Consider this thought experiment: In a spacecraft without a gravitational field, imagine moving yourself around a book rather than handling it. Since all positions are available, this illustrates that our shifting in space is intelligible in the same way as handling objects. All converge upon a singular underlying form.

This universal form extends to birds in flight, fish, bacteria, and white blood cells swimming, as well as leaves seeking light and roots seeking nutrients. Regardless of shapes or habitats, all organisms share this form. It is shared at all levels, from organs to groups of cells to single cells, organelles, and even their components. This form is deeply embedded in the fundamental architecture and functions of organic structures. To call this form intuitive only hints at the depth of our familiarity with it. Additionally, its universality enables organisms to make sense of one another through positions, movements, and object manipulation, suggesting it serves as a shared foundation for discernment among all organisms.

Before proceeding, consider an instance that is not a mathematical group: the set of times as time passes. Unlike a toy clock, a working clock does not satisfy the four criteria of groups; the hands never stand still (Identity) nor can they move backward (Invertibility). Thus, our experience of the progression of time is not a group. The form of groups applies when we calculate time to organize our year, week, day, or even the next few seconds. Planning is analogous to playing with a toy clock: Anticipating adheres to the form of a mathematical group, while our experience of time's progression does not. You plan, then life happens. This key difference constitutes a radical disjunction. To be alive is to continually cope with this disjunction, which forms the matrix of social life. Lulu can plan her next move, but the outcome remains uncertain.

To illustrate a non-group with a book: If flipping it to a new position made the previous position unavailable (no Inversion), or if it were impossible to keep the book still (no Identity), this underscores that the form of time's progression is radically unlike the form of objects and their movement.

15.7 Groups and Anticipation

Building on the previous elements, we can now examine anticipation. Imagine steering a sailboat at sea. In the distance, you spot another sailboat appearing to head toward you, leading you to wonder if you might be on a collision course. The trajectory of both boats is influenced by the wind's force and direction, as well as possible unseen water currents. Even so, the answer stares you in the face: A collision will occur if, and only if, the other boat remains fixed in your line of vision as it gradually grows larger.

You can verify this with a simple two-step experiment. Hold your right hand at arm's length from your face at eye level and point at yourself. Now, let your hand move not toward your nose, but toward an imaginary nose of a face next to yours on the left. Notice how your hand crosses your line of vision from right to left. For contrast, start again, but this time move your face, so your nose occupies the left position, allowing your right hand to reach your nose. Observe how the approaching hand remains still in your line of vision, ensuring the collision occurs by keeping your hand steady there. In fact, inadvertently, you actually ensure the collision occurs *by keeping your hand still in your line of vision.* We will return to this shortly.

This is what to watch for at sea. The relative position of the approaching boat either shifts or remains fixed. The shift is along a line analogous to the number line, or R+. Thus, this movement fulfills the four criteria of a group: The boat occupies a position along this line (Closure); parsing various shifts along the line does not affect the outcome (Associativity); there is a "no move" operation (Identity)—which indicates an imminent collision—and any movement along the line may be reversed (Inverse). This is how we anticipate collisions.

Catching a ball represents a desired collision. Consider a long pass from a football quarterback, a baseball flying toward the field, or a Frisbee thrown for a dog to retrieve. Do these actions involve complex calculations to determine the intersection of speeds and curves? No. Successful catchers keep the moving ball steady in their line of vision, ensuring the trajectory of collision. By managing the speed, position, and trajectory, they maintain the approaching object in a fixed view. The same principle applies to monkeys grabbing moving ropes, birds snapping up insects, and white blood cells seizing bacteria. This is how prey are caught and predators evaded.

It is worth pausing over the notion of "managing their own speed, position, and trajectory." The key is keeping an object still along the line of vision. How is this achieved in practice? You just did it to touch your nose with your finger, which likely zigzagged. This zigzagging allows organisms to reach directional goals. This is a form of oscillation. Another form of oscillation is speed management, alternating between slowing and accelerating. Trajectories are adjusted by veering side to side or up and down, which is also how we maintain balance. Generally, organisms manage steady states through oscillation which, as we have just seen, is intelligible in terms of mathematical groups.

15.8 Mathematical Groups in Displacement Strategies

We can now return to Lulu. Consider the combination of translations and rotations by which she navigates her territory. Her movements along beams and swings on ropes—akin to our displacement in a city—can be distilled into translations and rotations, making her navigation intelligible in terms of groups. Anywhere she goes is a place in the territory (Closure); regardless of how a sequence of moves is parsed, the outcome remains the same (Associativity); she can refrain from moving (Identity); and she can return to where she came from (Inverse).

Lulu's strategies also involve higher-level no-move options (Identity). She pauses to gauge her pursuer's next move and can delay until the last moment before swinging to the opposite side of the room. After swinging on a rope, she often keeps hold of it to prevent it from swinging back into a position from which it could be used by her pursuer.

Thus, her skills can be accounted for in terms of groups at various levels. At an elementary level, groups explain her basic management of speed and direction—her handling of herself as an object. On a higher level, they account for her navigation around the territory and her management of objects such as ropes. Additionally, they inform her planning—anticipating collision courses, whether it be catching ropes or avoiding her pursuer. Moreover, she organizes her multilayered skill, such as the "no move" of swinging in wide circles on a central rope to keep options open. This requires gauging gradual loss of momentum, which can also be understood in terms of groups (hanging from a nonmoving rope is a serious liability).

Grasping the sense of a territory does not always translate into understanding new situations. For example, we once had a squirrel persist in raiding a recently installed bird feeder. Her response to being chased away by us was to find new routes to approach the feeder—a brilliant use of navigation skills, but one that failed to solve her problem. Thus, while the dynamics between Lulu and her pursuer can be expressed in terms of territorial intelligibility, their skills extend beyond what mathematical groups can fully account for. Observing them is akin to being moderate chess players trying to grasp the moves in a match between masters. Their interaction forms a relationship in which both anticipate and implement strategies and counterstrategies in real time, actively interpreting each other's sense-making. We may struggle to predict their moves, but their sense becomes evident through the intelligibility of the territory. Interactions in relationships lead to higher orders of anticipation. For reasons explained in the final section, this touches on the deepest origins of discernment, a topic beyond the scope of this essay, which aims to establish the existence of organic intelligere or discernment.

The preceding discussion invites reflection on the nature and process of discernment in general. We will soon explore why this relates to the significance of mathematical groups. Their applicability across various fields indicates their potential as a framework for understanding discernment in its entirety, including how they account for the unfolding of Lulu's skill at different levels. I have argued that there is an "isomorphism between human cognitional structure-process, as we can recognize it in ourselves, and living distinctional structure-process, as we can recognize it in any living being, including ourselves." From this follows a heuristic principle: "The cognitional is a highly differentiated version of the distinctional" (Mayer-Foulkes, 2023, p. 5). In other words, all instances of discernment, from the highly differentiated to the deeply compacted, can illuminate one another. This isomorphism provides a key method of inquiry into the organic and the nature of aliveness.

15.9 The Two Blades of Discernment

Let us consider our own highly differentiated discernment. What insights can we glean from it regarding discernment in general? According to Lonergan, the heuristic structures of empirical methods operate like scissors: Two complementary blades, one representing empirical observation and the other mathematical abstraction, are at the core of scientific inquiry (Lonergan, 1992, p. 337). For example, Albert Einstein, just hours before his death, was occupied with mathematics as he sought to integrate subatomic and cosmological physics.

Are mathematical groups the upper blade in Lulu's skill? This is an extremely elementary form: Do something, then something more—the outcome is based on what you are doing (Closure); do something, then again something more—how these sequences are parsed is irrelevant (Associativity); refrain from action, and nothing has been accomplished (Identity); perform an action and then undo it is equivalent to not having acted (Invertibility).

This schema explains a spider's behavior when you try to pick her up. It is also how we make sense of her response. More significantly, it is how she makes sense of what we are doing to her. We make sense of each other through this shared schema because we both apply it in all our movements. That we share this schema in our movements suggests that it is at work in Lulu's skill, and moreover, that our account of Lulu's skill provides an upper blade for understanding organic movement generally: how organisms make sense of their environment, manage themselves as objects, and handle objects. How rigorously can this claim be substantiated?

Mathematical groups serve as an upper blade in three interlocking senses. First, they constitute an innate, preconceptual heuristic structure in Lulu, a spider, and us. This derives from the second sense: Lonergan's observation of mathematics in science as complementary to empirical observation. Third, they affirm that intelligere operates in all organisms as they discern the world.

These three interlocking senses ultimately lead to a fundamental question: Is mathematics discovery or invention? Is mathematics a sense the world *has*, or how we *make* sense of the world? *We* who—humans? Primates? Mammals? All organisms? What holds for mathematics also holds for mathematical groups.

While this is not an essay on the nature of mathematics, we will show how the foundation of mathematics aligns with the deepest nature of living beings. This is why these three senses interlock. It will be shown that mathematics is discovery, *given the invention of repetition by living beings*. Once living beings invent repetition, mathematics follows.

Mathematics, widely recognized as "the science of patterns" (Devlin, 2000, p. 3), prompts us to reframe the fundamental question: Is pattern—are patterns—discovery or invention?

The foundation of pattern is repetition, which arises with living beings. Consequently, once repetition is established, patterns can be discovered, leading to mathematics. Mathematical groups exemplify this. We will demonstrate that mathematical groups are based on and unfold repetition, functioning as an upper blade emerging from preconceptual organic discernment.

The three interlocking notions of an upper blade suggest a heuristic for inquiry into sense-making that is (1) framed by the foundations of discernment; (2) begins at the most rudimentary level of inquiry; (3) accounts for its unfolding; and (4) allows for comparative studies between compact and differentiated instances in terms of isomorphism.

15.10 Repetition: Sameness and Difference

Sameness and difference are foundational to the process of distinguishing; indeed, sameness-difference can define both distinction and repetition. This chapter draws from my argument that living beings are distinguishing distinctions (Mayer-Foulkes, 2023). Organisms function as distinctions that maintain their identity by distinguishing.

Distinguishing is achieved by establishing homogeneity (A=A). In principle, such homogeneity does not exist in the concrete world; it is made possible as organisms *neglect concrete residue*. Organisms arbitrarily establish differences that make a difference and those that do not (Mayer-Foulkes, 2023, p. 39). Differences that make a difference are defined as the fundamental unit of information by Bateson (1979).

For example, a bee seeking honey perceives flowers that provide nectar as homogeneous: A flower is a flower (A=A), and a source of honey is a source of honey (A=A), regardless of actual differences between them (A≠A). This capacity to intertwine sameness and difference (A=A, A≠A) forms the foundation for distinguishing, which originates from organisms and is crucial for their existence. The relationship (A=A, A≠A) underpins iteration. Repetition and distinguishing are intimately linked, representing two aspects of a fundamental capacity inherent to organisms. Roger N. Shepard's Theory of Generalization mathematically examines how such sameness and difference is established—his starting point for investigating perceptual-cognitive universals (Shepard, 2001). These serve as upper blades in sense-making. This chapter does not aim to specify universals, as Shepard has effectively done. Instead, it demonstrates that intelligere is always at work in organisms, rooted in mathematical universals grounded in repetition invented by living beings.

Self-regulation is rooted in (A=A, A≠A). For instance, this principle is illustrated in chemotaxis, where organisms move in response to chemical signals. Chemotaxis showcases how the bacterium *E. coli* navigates its environment by comparing chemical gradients using a two-component regulatory system (Mayer-Foulkes, 2023, p. 41–44). This system comprises a sensor (S) for detecting environmental differences and a motor (M) for executing responses, resulting in either "running" (straight swimming) or "tumbling" (random movement). While conditions detected by S remain constant, M continues to act.

What does it mean for "conditions to remain the same"? This (A=A, A≠A) is achieved by neglecting concrete residues. In actual environments, slight variations in attractant or repellent concentrations occur constantly. Only by ignoring these small differences can they be perceived as "the same" for S. Thus, the neglect of concrete residues is essential to establish binary states such as on/off or run/tumble. This neglect facilitates the transition from analog to digital states, leading to steady binary states A and Ā. Consequently, the distinguishing activity in all organic processes hinges on neglecting concrete residues to establish differences that make a difference.

This marks a fundamental distinction between living and nonliving processes. The establishment of (A=A) by organisms during their distinguishing processes transcends mere physical causality because the neglected concrete residues are arbitrary. Their determination is based on requirements for self-regulation within specific organisms. The arbitrary nature of this neglect is foundational to agency. In each distinction made by every organic process, differences that make a difference and those that do not are arbitrarily determined. Thus, the living world hinges on differences that make a difference established through the arbitrary neglect of concrete residues. This is the foundation for the gradual unfolding of agency.

The concepts involved in distinguishing activities—(A=A, A≠A), repetition, arbitrary neglect of concrete residues, two-element control systems, and differences that do and do not make a difference—relate to self-regulation and agency. This distinguishing activity, enabling organisms to exist and maintain their identities, operates from single cells to complex organisms composed of billions of semi-autonomous, self-regulating cells. Distinguishing occurs at all levels: by individual cells, tissue aggregates, functional tissue units, individual organs, and entire organisms. The distinction-distinguishing dynamic functions in physiology as much as in psychology. This observation clarifies why "botany and zoology represent a singular continuous-discontinuous field, illustrating the necessity for a re-foundation of biological concepts based on distinction-distinguishing" (Mayer-Foulkes, 2023, p. 43). Thus, at every level of living being, some degree of discernment and sense-making is inherently at work.

15.11 Sameness-Difference or Repetition in Mathematical Groups

Consider how mathematical groups are founded on, and unfold repetition, or sameness-difference. Let us go back to the definition of groups as illustrated by R+.

1. *Closure*: If a and b are members of R+, then $a + b$ is also a member of R+. The operation of addition ensures that the sum of two real numbers is another real number. As real numbers, a, b, exhibit sameness and difference, as does any other number $a + b$. Sets are, by definition, about sameness-difference. Closure signifies that any operation between members of a set produces further such sameness-difference. Therefore, groups fundamentally pertain to sameness-difference.
2. *Associativity*: The parsing of additions of numbers into subsets does not affect their sum. For any real numbers a, b, and c, it holds that $a + (b + c) = (a + b) + c$. In groups, sets of operation may be parsed differently but the final outcome is the same. Thus, Associativity establishes sameness-difference as it pertains to sets of operations.[4]
3. *Identity*: There exists an identity element for this group, specifically 0, because for any real number a, the equation $a + 0 = a$ holds true. The identity element expresses sameness-difference, linking the set to the operation. The operation of this element on any member of the set yields that member's value. It is the same value, and this repeats for every member of the set.

4. *Inverse*: Each element has an associated inverse that cancels its effect under addition. For every real number a, there exists a number a such that $a + -a = 0$. All members of the set have an opposite, which reflects sameness-difference. The operation applied to all members of the set is consistent and repeats; it and its opposite yield identity.

In summary, a group consists of elements (repetition) that are both the same and different, which together produce further same-different elements (repetitions). Different sequences of operations are equivalent (repetition). An identity element exists, as combining any other element with it yields that same element (repetition). Additionally, any element combined with its opposite yields the identity (repetition).

Clearly, repetition is the core of mathematical groups, just as it is of aliveness. As aliveness unfolds, so does repetition. So the scheme of groups can emerge and underlie discernment.

The same is true of logic. As Spencer-Brown (1969) illustrates, repetition is central to the axiomatic definition of distinction, from which a primitive arithmetic (Boolean arithmetic) is derived; this arithmetic serves as a calculus of distinctions. Boolean algebra emerges from this arithmetic, forming the primary non-numerical branch of mathematics we call logic (see Mayer-Foulkes, 2023, chapter 5). Since the two axioms defining distinction are expressed in terms of repetition, it follows that aliveness is a condition for the possibility of logic. From the algebra of distinctions (logic), Spencer-Brown derives second-degree equations, from which space and time are thence shown to be derived. Thus, repetition leads to distinction, Boolean arithmetic, logic, and ultimately, space and time, providing a framework for sense-making.

Sense-making, then, is founded on repetition unfolded through distinctions, which may be conceptual, preconceptual, or architectonic (Mayer, 2020). Living beings are not only distinguishers but also distinctions themselves. Just as a plurality of nested conceptual distinctions constitutes a calculus of distinctions, so a plurality of nested living distinctions calculates. This is *metabolism*, how sense-making begins.

However, a crucial caveat exists. A plurality of living distinctions—organisms and organic structures—indicates *relationships*. These are of a different order than logical relations. Discussing these is beyond the scope of this chapter. Unlike logical relations, such relationships involve paradox, a point we will revisit in the final section.

15.12 Repetition and Verification

A crucial foundational aspect of repetition for organisms is that it signifies affirmation and, by extension, verification (Mayer-Foulkes, 2023, p. 5). When individuals of a species act randomly, some behaviors are adaptive while others are not; all are repeated. Adaptive behaviors are verified and tend to persist, while nonadaptive actions are weeded out through natural selection. Those that endure are more likely to be passed on to future generations. This dynamic also occurs at the individual level: Adaptive actions are verified, while maladaptive behaviors gradually diminish as the individual ceases to perform them. Metabolisms are verified through repetition, just as neural networks are. This reflects Hebb's Law: "If it fires together, it wires together" (Hebb, 1949). At all levels, organisms verify through repetition. This brings us back to Lonergan's two blades: empirical observation and mathematical abstraction. The foundation of the upper blade is repetition. It plays a fundamental role in aliveness, mathematical groups, logic, and verification. Together, these elements form the core of rigorous, controlled inquiry—by organisms and in science. Scientific inquiry can be viewed as organic discernment par excellence, with organic discernment being inherently proto-scientific. Repetition underlies the isomorphism between compact and highly differentiated forms of verification, serving as the foundation of discernment.

15.13 Layers and Nesting in the Structure of Sense-Making

Beyond its foundation in repetition, how can this isomorphism be explained? We have addressed movement, collisions, and anticipation through mathematical groups. Chemotaxis illustrates how groups clarify the two-component regulatory systems common in living organisms. In these inherently analogical processes, organisms establish digital or binary states—such as on/off, start/stop, slow/fast, and hot/cold—and go on to oscillate between them. At higher levels, oscillation may occur between accelerating and decelerating, as well as shifting between warming and cooling. As previously noted, zigzagging is a form of oscillation used to verify directional goals. Correspondingly, discernment is continually practiced across various layers within the self-regulating processes of organisms. Thus, isomorphism can be explained in terms of layers. Nested layers enable increasingly higher levels of discernment, with each layer integrating and reframing lower ones.

Consider how we would account for the development of a skill such as Lulu's. We would do so in terms of insights, sequences, and nestings of insights. Except for their remarkable prehensile tails, the embodiment of woolly monkeys and humans is largely the same. We therefore share similar processes of learning. Certainly, intensive interaction with the monkeys over a decade left me in no doubt about this. Our deepest understanding of discernment comes from intimacy with our own development, a vastly complex process, as described here:

> The child would understand everything at once. It does not suspect that there is a strategy in the accumulation of insights, that the answers to many questions depend on answers to still other questions, that often enough advertence to these other questions arises only from the insight that to meet interesting questions one has to begin from quite uninteresting ones. There is, then, common to all men the very spirit of inquiry that constitutes the scientific attitude. But in its native state it is untutored. Our intellectual careers begin to bud in the incessant What? and Why? of childhood. They flower only if we are willing, or constrained, to learn how to learn. They bring forth fruit only after the discovery that, if we really would master the answers, we somehow have to find them out ourselves.
>
> Just as there is spontaneous inquiry, so too there is a spontaneous accumulation of related insights. For questions are not an aggregate of isolated monads. Insofar as any question is followed by an insight, one has only to act, or to talk, or perhaps merely to think, on the basis of that insight, for its incompleteness to come to light and thereby generate a further question. Insofar as the further question is in turn met by the gratifying response of a further insight, once more the same process will reveal another aspect of incompleteness, to give rise to still further questions and still further insights. Such is the spontaneous process of learning. It is an accumulation of insights in which each successive act complements the accuracy and covers over the deficiency of those that went before (Lonergan, 1992, p. 197).

Monkey learning is not mediated by language, nor are many aspects of our own learning. We discover how to touch our head and face, find our thumb for sucking, hold, sit, crawl, stand, walk, run, and climb. These foundational skills support further learning, such as carrying, opening, closing, pulling, pushing, throwing, and catching—none of which require language. From our earliest explorations, wonder at the world's mysteries flows forth, and each skill we acquire aligns with Lonergan's description.

Learning occurs in chunks, which I have termed "summing" and "lumping" (Mayer-Foulkes, 2024). We combine chains of chunks and summarize them into new units, illustrating how chunks nest. Anyone who drives a car recognizes the difference between their current skill and when they first learned to drive. This process mirrors our transition from letters to words to sentences to paragraphs while learning to read. Summing, lumping, and nesting demonstrate that learning is not merely aggregates of isolated monads. The ability to parse sequences is crucial, as these parsings break down and build up sequences. To discern is the spontaneous process of summing, lumping, and nesting, which is essential for skill development. Patterns are verified through action—discerning must be verified; otherwise, as shown, they are dismissed. Verification occurs through repetition, which is how we validate patterns throughout life. Whether explicitly or implicitly, to verify is to assent. Thus, repetition and verification underpin "the spirit of inquiry that constitutes the scientific attitude." In the specialized case of explanatory understanding, verification becomes systematic, as inquiry leads to the rational question, "Is this so, yes or no?" This invites the key self-reflective question, "Are there any further pertinent questions to ask?" This is the touchstone of the scientific attitude. Rooted in many nested layers of discernment unmediated by language, it unfolds the same organizing principles, including those of mathematical groups.

To visualize discernment in neural terms, think of an insight as a body fitting into an environmental niche. For example, as we navigate the constraints of a puzzle, neural patterns create a niche that elicits images into which a form fits or from which a form emerges. As this occurs, the constraints of the puzzle begin to be met, and this fit can be verified—or not. The correspondence between the image formed by neural patterns and the solution resembles an environmental niche where an organism fits snugly. This coupling, which constitutes discernment, unfolds in stages through summing, lumping, and nesting—over evolutionary timescales of hundreds, thousands, or millions of years, or in the case of a puzzle, days, minutes, or seconds.

15.14 Conclusion

In this chapter, discernment refers to how organisms, starting from the most rudimentary forms, actively make sense of the world based on principles ultimately grounded in repetition, as invented by living beings. Mathematical groups, which thoroughly unfold repetition, have been examined as an upper blade to uphold the notion of *intelligere* as inherent to aliveness.

We discern what Lulu discerns because, for both her and us, the principles of mathematical groups are invariant structures recurrently applied across various situations, integrated through summing, lumping, and nesting. This foundation, rooted in repetition, suggests that all organisms are deeply familiar with these principles, which guide their discernment, including awareness of what other organisms discern in terms of movement, collisions, and anticipation. These principles are rooted in the fundamental nature of living beings, distinguishing distinctions that gave rise to repetition and thereby verification.

How is the situation between Lulu and her pursuer resolved? They become tired, bored, or distracted. Something more interesting or urgent arises, causing them to forget each other. A short while later, they may sit peacefully side by side. As individuals who belong to a community made up of several families and friendships between them, the glue that bonds this social world is emotion. Beyond—or before—the mathematical relations they keenly discern, monkeys thrive in *relationships*. Their capacity to be discerning about relationships is as remarkable as their understanding of their territory.

Relationships inherently involve paradox—both/and. We are deeply familiar with paradox: We can simultaneously love and hate those closest to us; distance brings both relief and yearning; completing a long task yields happy satisfaction alongside sadness from a sense of loss; and a worthy opponent is both a rival and companion. *Relationships* allow relations to emerge, as articulated by mathematical groups (Mayer-Foulkes, 2023, chapters 5, 10).

This is so because living beings emerged in the plural. From the earliest single-celled organisms, plurality involved both competition and collaboration. The paradoxes of relationship have enabled the unfolding of ever more complex organisms and, with them, increasingly advanced discernment. Thus, discernment of the logical, mathematical, spatial, temporal, and movement is born of relationships and their paradoxes. Discerning social individuals make sense of each other making sense, managing and negotiating life together. Throughout evolution, a vast nesting of such social relationships has persisted (see Bourke, 2011). This is the realm of the ethical. Therefore, ethical discernment forms the foundation of all discernment (Mayer-Foulkes, 2023, chapter 10).

The world of relationships, which cannot be fully accounted for mathematically, accounts for the emergence of relations and mathematics. Making sense of relationships—that is a story for another day.

Notes

1. Note that Associativity differs from Commutativity, which states that (a + b) = (b + a), meaning the order of operations does not affect the outcome. This additional property obtains in a subset of 'Abelian' groups, of which R+ is an instance. Commutativity does not apply in some of the groups examined below.

2. Mathematical groups have long been used to model movement and perception (see Gromov, 1981; Jost, 2014; Kauffman, 1993; Laumond, 1998; Marr, 1982; Zucker, 1971). This discussion of mathematical groups does not imply that this is the only scheme by which organisms organize their experience. Significant research into such universal schemes has been conducted by Roger N. Shepard, discussed below.

3. In translation and rotation considered separately, commutativity applies (see note 1). Therefore, with a set of movements along a number line or a set of rotations of a toy clock, changing the order does not affect the outcome. However, when taken as a single group, only associativity holds. For example, the instructions "go North two blocks, turn right, go West three blocks" can be parsed into different subsets, but altering their order leads to different destinations.

4. Commutativity in Abelian groups adds further sameness-difference to sets of operations.

References

Bateson, G. 2002. *Mind and nature: A necessary unity.* Hampton Press.

Bourke, A. F. G. 2011. *Principles of social evolution.* Oxford University Press.

Devlin, K. 2000. *The language of mathematics: Making the invisible visible.* Holt Paperbacks.

Gibson, J. J. 1979. *The ecological approach to visual perception.* Psychology Press.

Gromov, M. 1981. *Metric structures for Riemannian and non-Riemannian spaces.* Birkhäuser.

Hebb, D. O. 1949. *The organization of behavior: A neuropsychological theory*. Wiley.

Hoffmeyer, J. 2008. *Biosemiotics: An examination into the signs of life and the life of signs*. University of Scranton Press.

Jost, J. 2014. *Mathematical methods in biology and neurobiology*. Springer.

Kauffman, S. 1993. *The origins of order: Self-organization and selection in evolution*. University of Pennsylvania Press.

Laumond, J. 1998. *Robot motion planning and control*. Springer.

Lonergan, B. 1992 [1957]. *Insight: A study of human understanding*. Collected Works of Bernard Lonergan, Vol. 3. University of Toronto Press.

Marr, D. 1982. *Vision: A computational investigation into the human representation and processing of visual information*. MIT Press.

Mayer-Foulkes, D. 2020. Significance. In D. Favareau, P. Cobley, and K. Kull (Eds.), *A more developed sign: Interpreting the work of Jesper Hoffmeyer* (pp. 259–262). De Gruyter.

Mayer-Foulkes, D. 2023. *The nature of living being: From distinguishing distinctions to ethics*. Springer.

Mayer-Foulkes, D. 2024. Review of Copenhagen Gathering: Reading and foundations. *Sign Systems Studies*, *52*(3–4), 611–620.

Partridge, E. 1958. *Origins: A short etymological dictionary of modern English*. Routledge and Kegan Paul.

Shepard, R. N. 2001. Perceptual-cognitive universals as reflections of the world. *Behavioral and Brain Sciences*, *24*(4), 581–601.

Spencer-Brown, G. 1969. *Laws of form* (1st ed.). George Allen and Unwin.

Spencer-Brown, G. 1973. *Laws of form* (limited ed.). Cognizer.

Weyl, H. 2017 [1952]. *Symmetry*. Princeton University Press.

Williams, L. 1978 [1971]. *Challenge to survival: A philosophy of evolution* (3rd ed.). Allison & Busby.

Zucker, S. W. 1971. Theoretical foundations of shape perception by humans and machines. *Perception*, *1*(1), 21–39.

16 Taking a Life Stance

Anton Markoš

The job of a translator, I wrote, is not just to transcribe from one language to another, but to transfigurate—to make a poem come alive in a new century, a new medium.
—J. P. Delaney (2018: 323)

16.1 Introduction

I start with some language analogies of life—beginning with a parable from a thriller by J. P. Delaney (2018) entitled *Believe Me*, whose book's cover graphics presents the title as BE*LIE*VE ME.

Does the title suggest a conundrum? Should the reader await that, in fact, the context covers three words—believe, lie, or belie? Not only this: Many book titles have also a subtitle explaining the content; this book, however, has a supertitle developing the conundrum further: Trust me. Love me. Just don't . . . BE*LIE*VE ME.

Forty characters (including spaces) conceal a rich bundle of meanings, often mutually opposing (though all will come up in the plot), not only in the ordering of basic characters but also in their arrangements (including spaces, upper-case letter, and italics)—all of this being achievable even in the English script, which is notoriously known as very poor as concerns diacritics.

Now, compare that with the title of the Czech translation, which bears the title *Věř mi*, cutting away all the brain teasers of the original work. In such a naked form, and translated, the title enters a new medium: If successful (read), it will acquire a new halo of meanings and contexts, comparable or not with the English book.

Written language is a considerable reduction of thinking and even talking, and the fact has been studied by many linguists. Let me mention here the Distributed Language Group (Love, 2004, 2007; or Cowley, 2017) coining the term "languaging" for the whole-body activity used to express a person's thoughts. The reverse process—that is, interpreting a written text—is also a complex process that depends not only on mere decoding; rather, it includes the construction of a narrative in the mind of the reader, dependent on his/her memory and experience (e.g., Eco, 1994a, b; Markoš and Cvrčková, 2013).

16.2 Clichés

Next, let us discuss some extreme, well-worn clichés concerning life that are too often encountered, sometimes even in the biosemiotic literature.

16.2.1 Discords

The first is the diatribe against "mechanistic," "deterministic," and "reductionist" beliefs that all characteristics of the living can be, and will be, understood in the framework, and by the methods, of physics (and chemistry). In other words, for such *physicalists*, "life" is a superfluous concept, an aggregate notion for the manifolds of chemical and physical processes. In contrast, according to the critics of such a scientific stance (let's call them here *vitalists*), life must represent something that is not palpable within the framework of science and is accessible only by the cenoscopic knowledge of "lived experience." The history of modernity is punctuated by the stubborn clichés that both of the above parties fell into, with the physicalist conviction usually winning. But this conviction has also had its evolution, as pointed out, for example, by Barbieri (2024, p. 141), as he illustrates with the concept of *mechanism*: "The first mechanistic model of the body was the clock-machine, then came the steam-engine-machine, and after that the computer-machine. Which amounts to saying that mechanism has introduced in biology first mechanical energy, then chemical energy, and finally information." This allows him to proclaim that (Barbieri, 2024, p. 141) "mechanism may well be able to change again and introduce in biology not only the concepts of energy and information, but also the last frontier, the concept of meaning." In the text below, I strongly argue *against* Barbieri's belief.

I suppose here that the reader is familiar with the rich history of splendid physicalist and vitalist endeavors, so I'll not analyze them further here, as they will not contribute anything to our biosemiotic investigation. Yet, in the next point, it is my duty to analyze efforts to introduce the concepts of meaning, information, and code into the area of physics.

16.2.2 Living Cosmos

Another view ignores the clinch between physicalism and vitalism: It is "hylozoic" (or "hylonomic") and will stress that *all* the universe is alive to some degree, its evolution involving parallel processes *negotiating* its next moves according to physics. It does so deterministicly, as characterized above, but always within the context of the memory and habits inherited from previous states.[1] Memory and rules, then, are involved in whatever we encounter in nature, analogous to phenomena we see optimally expressed in living beings.

Rivers depend on riverbed shapes formed in the past, as new stars depend on trends established by previous generations of stars, and geological processes mold the Earth's crust according to the results of its previous evolution. As in the previous discussion, life represents only an additional step to such overall processes. Quotations from the Czech phenomenologist Michal Ajvaz illustrate this view: "Any beginning points towards something that happened earlier, any new action is a result of past processes vaguely indicated" (Ajvaz, 2006, p. 18).[2] "The sedimentation (and transformation) of earlier states is an inseparable part of system's evolution, an endless game between extant situations; the force driving such a game [*onto-poietic tension*, see below], and unique situations brought about by previous history and contingency" (Ajvaz 2006, pp. 37, 43; also Ajvaz, 2023). Yes, here we encounter *memory* (of geological layers, of the countryside, of celestial bodies). But surprisingly, we come also upon *meaning*:

> The archaic layers of *meaning* do not represent just some relict contaminants that unfortunately cannot be cleared away and complicate or hamper our work of syntheses. . . . Such layers participate in the work of *creating meaning,* they have an important function. The meaning arises exactly from the game of all layers and layouts comprising it; the game of mutual challenges of both archaic and "higher" strata leads to a uniqueness and novelty of any form taken. (Ajvaz, 2006, p. 49, italics mine; see also 2017, p. 141; from the context, it is obvious that meaning is attributed to *all* complex systems with a history–that is, not life only.)

Here, the game is what counts: A static structure devoid of history is a mere abstraction. At the same time, the borderline between actuality and potentiality is also a mere abstraction (Ajvaz, 2006, p. 50). Such abstractions should be put aside in favor of self-structuration (Ajvaz, 2017, pp. 160; 2023, pp. 129–130). But "there is not, in the universe, any extra force or substance tending towards unity and organization, utilizing physical forces as described in physics to such ends. . . . It's *simply physical forces* with no addition expected from the side of philosophy. *Philosophy simply emerges from physics*, in that it takes the activity of such forces from a different angle, and designates them from a cohesive aspect of unifying, organization, and structuration. . . . We [philosophers] indicate only the general contours of such processes: in inanimate nature we observe some unifying *rhythm towards organization* (as well as a rhythm towards chaos) that later appears in a new form, also in living nature" (from a letter to the author, March 22, 2013, italics mine). Some scientists are not lagging behind this perspective: See, for example, the recent article by Wong et al. (2023) on evolution and selection in the universe, according to *universal* (even not yet formulated) general laws.[3]

The quotes above offer an important insight into the semiotic foundations of natural processes—that is, that *reflexing* is a universal cosmic movement and that "any order can arise only where there is an *intersection* with another organization or order, and only in the *interpretation* by that other order, which it itself interprets. Outside this doubled mutual *interpretation* there is no order or structure" (Ajvaz, 2017, p. 46, italics mine). But can we speak of interpretation, and even meaning, in these cases?

In the third book by Ajvaz (2023), subtitled *On the Construction of Meaning in the World*, we read, "The gushing of meanings [driven by *ontopoietic tension*] and their fulfillment is differentiated into streams of events, in a *different phase* compared to other events; moreover, such phases of fulfillment of unifying order come in clash with phases of sedimentation from other streams" (Ajvaz, 2023, p. 53).

Meanings enter the game everywhere here, and with them also *understanding*, and finally, *individual existence*—without specifying whether it is the living individual, or anything at all existing in the world: "Individual existence plays an important role in this game [of meaning extraction]; it was formed, or invented, in the very course of the game itself . . . as a solution of the riddle introduced in this or that stage of the ontopoietic game. . . . At stake for individual existence is its proper role in ontopoiesis according the rules of dynamic logos, rules and norms" (Ajvaz, 2023, pp. 335–336).

Enter our second author and critic of contemporary science, Michel Serres, a polymath who presents a similar vision of the dynamic world, but in a much broader scope, embracing mathematics, philosophy old and new, life, fiction and nonfiction, and many other branches of knowledge, all this taking place in a nonlinear space and time, and under "impure" conditions (i.e., the world is not a laboratory). Moreover, *locally* the spatial and temporal coordinates and velocities may differ substantially; we will revisit that topic in the last part of this paper. Serres is an author of more than fifty books in which he elaborates his multiple themes. Here, I take advantage of a book of interviews with Bruno Latour (Serres and Latour, 1995), of a selection of his essays (Serres, 1982), and of a monograph on Serres (Watkin, 2020), and I begin with his very impressive metaphor of the handkerchief:

> If you take a handkerchief and spread it out in order to iron it, you can see in it certain fixed distances and proximities. If you sketch a circle in one area, you can mark out nearby points and measure far-off distances. Then take the same handkerchief and crumple it, by putting it in your pocket. Two distant points suddenly are close, even superimposed. If, further, you tear it in certain places, two points that were close can become very distant. This science of nearness and rifts is called topology, while the science of stable and well-defined distances is called metrical geometry. Classical time is related to geometry, having nothing to do with space, as Bergson pointed out all too briefly, but with metrics. On the contrary, take your inspiration from topology, and perhaps you will discover the rigidity of those proximities and distances you consider arbitrary. (Serres, 1995, p. 60)

Several pages later, we read,

> Suddenly, two or three objects separated by great distances, with no previous link between them, belong to the same family. . . . The advantage that results from it is a new organization of knowledge; the whole landscape is changed. In philosophy, in which elements are even more distanced from one another, this method at first appears strange, for it brings together the most disparate things. People quickly criticized me for this—for bringing together turbulence theory and Lucretius' poetry, thermodynamics and Zola's novels, and so on. But these critics and I no longer have the same landscape in view, the same overview of proximities and distances. With each profound transformation of knowledge come these upheavals in perception. (Serres, 1995, pp. 70–71)

And, for Serres, the same holds for the topology of time, when traits of different origin come into close interactions:

> Time is paradoxical; it folds or twists; it is as various as the dance of flames in a brazier—here interrupted, there vertical, mobile, and unexpected. . . . It's not always laminar. The usual theory supposes time to be always and everywhere laminar, with geometrically rigid and measurable distances—or at least constant. Someday it will be said that that is eternity! It is neither true nor possible. No, time flows in a turbulent and chaotic manner; it percolates. All of our difficulties with the theory of history come from the fact that we think of time in this inadequate and naïve way, . . . every historical era is likewise multitemporal, simultaneously drawing from the obsolete, the contemporary, and the futuristic. (Serres, 1995, pp. 58–60)[4]

Two more parables illustrate Serres's approach to understanding nature. The first is that of the Harlequin, a character in a patchwork coat, and is similar to the parable of the handkerchief above, but with an added value. If you rotate him swiftly, suddenly you get another character: the snow-white Pierot; as the white light is an assembly of many colors, so is Pierot. The colorful world became—in hands of Modern science—an ironed, linear model shining "white."

The second parable concerns the *Parasite.* Apart from established international meanings, in French, the word is also close to the name for "white noise." Serres, however, takes the word in a broader sense, including phenomena like background music or babbling; that is why the word "parasite" remains untranslated when his text is put into foreign languages. It follows that novelty, new solutions or paths, are impossible without the presence of such *parasites*: If science tries to get rid of them, it becomes static, nontemporal, a pale reflection of reality. In the editorial preface to Serres (1982, p. xxvii), we read, "Order is not the law of things but their exception. [And quoting Serres] 'Consequently, it is necessary to rethink the world not in terms of its laws and its regularities, but rather in terms of perturbations and turbulences.'" Further, we read Serres's own words: "To isolate an ideal form is to render it independent of the empirical domain and of noise. Noise is the empirical portion of the message just as the empirical domain is the noise of form" (Serres, 1982, p. 70).

16.2.3 Information

One more cliché, and really a riddle for me. We often encounter a strange statement, or allegation, that world has at its basis three components: matter, energy, and information—an allegation, because to my knowledge, no physicalist statements mention/define information as a fundamental principle. I would argue that the statement is nonsense, because, first, information must be defined *for* a given system—that is, a particular living being—or to its extended artificial sensors, such as a thermometer. In other words, we study the color, structure, or metabolism of an insect, of an ecosystem, of history thereof, and at a particular accuracy set (resolution, or size of the "pixels"). Last but not least, it must be always stated *for* or *to whom* such "information" is aimed. In other words, information is always bound to a context that *necessarily* involves a living being—it is not an independent quantity (Hoffmeyer, 1995). Only living beings (often specially trained—for example, a physician or a dog) can detect, extract, and evaluate information (often with the help of special detectors and software). This does not mean that we are going to abandon thinkers like Ajvaz or Serres above, but from now on, their inferences would hold only for life and living systems proper. The inorganic world, in contrast, will be left only with strata of the past ("memory") accumulated, converted (metamorphosis), or put into different constellations, without any reflections or purposeful activity from their part. And without playing any game. With this statement, we enter the second part of my investigation into the extraordinary phenomenon of life.

16.3 On Interpretation: Relations, Signs, Codes, and Meanings

Let us concentrate on life only and take interpretation of the world as a distinctive property of living beings (and in the case of humans, equipped not only with body sensors but also with special instruments, and with language and mind). Doing so, we approach a very important issue: It is the living being who decides *what* is information *for it*, and *how* to interpret it. Hofstadter (1979, p. 166) recognizes three levels of recognizing/interpreting information *as* a message: (1) The *frame message* announces, "I am a message, decode me!" (e.g., a Czech addressee receives a letter written in the Latin alphabet, yet in an incomprehensible language); (2) the *outer message* suggests (by a set of triggers) how to decode it. (The letter bears an Estonian stamp, presumably it is in Estonian. The receiver looks for an Estonian speaker, who, however, tells them that the letter is actually in Finnish.); and finally, at the level (3), we come to the message itself. (The receiver finds a bilingual Finn who translates the message, and the receiver finally is able to extract its *meaning*.) The whole process is, of course, facilitated if the receiver knows the intentions of the sender. Consider these three levels of message recognition in these examples from biology: (1) a "footloose" female bird detecting (in breeding season) the trills of a conspecific male; (2) a male insect attracted by female pheromone(s); or, (3) in flowering plants, the pollen tube (grown from a pollen grain if it landed on a conspecific stigma) will be led to the embryonal sac. It follows that to detect a message, the receiver must be equipped by appropriate sensors and must be able to process it and translate it into a right context with its biological endowment. It goes without saying that unintended receivers, such as, for example, parasites or predators, may also recognize such messages and interpret them in their own way.

16.3.1 Codes

The word *code* is derived from the Latin *codex*, the book; later it also took on a meaning of unbreakable rules as they are often written down in such a book. Unbreakable, but not always watertight—consider, for example, the Penal code with cohorts of lawyers endlessly interpreting its meaning. Because the language of the lawmaker might be ambiguous, the language itself has developed, as has the society and its sensitivity to various kinds of crimes and sentences. The concept of code (more or less exposed to interpretation) was later accepted into many other branches of human activity (e.g., bar code, access code, machine code) as well as in biology, as discussed below.

In biology, to recognize and internalize information into wetware's workings, a series of more and less hardwired codes must necessarily exist. Here, we enter the realm of the living that cannot be inferred from biochemistry; in other words, the wetware and its rules are a product of history and evolution in a given lineage. Marcello Barbieri (see 2024 as pars pro toto of his many earlier books and articles) has devoted most of his scientific life to this fact that still did not occur to many biologists (who will still stubbornly propose the slogan "Life is chemistry").

An obvious instance of codes is the genetic code, the rules of translation of genetic information (sequences of nucleotides in RNAs) into sequences of amino acids in proteins. Here, a principal role is played by proteins, as set by aminoacyl-tRNA synthases, recognizing a particular triplet on the RNA, and coupling it—according to the code—to a "prescribed" amino acid. It can be said that any enzyme recognizing its substrate and canalizing its reshaping according to strict rules may serve as an example of such code use. Some proteins may bear also multiple "affiliations" in different contexts—for example, a banal metabolic enzyme (transketolase in the cytoplasm) plays a role as a regulator of gene expression in nuclei of neurogenic cells (Pinson et al., 2022).

The code, as said above, may be a product of the "negotiation" and acceptance of a set of universally recognized, and binding in a given community, rules; other such fixed codes can be found in metabolic pathways, in embryonal development, and in ethology.

Binding, but not fixed: Take, for example, the code systems of any language, with its lexicon, grammar, orthography, idioms, and ways of usage in different contexts. But conversation will not stick *absolutely* to such rules. As already mentioned, similarly as in other instances, with codes (especially those written down or frozen), a rule represents the *reduction* of a much richer world—the "body language," as it were, of all domains of life, from molecules to communities. Conversely, message exchanges within a group of people, especially those who know each other, employ such rules just as *one of many* means of coming to a mutual understanding. Stephen Cowley, an eminent member of the Distributed Language Group (since 2007) coined the concept of *distributed language*, characterizing it as follows: "Far from being housed in the skull, language activity links people across time-scales. Not only does articulation give rise to speech but, together with gesture, this is partly anticipatory, uses long term and working memory, and links with both previous interactions and linguistic history" (Cowley, 2012, p. 17). Moreover, "the distributed perspective challenges [the idea that] language is essentially 'symbolic', [calling into question] two dominant orthodoxies. First, it denies that verbal patterns are represented inside minds (or brains). Rather, language is, at once, collective, individual, and constitutive of the feeling of thinking. It is distributed between us" (Cowley, 2011, p. 1). The action of communicating is, then, also a process of "languaging." Moreover, language is to a great extent plastic, and—what is important—evolving, and so are its various codes (e.g., Barfield, 1967; Trask, 2010).

In a similar spirit, Donald Favareau (2008) shows how talk-in-interaction casts doubt on the idea of language as a system of codes only. Such "purified" codes are products of "freezing" the language, isolating it from the real world.[5] He states examples and cites empirical evidence (Favareau, 2008, p. 180), that, in naturally occurring conversational interaction, the "majority of utterances are not actual grammatically complete sentences, and many are not even words," and "for unlike with the written word, speech never comes to us complete and ad hoc, but piecemeal and in real time" (Favareau, 2008, p. 183). "Listeners' moment-to-moment gestures, body torque, eye movements and even breathing can be, and often are, causally influential factors in how a spate-of-talk 'in progress' will alter it's very content, as well as its form" (Favareau, 2008, p. 186). The meaning of a talk (discussion) is cocreated by the group, and the results are in no way prestated in advance. And a final quote: "The semiotic constraints and possibilities that are opening up and closing down in each immediately actualizing 'adjacency' moment—is common to everyday experience upon a few moments reflection, but lost utterly to an over-formalized theory of language (and thought) as computation and as an exchange of messages that are first fully pre-formulated in the brain, translated into words, delivered into the public environment via 'the minimal meaning-bearing units.'" (Favareau, 2008, p. 187).

I believe that analogies from such linguistic studies can, and should, also be used for the code systems in everyday use (i.e., not those put on the record or into the codex). This way, we could get a much more plastic understanding of life's doings instead of counting on fixed, hardwired codes—such as depicted in grammar and biology textbooks.

Importantly, the evolution of codes may proceed at considerably different speeds: The genetic code (or codes) employed in early animal embryogenesis are today almost fixed across the whole biosphere, whereas the codes of dialects in language, say, or of birdsong, postures, or immunoglobulin specificity may change within an individual's lifetime. Yet, due to their *relative* stability, codes can and are being studied by standard methods of biology. We can therefore conclude that the overall behavior of living beings (and communities thereof) can be characterized and studied as an endless buzz of information processing via a variety of preestablished codes that differ widely in their duration and plasticity.

16.3.2 Codes Are Not Signs, but They Are Prone to Produce Them

The working of codes radically transforms the initial inputs both external (light, sounds, hormones, nutrients) and internal (presence or absence of particular proteins, settings, metabolites). They may or may not become sign vehicles, and signs—in their multiple contexts—may or may not point toward particular objects or actions, resulting in meaning extraction for the given critter or community. In our explanations, we will rely on the explanatory power of triadic semiotics of Charles S. Peirce, as elaborated further by John Deely (2009). What follows is an attempt to "zoom out" from the system developed for humans to all forms of life.

Processing (putting into context via code systems) of external and internal cues (signs) will bring them into the manifold of relations that finally will be interpreted by a living being in a way to extract *meaning(s)* to/for it. Such relations are *suprasubjective* in that they cannot be pointed at by a finger or analyzed in a lab. They are, rather, specific for a given living being in a given time and/or situation. This is obvious from the fact that an identical sign vehicle (word, posture, hormone, season) will elicit different reactions in different life forms (and their communities), according to different contexts put together by their respective memory, experience, and momentary settings. In this way, *all* inhabitants of the biosphere forge meaning from their momentary position in the world (or, in other words, stipulate their coordinates within their umwelt).

Hence, sign relations belong to the most basic characteristics of life. Therefore, we must now turn to life itself and analyze the preconditions that enable living beings to behave in this meaning-driven way—a way that differentiates them from nonlife (and that will definitely free us from hylozoic ruminations).

16.4 Closures

In our recent book (Horský and Markoš, 2025, forthcoming),[6] we maintain that for a living being, the principal precondition for evaluating and representing (interpreting) the world is its *innerness*—an internal domain isolated from the surroundings, thus distinguishing between self and non-self. Stuart Kauffman (2000; see also Kauffman, this volume) named this inner realm *closure*. Closure (*clausura*) is also the name for a regime of most types of monasterial life. Monks or nuns lead a life bound by strict rules (*regulae*) that is noticeably different from the lives of people living in the external world. The boundary between both worlds is, of course, "semi-permeable," allowing the exchange of goods and information, but again under strictly defined (i.e., delimited) conditions. The Kauffmanian analogy with such situations in understanding living beings (*autonomous agents* in his terminology) is felicitous. Horský and I further elaborated on the concept by recognizing two types of closures, both inseparable from life since its very beginning (see also Švorcová and Markoš, 2023; see also Gilbert, this volume). We named them *Closure 1* and *Closure 2*.

16.4.1 Closure 1

Closure 1 is represented by all types of cells, and cells only, and "constitutes an unconditional prerequisite of life" (Horský and Markoš, 2025). Extant cells are all descendants of preexisting cells, all the way back to the origins of life (there is no evidence of more than one event of life's origins). This suggests that Closure 1 has never ceased to exist and keeps memory, experience, and structures of the past (albeit set up differently in all lineages or even in individual cells). Hence, all types of cell lineages that appeared in evolution share common origins, and also many rules (or their modifications) established at that time. In this way, lineages (even those differentiated in the remote past) may communicate to this or that extent, exchange information, metabolites, and even form nonrandom chimeric assemblages such as multicellular bodies or ecosystems bound on a mutual understanding (see more in Markoš and Švorcová, 2019). Such mutual communication of cells is a precondition for a second type of closure.

16.4.2 Closure 2

Closure 2, in contrast, is represented by multicellular assemblages of various forms that arise de novo, persist for different time periods (from minutes to billions of years), and perish. In the case of humans, examples of such assemblages may include a multicellular body, a holobiont, a religious community, a fan club of this or that team, a community speaking the same language or dialect, an army, a family, cultures, and civilizations; as a counterexample, we may think of an overfilled tram or visitors in the fast-food restaurant. For assemblages not including humans, let's name, for example, a forest, a bacterial mat, or a tree with its symbionts.

Closures 2 are necessarily based on cells, but such cells can be simultaneously members of many different assemblages, of different duration. A human cell is a member of a multicellular body, a holobiont, a member of a motorcycle gang; of course, it belongs to the species *H. sapiens*, to vertebrates. Closures 2 also may become members of other Closures 2 (just replace, in the previous sentence, the word "cell" with "human being"). In Horský and Markoš (2025, forthcoming), we used a metaphor of the mikado game (see, for example, https://en.wikipedia.org/wiki/Mikado_(game)) with different sticks representing different Closures 1 and 2, united in more complex Closure 2. Some sticks can be easily removed (an officer stops visiting a motor gang), others are not so easy (one can defect to a different nation), and some cannot be left at all (e.g., "membership" in the class of vertebrates).

Of course, both types of Closures cooperate and influence each other in trillions of ways; we will not be pursuing that discussion here. But for the next section, we turn to history.

16.5 History

In our abovementioned book (Horský and Markoš, 2025), we claim that both historians and evolutionary biologists interpret extant traces from the past and spin narratives interpreting past events (i.e., abductions based on such traces). Further, we say that should biosemiotics be taken seriously. Then it must be recognized that *all* living beings are also endowed with such capacity for interpreting the traces endowed to them by their evolution. As mentioned above, Markoš and Švorcová (2019) have outlined the preconditions for such interpretative abilities.

The "general" evolution of the world has been outlined by authors like Serres or Ajvaz, as referred to in part one above. It reaches, however, a new dimension with the arrival of life. Even if many facets of spatiotemporal evolution can be recognized in a lifeless world as well—as we have seen—only with the ability of interpreting hints and traces indicating a particular history did *narrativity* became possible, so as to weave a story, a narrative, helping—among other thing—to decide, or better, negotiate, a path toward one of many possible futures (in the sense of Kauffman's *adjacent possible*). To achieve this, continuity must be achieved at least to the extent of Closure 1, enabling a memory consisting of accumulated, and also lost or transformed, traces. What follows then is the ability to evaluate of such traces (i.e., sign vehicles) and to elaborate them in a way to assign them a meaning that will be appropriate for the situations here and now—in other words, to confirm the older narrative or create a new version of it. Of course, external changes, as well as mutations at all levels of signaling and signal processing, also enter the game.

At the same time, intercellular relations—direct or via cell assemblages—allow the weaving together with other narratives based on different traditions, as well as with novelties—for example, traditions of different duration and/or importance. Even the settings of the biosphere (or Gaia) can be considered in this way.

To get an idea of such processes and events, let us discuss but one kind of long-lasting memory—a linear script constructed of a final (and finite number of) alphabetical characters.[7] At the level of many human civilizations, this presents a reliable and lasting way of reduction (and freezing) of ever-evanescent oral utterances. But the slogan *littera scripta manet* is not very reliable: Interpretative efforts of readers will create again a new inflation, a new halo of meanings, that may or may not correspond to the intents of the original speaker (concerning the process of such reduction–inflation, see Markoš and Cvrčková, 2013).

To return to our introductory example with the book title: Undoubtedly it will arouse different expectations in different readers—albeit written in a standard English, with no usage of dialects, idioms, or argot, and no borrowings from other languages/cultures. In the flow of reading the book, the text may add—to the reader—nuances to the title, depending on its language (including national and familial) anchoring. Even the laconic Czech title will acquire, this way, a halo of semiotic supplements—which immediately opens new spaces of meaning (not necessarily identical with that of native speakers reading the original), supported by the fact that the translator is not an AI. Moreover, the whole original story under discussion has as its background the work of the nineteenth-century French poet Charles Baudelaire (translated into contemporary English)—and the Czech translation of the same poems may accent different aspects of the work, and, of course, employ different usages of words.

Analogies of what was said in the former paragraph can be found in lineages of living beings, whenever genetic scripts are transferred into radically new background. Ordinary sexual processes (crossing) may serve as examples, but such transfers can take place even between distant lineages via *horizontal gene transfer*. The prokaryotic (bacterial and archaeal) biosphere is interconnected to an extent that some authors speak about the "prokaryotic pangenome"—referring either to the strains in the same species, or more generally across the whole biosphere (e.g., Rosconi et al., 2022; Tettelin et al., 2008) (with "species" defined loosely as a set of core genes that do not participate in such gene flow).[8] Any gene, save the core, contained anywhere in such a pangenome, can be theoretically called up and "googled out" by a cell that happens to be in a need of it—such as in the case of spread of resistance to antibiotics and their derivatives that are used in medicine. Thus, genomes of bacteria and archaea are submitted to endless "creolization," with only "kernel" parts being protected from such incessant exchange. In the world of eukaryotes, such horizontal transfers are not so frequent, yet existing. For example, key genes of mammalian placenta were incorporated from retroviral genomes (Chuong, 2013); similarly, plant hemoglobins came from bacteria or animals (Vázquez-Limón et al., 2012). Such transfers may result in new, innovative usage of foreign genes. Thus, the concept of *semiosphere*, coined by Lotman (1990) for human global civilization, can be easily taken as a special case of *biosphere*.

16.6 Plasticity and Elasticity Model of Evolution

The evolution of culture is often modeled as long periods of stasis, punctuated by sudden bumps of novelty ("explosions" *sensu* Lotman, 2009); a similar model was suggested for evolution of life (Gould and Eldredge). Flegr (2010, 2021) developed the evolution model further, suggesting that the evolution of a species consists of two phases. The plastic phase constitutes about 5 percent of species' lifetime, whereas the rest is the *elastic phase*. In plastic phase, the species is highly malleable by both external and internal factors; thus, its fate is hard to assess. Later, the species becomes elastic—that is, highly resilient to such factors: If a deviation occurs, there is a tendency to return to original "typus."

In Horský and Markoš (2023), we developed the scheme further by replacing "species appearance" by "phenotypic traits." This deserves a note: The appearance of an individual—and subsequently of generalization to a whole species—is a sum of many phenotypic traits typical for that individual (strain, species). So, too, an appearance of a human being can be specified by traits such as his/her eye color, height, behavioral complexity, skeletal morphology, and inclination to obesity. The scientific jargon mostly takes "phenotype" and "appearance" for synonyms; here, however, I suggest distinguishing them.

In figure 16.1, several such graphs are laid over each other; note that the axes (and curves) have no beginning: There has always been something before. Curve a shows the development of a taxon-specific trait, such as, for example, the evolution of forelimbs in cetaceans. For curve b, the wavy, nearly horizontal line indicates some ancient, elastic

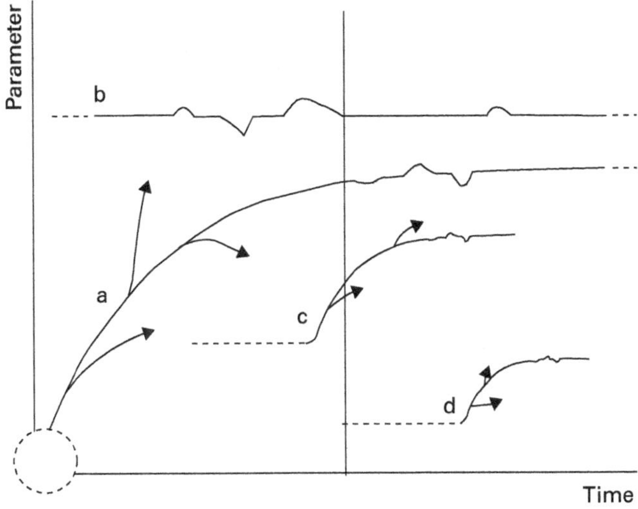

Figure 16.1
Plastic and elastic phases in the evolution of phenotypes. Curves explained in the main text. Note that the plastic phase is disproportionally extended for the sake of clarity. (After Horský and Markoš, 2023; see also Švorcová and Kleisner, this volume).

feature (e.g., early vertebrate embryogenesis). Curves c and d are two examples (out of possible trillions of similar curves for any species) of "explosions" (e.g., a new symbiosis within the holobiont, adaptive radiation, hybridogenesis, or the sudden discovery of a new trait, such as mind or speech). The *appearance* of the individual or species is then given, at a given moment, by a perpendicular cut through the tangle of all those curves structuring the overall appearance, with each of its traits having a different dynamic at a given moment. This actualized appearance is a result of a tradeoff of all phenotypic traits and codes, in different phases of their evolution, and negotiated at a given time. Take it as an analogy of a handkerchief model of Serres, described above.

16.7 Conclusion

Generating and interpreting signs is the *defining property* of life, and life *only*. Any signs used for communication are eligible to this set, but in order to explain the idea, I kept here mainly to strings of lasting discrete units ("characters"), such as those used in the genetic script, and to the scripts in languages that use alphabetic notation. I started with the analogy between both ways of recording and interpreting, continued with the need for abductive processes for the interpretation of the world, and shown that two types of closures are required for the process to be feasible. The role of memory and experience of a given living being is a key factor for interpreting—in both types of closures—internal and/or external cues. Interpretation, then, is highly dependent on the history of an individual or its lineage (or other types of community) and is prone to the endless whirl of evolutionary games.

I proceeded through several "Yes" and "No" statements, and I will now try to make a list of them:

No to the physicalization of life. Hylozoic attempts illustrated here (by Ajvaz, Serres, Wong et al., or Prigogine and Stengers) provide astonishing descriptions of the dynamics of our universe but cannot be automatically extended to living beings (however, they can be successfully used for describing some of the dynamic processes therein).

Yes to the definitory property of all living beings is two types of closures.

No to rigid codes set in advance.

Yes to a belief that life is a special quality in the universe. Despite this, it follows:

No for external cues operating on life as if "clicking" on a computer icon. Hence,

Yes for the view of codes negotiated in evolution of life, and life only, but

No to their rigidity, reducibility to a kind of entries in codices and Excel tables.

Rather, semiotic phenomena experienced by us, living beings, especially language, should be taken as an analogy for extending the language metaphor to many of life's doings.

Notes

1. In the area of language, we often encounter similar statements (e.g., Barfield, 1967; Trask, 2010). Here, however, I examine only claims of a more general drive in the universe.

2. This and following quotations from Ajvaz are my own translations.

3. It is noteworthy that the article does not cite older approaches to the problem—not even those of Ilya Prigogine, who formulated the theory of open systems and their self-formation (e.g., Prigogine and Stengers, 1984).

4. Historiographers long ago discovered this mutual interdigitations of different strata of historical time. I consider especially impressive the concept of *hypolepsis* (Assmann, 2011), the possibility to read, further develop, or disagree with written documents of different times, origins, and languages. Same for myths, religions, or morals and—as we will see below, also for biological evolution. Serres, however, takes the liberty to include the whole world.

5. Recall the parasite in Serres, as discussed above.

6. Jan Horský is a historian, and I am a biologist: our common goal, in the book, was to bring historiography and evolutionary biology under a common denominator.

7. The diversity of such alphabetical signs can often be considerably amplified with various kinds of diacritics—in both written texts and genomic sequences (see, for example, the introductory frame at the beginning of this paper), but here I am not going to complicate the text further; see Markoš and Švorcová, 2019, for further discussion of this matter.

8. Read: "We have not detected yet such a flow of these genes."

References

Ajvaz, M. 2006. *Příběh znaků a prázdna* [The tale of signs and emptiness]. Druhé město.

Ajvaz, M. 2017. *Kosmos jako sebeutváření* [Cosmos as self-organization]. Pavel Mervart.

Ajvaz, M. 2023. *Dynamický logos: O významové stavbě světa* [The dynamic logos: On the construction of meaning in the world]. Filosofia.

Assmann, J. 2011 [1997]. *Cultural memory and early civilization: Writing, remembrance, and political imagination.* Cambridge University Press.

Barbieri, M. 2024. *Codes and evolution: The origin of absolute novelties.* Springer.

Barfield, O. 1967 [1926]. *History in English words.* Lindisfarne Books.

Chuong, E. B. 2013. Retroviruses facilitate the rapid evolution of the mammalian placenta. *Bioessays, 35*, 853–861.

Cowley, S. J. 2011. Distributed language. In S. J. Cowley (Ed.), *Distributed language* (pp. 1–14). John Benjamins Publ. Co.

Cowley, S. J. 2012. Mimesis and language: A distributed view. *Interaction Studies, 13*, 17–40.

Cowley, S. J. 2017. Changing the *idea* of language: Nigel Love's perspective. *Language Sciences, 61*, 43–55.

Deely, J. 2009. *Purely objective reality.* Mouton de Gruyter.

Delaney, J. P. 2018. *Believe me.* Ballantine Books.

Eco, U. 1994a. *Six walks in the fictional woods.* Harvard University Press.

Eco, U. 1994b. *The limits of interpretation.* Indiana University Press.

Favareau, D. 2008. Collapsing the wave function of meaning: The epistemological matrix of talk-in-interaction. In J. Hoffmeyer (Ed.), *A legacy for living systems* (pp. 169–211). Springer.

Flegr, J. 2010. Elastic, not plastic species: Frozen plasticity theory and the origin of adaptive evolution in sexually reproducing organisms. *Biology Direct, 5*, 2.

Flegr, J. 2021. *Frozen evolution: A farewell to selfish gene.* Andrias Press.

Gould, S. J., and Eldredge, N. 1977. Punctuated equilibria: The tempo and mode of evolution reconsidered. *Paleobiology, 3*, 115–151.

Hoffmeyer, J. 1995. *Signs of meaning in the universe.* Indiana University Press.

Hofstadter, D. R. 1979. *Gödel, Escher, Bach: An eternal golden braid.* Vintage Books.

Horský, J., and Markoš, A. 2025 (forthcoming). *Movements of seeds and forms.* Springer.

Lotman, J. 2009. *Culture and explosion.* De Gruyter Mouton.

Lotman, Y. M. 1990. *Universe of the mind: A semiotic theory of culture.* Tauris.

Love, N. 2004. Cognition and the language myth. *Language Science, 26*, 525–544.

Love, N. 2007. Are languages digital codes? *Language Science, 29*, 690–709.

Markoš, A. 2016. The birth and life of species–cultures. *Biosemiotics, 9*, 73–84.

Markoš, A., and Cvrčková, F. 2013. The meaning(s) of *information, code . . . and meaning. Biosemiotics, 6*, 61–75.

Markoš, A., and Švorcová, J. 2019. *Epigenetic processes and the evolution of life.* CRC Press.

Markoš, A., and Švorcová, J. 2020. Cutting the Porfyrian tree: Objective reality as created by the innerness of living beings. In Ľ. Lacková, C. J. Rodríguez, and K. Kull (Eds.), *Gatherings in biosemiotics XX* (pp. 24–36). University of Tartu Press.

Pinson, A., et al. 2022. Human TKTL1 implies greater neurogenesis in frontal neocortex of modern humans than Neanderthals. *Science, 377*, eabl6422.

Prigogine, I., and Stengers, I. 1985 [1979]. *Order out of chaos.* Flamingo.

Rosconi, F., et al. 2022. A bacterial pan-genome makes gene essentiality strain-dependent and evolvable. *Nature Microbiology, 7*, 1580–1592.

Serres, M. 1982. *Hermes: Literature, science, philosophy.* J. Hopkins University Press.

Serres, M., and Latour, B. 1995 [1990]. *Conversations on science, culture, and time.* University of Michigan Press.

Švorcová, J., and Kleisner, K. 2018. Evolution by meaning attribution: Notes on biosemiotic interpretations of Extended Evolutionary Synthesis. *Biosemiotics, 11*, 231–244.

Švorcová, J., and Markoš, A. 2023. Closures as a precondition of life, agency, and semiosis. *Biosemiotics, 16*, 45–59.

Tettelin, H., et al. 2008. Comparative genomics: The bacterial pan-genome. *Current Opinion Microbiology, 12*, 472–477.

Trask, R. L. 2010. *Why do languages change?* Cambridge University Press.

Vázquez-Limón, C. et al. 2012. The evolution of land plant hemoglobins. *Plant Science, 191/192*, 71–81.

Watkin, C. 2020. *Michel Serres: Figures of thought.* Edinburgh University Press.

Wong, M. L., et al. 2023. On the roles of function and selection in evolving systems. *PNAS, 120*, e2310223120.

Index of Names

Index of Terms

Publisher contact:
The MIT Press
Massachusetts Institute of Technology
77 Massachusetts Avenue, Cambridge, MA 02139
mitpress.mit.edu

EU Authorised Representative:
Easy Access System Europe, Mustamäe tee 50,
10621 Tallinn, Estonia
gpsr.requests@easproject.com

Printed by Integrated Books International,
United States of America